# Influence Function Approach

## Selected Topics of Structural Mechanics

WIT*PRESS*

WIT Press publishes leading books in Science and Technology.
Visit our website for the current list of titles.
www.witpress.com

WIT*eLibrary*

Home of the Transactions of the Wessex Institute, the WIT electronic-library provides the international scientific community with immediate and permanent access to individual papers presented at WIT conferences. Visit the WIT eLibrary at
http://library.witpress.com

# Influence Function Approach

## Selected Topics of Structural Mechanics

**Yuri A Melnikov**
*Middle Tennessee State University, USA*

WIT*PRESS* **Southampton, Boston**

**Y. A. Melnikov**
*Middle Tennessee State University, USA*

Published by

**WIT Press**
Ashurst Lodge, Ashurst, Southampton, SO40 7AA, UK
Tel: 44 (0) 238 029 3223; Fax: 44 (0) 238 029 2853
E-Mail: witpress@witpress.com
http://www.witpress.com

For USA, Canada and Mexico

**WIT Press**
25 Bridge Street, Billerica, MA 01821, USA
Tel: 978 667 5841; Fax: 978 667 7582
E-Mail: infousa@witpress.com
http://www.witpress.com

British Library Cataloguing-in-Publication Data

A Catalogue record for this book is available
from the British Library

ISBN: 978-1-84564-129-0

Library of Congress Catalog Card Number: 2007932025



Printed in Great Britain by Athenaeum Press Ltd.

to Sasha with love

# Contents

# Preface

Structural mechanics is the study of the effects that forces of different physical origin (mechanical, thermal, magnetic and so on) produce on elements of structures such as cables, pillars, beams, plates and shells. In presenting the material in this text, it was presumed that the reader's background is equally solid in undergraduate mathematics and mechanics. The reader is assumed to be relatively fluent in differential and integral calculus and to possess, at the same time, a workable knowledge of the fundamental principles of statics and dynamics. Knowledge of the basics in mathematics is critical. It provides the reader with an understanding of mathematical models, qualitative and quantitative analysis, which helps to estimate the load-carrying capacity of elements of structures.

This book covers only a limited number of topics from the undergraduate course of structural mechanics and cannot, therefore, be considered for the principle text. The objective in designing this volume was the development of a supplementary text for the course. It can also be helpful as such for other core courses in the mechanical/civil engineering curriculum that deal with elements of structures. On the other hand, since mathematical aspects of the discussion were always in the author's mind, the chosen language of presentation gives hope that this book could also attract mathematic's majors in the curriculum of applied or industrial mathematics. As to the applied mathematics curriculum, the book can be adapted as a graduate text for a course on computational mechanics where a student could use strong mathematical background in modelling and solving different problems from mechanics.

Given that dozens of texts on subjects related to mechanical engineering are widely available, the intention to work out another does not probably look reasonable and well grounded. The author believes, however, that the content specificality of the present book can justify the desire to add another volume to the tall pile of existing well-established texts. The specific feature of the present text becomes clear just from examining its table of contents. Indeed, this text represents the first ever attempt to include in book format a number of standard problems from structural mechanics, which are treated by means of a single mathematical approach that is novel in the field. The *influence* (*Green's*) *function method* constitutes the basis for this approach.

In the preparatory sections of this book, effective procedures are proposed for the construction of Green's functions to a variety of boundary-value problems that are stated for ordinary and partial differential equations that simulate static equilibrium of beams, thin plates and shells. Then, using the analogy between Green's functions and influence functions of a point force, we are analyzing the behavior of structural elements undergoing point concentrated and distributed loads. An extensive set of influence functions of a point force is obtained for beams, plates and shells, and computer friendly algorithms are developed that use the influence function method for solving some natural vibrations and buckling problems. With the aid of these algorithms a vast number of different problem settings is considered for elements of structures.

Note again that designing a primary text for a course on structural mechanics has never been in the author's mind. The book was planned, instead, to cover only a selected set of the traditional topics in the course. At the same time, the method of presentation that was developed and implemented here is a complete departure from those standard and well-established approaches to mathematical models of the typical problems that are usually considered in structural mechanics. This specific feature could make the book a very convenient source of an alternative supplementary reading on the subject. Supplementary texts are especially important nowadays in light of the necessity to treat the subject with today's level of breadth, generality and rigor.

In the process of preparation of this text, an attempt was made to be as close to the student's viewpoint as possible. While teaching a number of courses within the engineering and applied mathematics curricula, the author came to the conclusion that the principal difficulty that students who major in mechanics usually experience is a lack of mathematical thinking while going through mechanical topics. Whereas for students majoring in mathematics, the basic hurdle is a lack of understanding and proper interpretation of mathematical models. This is a result of the shortage of the so-called bridging courses in both mathematics and mechanics curricula, courses that fill out the gap between applied mathematics and engineering science. The author hopes that the present work could be considered as a text for such a bridging course in either curriculum.

When launching this project the author had stated the *number one goal* for the presentation was that the text ought to be *for students* in the first place. There is no doubt that the failure to achieve this goal would dramatically reduce the overall value of the work, since its context is unique, and it is hard (if not impossible) to find another alternative text on influence functions that could address the readers' concerns, clarify their hesitations and answer arising questions. Thus, the presentation must be as self-contained and comprehensive as possible within the scope of the selected limited volume.

Such a demanding goal required a specific methodology in the presentation where it was assumed that not every concept or statement, which is obvious to the instructor, is that clear to the student. So, a specific effort is required to attract a wider readership, to gain the readers' respect and to make this book a pleasant and, at the same time, useful read after all. This implies that the presentation must contain clear definitions, straightforward constructive and easy to follow proofs, a considerable number of convincing illustrative examples, and, on the top of that, maximum clarity in explanations.

The author hopes that decades of his experience, first in learning and then teaching the subject, and years of a challenging work on this book has helped him to achieve the *number one goal* as stated earlier. But, this being said, it is evident that the best and perhaps the only unbiased referee on whether or not the goal is achieved is, however, the reader of this manual who can and will ultimately come up with a well-grounded judgement.

To pay tribute to people who either explicitly or implicitly contributed to this project, it is with great pleasure that the author acknowledges his collaboration, over many years, with Drs V.A. Boborykin, Ye.A. Bobylyov, I.M. Dolgova, V.B. Govorukha, V.A. Koshnarjova, R.D. Krasnikova, A.V. Krasnikov, V.V. Loboda, N.V. Polyakov, T.V. Rydvanskaya, V.V. Shubenko, S.A. Titarenko, E.Ts. Tsadikova and V.L. Voloshko (at Dnipropetrovsk National University, Ukraine) and Drs M.Y. Melnikov, J.Q. Powell and K.L. Shirley (at Middle Tennessee State University, USA). The author wishes to also acknowledge his students at MTSU (A.S. Arman, M.T. Hall, S. Hughes, S. McDaniel, P.L. Roubides and T. E. Slowey), whose research projects enhanced the quality of this manual. Each of the talented individuals named above can find at least a single spot in the text that reflects our fruitful collaboration.

The production of this book required the help of a number of individuals from WIT Press where the author is especially grateful to the production editor Isabelle Strafford for her professional work that made the book a much better read.

During his involvement with this project for over three years, both the Department of Mathematical Sciences and the Office of Graduate Studies at Middle Tennessee State University supported the author. This support has significantly hastened the work and is very much appreciated.

**Yuri A. Melnikov**

# Introduction

Presentation of the material in this volume is based on the implementation of two important notions taken from different sciences. One of them (the *influence function of a point concentrated force*) is brought from structural mechanics, while another (the *Green's function of a boundary-value problem*) is taken from mathematics. They are closely related to each other and their relation represents the keystone in this text. It appears that bringing these notions together allows us to create a single methodological approach to a variety of problems in structural mechanics, makes their analysis easier and builds up a solid foundation for some further developments in the field.

The notion of Green's function is traditionally playing a significant role [8, 11, 17, 18, 20, 25, 26, 37, 38, 53, 54, 57, 60, 61] in the qualitative theory of ordinary and partial differential equations that simulate phenomena and processes studied in natural sciences. Students of the undergraduate mathematics curriculum are exposed to this notion as early as in the course of differential equations. Intensive studies of recent decades (see, for example, [9, 10, 14, 24, 29, 31–35, 40–52, 59]) have also revealed a remarkable computational potential of Green's functions in applied differential equations, making these functions an indispensable tool in both qualitative and quantitative aspects of this area of research. It is evident that in nowadays Green's functions cannot any more be considered as an apparatus for exclusively 'proof and derivation' type research.

The notion of influence function of a point force for an element of structure, on the other hand, is important in structural mechanics. It represents an effect that a concentrated force applied at an arbitrary point produces on the element. It is known in mechanics [4, 12, 27, 31, 37, 59, 65] that when possessing the influence function of a point force, one can readily evaluate the equilibrium state of the structural element if the latter undergoes a distributed load or even a combination of loads of different kinds. This integrative property of influence functions is unique and raises them to the level of a universal tool in solving many problems in mechanics.

**Influence function–Green's function relation:** In what follows, a correspondence will be drawn between these two notions. This correspondence is of primary importance in grasping the conceptual idea of the approach used in this text to a broad variety of problems in mechanics. The Green's function of a certain boundary-value problems in ordinary or partial differential equations can, in fact, be identified with the influence function of a point force in the phenomenon of mechanics, for which the boundary-value problem serves as a mathematical model. This allows us to build up a bridge between mathematics and mechanics and formalizes solution procedures for many standard problems in mechanics which are being discussed in this book.

To reveal the relationship between the influence function of a point force for a structural element and the Green's function of a boundary-value problem that models the equilibrium of that element, let us revisit particular settings in mathematics and recall phenomena in physics which are simulated by those settings. In doing so, we take a look at a few illustrative examples.

**Example 1:** Let the linear second order differential equation

$$\frac{d}{dx}\left(m(x)\frac{dy(x)}{dx}\right) = f(x), \quad x \in (0, a) \tag{1}$$

be subject to the boundary conditions

$$y(0) = 0, \quad y(a) = 0 \tag{2}$$

This boundary-value problem could be regarded as a model for several phenomena of different physical nature. It could, for example, be identified as a model for a one-dimensional steady-state heat conduction in a rod of length $a$ with insulated lateral surface. The rod is made of a material whose conductive properties depend on the $x$ coordinate, and is subject to internal heat sources.

Another interpretation of the boundary-value problem that appeared in eqns (1) and (2) can be associated with the static equilibrium of a cable of length $a$, having a variable cross-section, fixed at the end-points and subject to a distributed transverse load $f(x)$.

These different interpretations of a single boundary-value problem in eqns (1) and (2) represent just two of many other possible physical models that can be related to this mathematical setting.

In the case of heat conduction, the function $y(x)$ in eqns (1) and (2) is interpreted as the temperature at point $x$ in the rod, made of a nonhomogeneous material, with a variable *thermal diffusivity* $m(x)$. Presence of the right-hand side function $f(x)$ in the governing differential equation presumes an internal heat acting in the rod. As to the boundary conditions imposed by eqn (2), they specify zero temperature at the end-points $x = 0$ and $x = a$.

If the setting in eqns (1) and (2) is regarded as the cable problem, then $y(x)$ is interpreted as the displacement of the cable, caused by a transverse distributed load that is proportional to $f(x)$. The cable has variable mass density $m(x)$, and according to the boundary conditions in eqn (2), the end-points $x = 0$ and $x = a$ of the cable are fixed.

Regardless of a physical interpretation of the boundary-value problem posed by eqns (1) and (2), if the latter has a unique solution, then, as we show in Chapter 2, there exists a *Green's function* $g(x, s)$ for the corresponding homogeneous equation

$$\frac{d}{dx}\left(m(x)\frac{dy(x)}{dx}\right)=0 \tag{3}$$

subject to the boundary conditions in eqn (2).

The reader will soon find out that Green's functions of linear ordinary differential equations are defined, in compliance with their properties, in two pieces. If the coefficient $m(x)$ is, for example, defined as $m(x)=\beta^2x^2+1$, where $\beta$ is a fixed parameter, then the Green's function to the boundary-value problem in eqns (2) and (3) appears in the form

$$g(x,s)=\frac{1}{\Delta}\begin{cases}\arctan\beta x(\Delta/\beta-\arctan\beta s), & \text{if } x\le s\\ \arctan\beta s(\Delta/\beta-\arctan\beta x), & \text{if } x\ge s\end{cases}$$

where $\Delta=\beta\arctan\beta a$.

If the setting in eqns (1) and (2) is viewed as the cable problem, then $g(x,s)$, as a function of $x$, represents the displacement of the cable caused by the transverse unit force applied to an arbitrary point $s$. Due to this physical interpretation, $g(x,s)$ is called in mechanics the *influence function of a point concentrated force*. The variables $x\in[0,a]$ and $s\in[0,a]$ in $g(x,s)$ are usually referred to, in mechanics, and often times in mathematics, as the *observation (field) point* and the *source point*, respectively.

Viewing the setting in eqns (1) and (2) as the heat conduction problem, we interpret $g(x,s)$, as a function of $x$, as the temperature distribution in the rod generated by a unit thermal source acting permanently at an arbitrary point $s$. That is why, in thermal sciences, $g(x,s)$ is called the *influence function of a unit heat source*.

These two different physical interpretations of the Green's function of the boundary-value problem posed by eqns (2) and (3) reveal an important fact that the Green's function–influence function correspondence is not necessarily a one-to-one relationship, because the same single boundary-value problem might have different readings in different areas of physics. ▲

**Example 2:** As another example of the Green's function–influence function correspondence, let us consider a boundary-value problem where the linear ordinary fourth order differential equation

$$\frac{d^2}{dx^2}\left(EI(x)\frac{d^2w(x)}{dx^2}\right)=q(x),\quad x\in(0,a) \tag{4}$$

is subject to the boundary conditions

$$w(0)=\frac{d^2w(0)}{dx^2}=0,\quad w(a)=\frac{dw(a)}{dx}=0 \tag{5}$$

This setting simulates, in mechanics, the lateral deflection $w(x)$ of an elastic beam of length $a$, caused by the transverse distributed load $q(x)$. According to the conditions in eqn (5), the left end-point, $x = 0$ of the beam is assumed to be hinged or, as we will refer to in this text, *simply supported*, while the beam is built in a wall (we will say *clamped*) at its right edge, $x = a$.

Specific boundary conditions in eqn (5) are chosen just for the sake of certainty. Indeed, a number of other physically feasible boundary conditions can be imposed instead, and a wide set of such conditions can be found in problems of Chapters 3 and 4 where bending of beams is discussed in detail. The term $EI(x)$ represents the so-called *flexural rigidity* of the beam. It is a product of the *elasticity modulus* $E(x)$ of the material of which the beam is made and the *moment of inertia* $I(x)$ of the beam's cross-section.

The Green's function $g(x, s)$ of the boundary-value problem in eqn (5) for the homogeneous equation

$$\frac{d^2}{dx^2}\left(EI(x)\frac{d^2w(x)}{dx^2}\right) = 0, \quad x \in (0, a)$$

is associated with the *influence function of a unit transverse point force* applied to the beam at a point $s$. Thus, $g(x, s)$ represents the deflection of the beam at point $x$, in response to a unit transverse force applied to an arbitrary point $s$.

For a beam that is clamped at $x = 0$, while the edge $x = a$ is free (we call such a beam *cantilever*) and the flexural rigidity is a linear function $EI(x) = mx + b$ of $x$, the influence function has been constructed in this text as

$$g(x, s) = \frac{1}{2m^3}\begin{cases} mx[mx + 2(ms + b)] \\ \qquad + 2(mx + b)(ms + b)\ln\dfrac{b}{mx + b}, & \text{if } x \leq s \\ ms[ms + 2(mx + b)] \\ \qquad + 2(mx + b)(ms + b)\ln\dfrac{b}{ms + b}, & \text{if } x \geq s \end{cases}$$

Influence functions of a point concentrated force for many other beam problems with a wide variety of edge conditions are available in this text. A number of them for single-span beam can be found in Chapter 3. Some of these influence functions are presented in a book format for the first time. ▲

**Example 3:** Another illustration of the analogy between Green's functions and influence functions can be found in the classical Poisson–Kirchhoff plate theory (e.g., [35, 43, 45, 51, 59, 65]). According to this theory the biharmonic nonhomogeneous!equation written in Cartesian, for example, coordinates

$$\frac{\partial^4 w}{\partial x^4} + 2\frac{\partial^4 w}{\partial x^2 \partial y^2} + \frac{\partial^4 w}{\partial y^4} = \frac{f}{D}, \quad (x, y) \in \Omega \tag{6}$$

posed on a simply connected region $\Omega$ bounded with a smooth contour $\Gamma$, and subject to the following boundary conditions

$$w(x, y) = \frac{\partial w(x, y)}{\partial n} = 0, \quad (x, y) \in \Gamma \tag{7}$$

with $n$ representing the normal direction to $\Gamma$, models the lateral deflection $w = w(x, y)$ of a thin elastic plate whose middle plane occupies the region $\Omega$. The edge of the plate is, according to the conditions in eqn (7), clamped. The plate undergoes a transverse distributed load $f = f(x, y)$ and the constant $D = Eh^3/12(1-\sigma^2)$ is defined through the elasticity modulus $E$, the Poisson ratio $\sigma$ of the material of which the plate is made and the thickness $h$ of the plate. The parameter $D$ is traditionally referred to, in mechanics, as the *flexural rigidity* of the plate.

In the plate theory, it is shown that the *Green's function* $G(x, y; s, t)$ of the homogeneous equation corresponding to (6) subject to the boundary conditions in eqn (7) can be interpreted as the deflection of the plate at a point $(x, y)$, caused by a unit transverse force applied at an arbitrary point $(s, t)$. That is why $G(x, y; s, t)$ is called, in mechanics, the *influence function of a point force* for the plate. Analogously to the beam theory, we will call $(x, y)$ the *observation (field) point*, while $(s, t)$ will be referred to as the *force application point*.

For a clamped, for example, circular plate of radius $a$, the influence function of a point force is well known [38, 45, 65]. It appears in polar coordinates as

$$
\begin{aligned}
G(r, \varphi; \varrho, \psi) &= \frac{1}{16\pi}\left[\frac{1}{a^2}(a^2 - r^2)(a^2 - \varrho^2)\right. \\
&\qquad \left. - (r^2 - 2r\varrho\cos(\varphi - \psi) + \varrho^2)\ln\frac{a^4 - 2a^2 r\varrho\cos(\varphi - \psi) + r^2\varrho^2}{a^2(r^2 - 2r\varrho\cos(\varphi - \psi) + \varrho^2)}\right]
\end{aligned}
$$

with $(r, \varphi)$ and $(\varrho, \psi)$ representing the observation point and the force application point, respectively.

Later in this text the reader is exposed to a variety of explicit readily computable representations of influence functions for different plate problems. Poisson–Kirchhoff plates and plates resting on elastic foundation are considered in Sections 5.2 and 5.3, while Section 5.4 deals with Reissner plate problems. ▲

While considering boundary-value problems of applied mechanics, one can adopt either the language of mathematics (the Green's function terminology) or the influence function terminology. Clearly, the Green's function language is more suitable and recommended in discussing mathematical issues, whereas in talking mechanics the influence function terminology will be used instead. But this is not a dogma and it will not keep us, in this text, from a more flexible use of these terms, once the influence function–Green's function correspondence is set up within the scope of a certain topic.

Note that not every standard undergraduate text on differential equations introduces the Green's function concept. And even when a text does so (as, for instance, [15, 23]), the discussion is usually superficial and does not cover the many details required for comprehending the material of the present text. This concept is usually discussed in detail in graduate texts on differential equations, which cannot, of course, be considered as a prerequisite to our book if it is adopted as a text in the mechanics curriculum. This observation stimulated our discussion in Chapter 2, where the Green's function concept receives the most serious study to prepare our reader to a productive use of the influence function formalism in later sections of the text.

Two stages can be distinguished in the influence function method. The first stage focuses on the construction of a computer implementable representation of the required influence function of a point force. To obtain the influence function, we recommend the construction of the Green's function to the corresponding boundary-value problem instead, because such procedures are better described in literature. This stage is mostly analytic and requires rigorous mathematical developments. At the second stage in the influence function method, an influence function based algorithm is designed to obtain components of the solution that are required in practice. The two stages represent actually parts of a single influence function procedure and, in this text, we will pay a significant attention to both of them.

**Text organization:** Familiarity with text organization is always important to the reader, because it allows a systematic study of particular topics of the material while keeping in mind an integral sense of the text as a whole. The reader is therefore advised to closely examine this section of the Introduction.

To introduce the reader to the type of the text organization that we accepted, note that a two-integer numbering is used for the separate line equations as well as for theorems, figures and tables. That is, the first numeral in an object's number indicates the chapter number while the second stays for the object number in the chapter. A similar two integer numbering is used for the section examples, but in contrast to the equations numbering, the first numeral in an example's number specifies the section to where the example belongs.

Preparing for a review of the contents in this book, note that the material in the first two chapters is critically important for preparing the reader to the study of the influence function method. That is why it is strongly recommend that the information from Chapters 1 and 2 is necessarily presented to the student helping to create an appropriate studying environment later in the text.

Chapter 1 is specifically designed to make some classical topics of mathematics easily affordable by students who major in mechanics. A score of such topics is included in the chapter as they are crucial for the further study in the text. First, elements of high order linear ordinary differential equations and basic concepts of linear algebra associated with this topic are all covered in Section 1.1. Cramer's rule that is convenient for analytic solution of systems of linear algebraic equations is briefly reviewed in Section 1.2.

Section 1.3 brings an introductory discussion on trigonometric Fourier series. Some series summation formulae are derived in Section 1.4. Note that the material of these sections helps to create a comfortable environment for the reader when proceeding through Chapter 5 dealing with plates and shells problems. Some numerical procedures recommended for approximate solution of ordinary differential equations are reviewed in Section 1.5. A brief introduction to linear integral equations is presented in the concluding section of Chapter 1.

Chapter 2 is devoted to the notion of Green's function for linear ordinary differential equations. In Section 2.1 the definition of Green's function is given and the existence and uniqueness theorem is formulated and proved. The proof is constructive in nature, which provides us with a method that is traditionally used for the construction of Green's functions. An extensive set of illustrative examples is presented highlighting particular stages of the construction procedure. Section 2.2 discusses the self-adjointness of boundary-value problems and a special issue of the symmetry of Green's functions. An alternative method for the construction of Green's functions is discussed in Section 2.3. It is based on the classical method of variation of parameters.

In contrast to the first three sections of Chapter 2, which represent just a compilation of issues that are traditionally related to the theory of Green's function, Sections 2.4 and 2.5 contain novel material that has not been included in standard texts on differential equations. An extension of the Green's function concept is proposed in Section 2.4 that widens the sphere of the Green's function applications in mechanics. The so-called multi-point posed boundary-value problems are introduced and discussed in detail. The notion of *matrix of Green's type* replaces for these problems the notion of Green's function. We formulate and prove the existence and uniqueness theorem and show that the construction procedures for matrices of Green's type can be naturally derived from the routines traditionally used for Green's functions. The material in these sections is supported with a number of helpful examples.

The applicability of the influence matrix formalism developed in Section 2.4 is limited to a "sandwich type" assembly in which the material is piecewise homogeneous. To broaden its application range, the formalism is extended, in Section 2.5, to a more general type of multi-point posed boundary-value problems. These problems, of a more complex type, cover a wider variety of situations in applied mechanics. Framework of graph theory is used for that purpose. Sets of linear ordinary differential equations are considered, formulated on finite weighted graphs in such a way that every equation in the set governs a single unknown function and is stated on a single edge of the graph. The individual equations in the set are put into a system form by imposing contact and boundary conditions at the vertices and end-points of the graph, respectively. Based on this setup, a new definition of the matrix of Green's type is introduced. Existence and uniqueness of such matrices are discussed, two methods for their construction are proposed and some particular examples are analyzed.

Bending problems for single-span Poisson–Kirchhoff beams are considered in Chapter 3. Statement of basic beam problems is reviewed in Section 3.1.1. The influence function–Green's function relation for beams is highlighted in Section 3.1.2. A number of influence functions of a point force for beams of uniform flexural rigidity are constructed in Section 3.1.3. A variety of physically natural sets of edge conditions is considered.

Section 3.2 shows how the response to transverse point forces or bending moments of different intensity can be expressed for a beam in terms of the influence function of a unit transverse point concentrated force. Combinations of loads are treated by means of this approach. Analytic expressions for the deflection function as well as for the stress-related components are derived for many different problem settings. A numerical influence function approach is developed for complicated loads that do not allow analytic solution.

The influence function approach to beams resting on the simple (single-parameter) elastic foundation is developed in Section 3.3. Analytic expressions of influence functions are derived for infinite and semi-infinite beams. Different boundary conditions are considered and beams undergoing a variety of transverse loads are examined. A compendium of influence functions for single-span beams of uniform flexural rigidity concludes Chapter 3.

Section 4.1 of Chapter 4 illustrates the practical solvability of problems for single-span beams of variable flexural rigidity by means of the influence function method. It is shown that analytic expressions of influence functions for such beams are possible when the flexural rigidity represent either linear or exponential function. Numerical algorithms are recommended for more general cases of flexural rigidity when an exact solution of the governing differential equation is either impossible or inconvenient.

Some indirect applications of the influence function method are proposed in Sections 4.2 and 4.3, where it is shown how one can benefit from the knowledge of the influence function of a unit transverse point concentrated force in solving a variety of other beam problems. Frequencies and mode shapes of natural vibrations of a beam are analyzed in Section 4.2. In Section 4.3, influence functions of a unit transverse point force are used to find such *critical* values of axial forces applied to a beam, which cause the loss of stability of the original equilibrium state. The classical Euler formulation of buckling problems is considered.

Section 4.4 deals with multi-span beam problems, which reduce to the so-called multi-point posed boundary-value problems for systems of linear ordinary differential equations. These are not, however, boundary-value problems for systems of equations in the conventional sense, where several unknown functions are supposed to have a common domain for an independent variable, and at least one of the equations in the system involves more than one unknown function. Instead, each equation in the systems that are considered in Section 4.4 governs a single unknown function and is formulated over an individual domain. The system is actually formed by letting those single domains interact with each other at their end-points. End-points for the individual equations become contact points in the system at which appropriate contact conditions are formulated.

The notion of *matrix of Green's type* is introduced as appropriate for a particular type of multi-point posed boundary-value problems stated for a "sandwich" type media. Based on that, in Section 4.4.1 we apply the notion of matrix of Green's type to multi-span Kirchhoff beams. Several particular examples are considered where we do not only construct influence matrices but also show, in Section 4.4.2, how they can be used in computing components of stress-strain states for particular multi-span beams undergoing transverse loads.

In Chapter 5 we turn to thin elastic plates and shells where the list of influence functions available in literature is very limited. Section 5.1 gives the definition of the influence function of a transverse point force, which represents the Green's function (matrix) to a partial differential equation (system) that governs the elastic equilibrium of the plate or shell. A technique is presented in Section 5.2 for obtaining some new influence functions for rectangular and circular Poisson–Kirchhoff plates with a variety of edge conditions imposed. The technique is based on the so-called method of eigenfunction expansion [54] that has proven earlier [43, 45, 47, 51] to be especially effective for a number of problems in computational continuum mechanics.

In this technique, an influence function is first expanded in terms of its Fourier series with respect to one of the independent variables. This consequently results in the construction of the Green's function for an ordinary differential equation in the coefficients of the Fourier expansions (the first stage of the technique). This construction can be done by using either the method based on the defining properties of Green's functions (Section 2.1), or by the method of variation of parameters, as described in Section 2.3. The influence function of interest is then constructed by the complete or partial summation of the Fourier series (the second stage of the technique).

In Sections 5.3 and 5.4, we extend the influence function formalism to the Poisson–Kirchhoff plates resting on a simple (single parameter) elastic foundation and to the Reissner plates model [56] where the effect of transverse normal stress and transverse shear deformation is accounted for. Influence functions of a transverse point force are constructed for plates of different shape and the use of these functions is explained in solving for the stress-strain state in plates subject to a variety of transverse loads.

Thin shells of revolution are under discussion in Section 5.5. A general procedure is described for the construction of influence matrices of a point force for shells with an arbitrary shape of the meridian. Different problem settings are considered for circular cylindrical shells. Analytic expressions for the entries of the influence matrix are, for example, obtained in the case of a simply supported section of the cylindrical shell. A special attention is paid to a fundamental set of solutions to a system of ordinary differential equations governing an axially symmetric stress-strain state of a cylindrical shell.

Since this book was intended as a course text in the first place, a strong emphasis was laid on the practical part of the student's work. To reduce to a minimum the hardship of comprehending the material by the reader, each section of the text is supplied with carefully designed illustrative examples that represent the subject in full terms and reach necessary depth and completeness in the presentation. An extensive set of the End Chapter Exercises is specifically developed for each chapter. Answers and comments to most of the End Chapter Exercises are available in the Appendix.

# Chapter 1

# Mathematics Behind Influence Function

As it has earlier been indicated in the Introduction section, the influence function of a point concentrated “source” (which is either an actual source in the case of a conduction type phenomenon, or represents a force, if the phenomenon is mechanical in nature) could be associated with the Green’s function of a boundary-value problem for a differential equation that simulates the phenomenon. This chapter is designed to briefly review some topics from mathematics that are of a great importance in understanding the notion of Green’s function and in supporting the notion of influence function.

In making this text as self-descriptive as practically possible, Section 1.1 presents a brief but quite comprehensive review, of some mathematical issues that directly relate to elements of linear algebra and high-order linear ordinary differential equations. This is the author’s belief that such review lays down a background for the reader’s productive journey through the present text. The review could definitely assist the reader (whose background in mathematics has not to extend beyond standard undergraduate Calculus) in being comfortable as to the concepts of mathematics which stay behind and support the basics of the influence function method.

A brief review of the Cramer’s Rule for well-posed systems of linear algebraic equations is available in Section 1.2. This method for solving linear systems is given a special consideration due to its wide employment in this text. The point is that it is convenient for computer-algebra developments in dealing with a low-dimension system whose coefficient matrix and the right-hand side vector are not numeric but rather algebraic.

Another standard topic from mathematics is reviewed in Section 1.3, where a brief introduction is offered to trigonometric Fourier series. This is helpful in studying the material in the later parts of this text, where boundary-value problems for partial differential equations are considered. Such problems are tackled by a series expansion which is used for the construction of Green’s functions that are identified as influence functions in mechanics.

In Section 1.4 we show how some series summation formulae can be derived that make it possible to radically increase the convergence rate and enhance, therefore, the computability of series representations of influence functions. The material of this section is important for the development of productive computer algorithms in solving a variety of plate and shell problems. A superficial acquaintance with complex variables is desirable for the reader's comfort in reviewing the material of this section.

Section 1.5 reviews basic numerical methods widely used in solving either initial-value, or boundary-value, or eigenvalue problems for a single ordinary differential equation and systems of such equations. This review can help the reader to closer follow our presentation in those parts (see Chapters 3, 4 and 5) where numerical tackling is necessary for governing differential equations.

The use of integral equations is very limited in this text. But, since this topic is touched upon herein and is rarely covered in standard undergraduate texts, it will be to the readers advantage that we review in Section 1.6 some essentials of integral equations to make this material handy for our reader.

## 1.1 Elements of Ordinary Differential Equations

A broad variety of physical phenomena that we are dealing with within the scope of this text are simulated with high-order ordinary differential equations. The author believes that a brief review of the qualitative theory of such equations, which is available in this section, will help the reader in studying the method of influence function for mechanical problems.

Our major emphasis in this text is on linear differential equations. In many cases we avoided complicated and lengthy proofs focusing more on practical issues. For a more complete and justified presentation of the theory of linear ordinary differential equations, the reader is advised to visit standard texts [2, 15, 23, 60, 68] on this topic.

Also in this Section, one can find a brief description of some essentials of linear algebra that are critical for our presentation of linear differential equations and are actively employed in this text.

### 1.1.1 High order linear equations

First, let us touch upon some issues that are of a terminological and classification matter. The following

$$L[y(x)] \equiv p_0(x)\frac{d^n y}{dx^n} + p_1(x)\frac{d^{n-1} y}{dx^{n-1}} + \cdots + p_n(x)y = f(x) \qquad (1.1)$$

is said to be linear nonhomogeneous differential equation of order $n$ in $y = y(x)$, with $L$ being called a linear ordinary differential operator, $p_i(x)$, with $i = \overline{0, n}$ (this is a short-handed notation that we will be using in this text for $i = 0, 1, 2, \ldots, n$), are referred to as the equation coefficients, while $f(x)$ is called the right-hand side term. If $f(x)$ is identically zero in an interval $(a, b)$, then we call (1.1) a homogeneous equation. The functions $p_i(x)$ and $f(x)$ are supposed to be continuous on $(a, b)$, with $p_0(x)$ being non-zero for every $x$ in $(a, b)$.

To avoid possible confusion that may occur in regard to the term *homogeneous*, the reader must discern different meanings of this term in mathematics and mechanics. Indeed, in mathematics we usually say *homogeneous* boundary-value problem, *homogeneous* equation, or *homogeneous* boundary condition, when the right-hand side in the corresponding equality is zero. In mechanics, on the other hand, when specifying properties of materials, we usually use the term *homogeneous* to indicate that an object under consideration is composed of a material whose properties do not vary with space coordinates. Within the present text, this term will be used in both senses.

If each of functions $y_1(x)$, $y_2(x)$, ..., $y_k(x)$ represents a particular solution to the homogeneous equation

$$p_0(x)\frac{d^n y}{dx^n} + p_1(x)\frac{d^{n-1} y}{dx^{n-1}} + \cdots + p_n(x)y = 0 \tag{1.2}$$

associated with (1.1), then any linear combination

$$\sum_{j=1}^{k} C_j y_j(x)$$

of $y_j(x)$ with arbitrary constant coefficients $C_j$ is also a solution to (1.2). This statement immediately follows from the linearity of eqns (1.1) and (1.2).

A crucial notion from linear algebra needs to be introduced before we go any further with the analysis of differential equations. That is the notion of *linear dependence* of functions. From standard undergraduate texts on linear algebra we learn that a set of functions $g_1(x)$, $g_2(x)$, ..., $g_k(x)$ is called *linearly independent* on an interval $(a, b)$ when the linear combination

$$\sum_{j=1}^{k} C_j g_j(x)$$

identically equals zero on $(a, b)$ if and only if all the coefficients $C_j$ are zero. Otherwise, the set is said to be *linearly dependent*.

From the definition of linear dependence, it in fact follows that a set of functions is linearly dependent on an interval if at least one function from the set can be expressed as a linear combination of the remaining functions in the set. In the case of two functions, this implies that when neither of the two function is a constant multiple of the other one on an interval, then the functions are linearly independent on that interval.

**Example 1.1:** Just by inspecting graphs of the functions $g_1(x) = x$ and $g_2(x) = |x|$, one concludes that these functions are linearly independent on the entire $x$-axis. Indeed, neither of them is a constant multiple of the other on $(-\infty, \infty)$. Whereas, these functions are linearly dependent on both $(-\infty, 0)$ and $(0, \infty)$. ▲

**Example 1.2:** The functions $g_1(x) = \cos 2x$, $g_2(x) = 3\sin^2 x$, and $g_3(x) = -2\cos^2 x$ are linearly dependent on any interval, because the linear combination

$$C_1 \cos 2x + C_2 3\sin^2 x + C_3(-2\cos^2 x)$$

of them equals identically zero when $C_1 = 1$, $C_2 = -1/3$ and $C_3 = -1/2$, since a double-angle identity in trigonometry states that $\cos 2x = \cos^2 x - \sin^2 x$. ▲

**Example 1.3:** The functions $g_1(x) = x^2 + x$, $g_2(x) = 2x^2 - 3x$, and $g_3(x) = x^2$ are linearly dependent on the interval $(-\infty, \infty)$, because it can be easily seen that one of them is a linear combination of the others. Indeed

$$g_3(x) = \tfrac{3}{5} g_1(x) + \tfrac{1}{5} g_2(x) \qquad \blacktriangle$$

There exists [15, 23, 60, 67] a simple and straightforward criterion that allows us to find out if a set of functions is linearly independent. Suppose each of the functions $g_1(x), g_2(x), \ldots, g_k(x)$ is at least $n-1$ times differentiable on an interval $(a, b)$. Then, if the determinant

$$W(g_1, g_2, \ldots, g_k) \equiv \begin{vmatrix} g_1(x) & g_2(x) & \cdots & g_k(x) \\ g_1'(x) & g_2'(x) & \cdots & g_k'(x) \\ \cdots & \cdots & \cdots & \cdots \\ g_1^{(k-1)}(x) & g_2^{(k-1)}(x) & \cdots & g_k^{(k-1)}(x) \end{vmatrix}$$

is not zero for at least one point in $(a, b)$, then the functions $g_1(x), g_2(x), \ldots, g_k(x)$ are linearly independent on $(a, b)$. We can also say that if the functions are at least $n-1$ times differentiable and are linearly dependent on $(a, b)$, then $W(g_1, g_2, \ldots, g_k)$ is identically zero in $(a, b)$.

The determinant $W(g_1, g_2, \ldots, g_k)$ plays an important role in linear differential equations. It is called the Wronskian of the functions $g_1(x), g_2(x), \ldots, g_k(x)$ and named after a Polish mathematician J.M. Wronski (1778–1853).

**Example 1.4:** The functions $g_1(x) = 1$, $g_2(x) = x$, $g_3(x) = x^2$, $g_4(x) = x^3$ are linearly independent on $(-\infty, \infty)$ since their Wronskian

$$W(1, x, x^2, x^3) = \begin{vmatrix} 1 & x & x^2 & x^3 \\ 0 & 1 & 2x & 3x^2 \\ 0 & 0 & 2 & 6x \\ 0 & 0 & 0 & 6 \end{vmatrix} = 12 \neq 0 \qquad \blacktriangle$$

**Example 1.5:** Wronskian criterion can be applied to the functions from Example 1.3. Indeed, the functions in there are linearly dependent on $(-\infty, \infty)$, because simple algebra shows that

$$W(x^2 + x, 2x^2 - 3x, x^2) = \begin{vmatrix} x^2 + x & 2x^2 - 3x & x^2 \\ 2x + 1 & 4x - 3 & 2x \\ 2 & 4 & 2 \end{vmatrix} = 0 \qquad \blacktriangle$$

We revisit again differential equations and introduce another key notion for linear $n$-th order equations. Any set $y_1(x), y_2(x), \ldots, y_n(x)$ of $n$ linearly independent particular solutions of eqn (1.2) on $(a, b)$ is said to be the *fundamental set of solutions* to (1.2) on the interval $(a, b)$.

It can be shown that, due to the linearity of equation (1.2), if the set of functions $y_1(x), y_2(x), \ldots, y_n(x)$ represent its fundamental set of solutions on $(a, b)$, then any solution of (1.2) can be written as the linear combination

$$Y(x) = \sum_{j=1}^{n} C_j y_j(x)$$

where $C_j$ are arbitrary constants. Hence, by linearly combining the components $y_j(x)$ of the fundamental set of solutions, we can readily construct a linearly independent set of functions $Y_1(x), Y_2(x), \ldots, Y_n(x)$ representing another fundamental set of solutions to eqn (1.2). In other words, the fundamental set of solutions of a linear homogeneous differential equation is not unique. To address this point, we consider just an illustrative example.

**Example 1.6:** It is evident that the exponential functions $e^x$ and $e^{-x}$ represent a fundamental set of solutions to the equation

$$y''(x) - y(x) = 0 \tag{1.3}$$

Indeed, it can be shown by inspection that either $y_1(x) = e^x$ or $y_2(x) = e^{-x}$ is a solution to this equation. In addition, these functions are linearly independent on $(-\infty, \infty)$, because their Wronskian is non-zero

$$W(e^x, e^{-x}) = \begin{vmatrix} e^x & e^{-x} \\ e^x & -e^{-x} \end{vmatrix} = -2 \neq 0$$

But, at the same time, two other functions

$$Y_1(x) = \sinh x = \tfrac{1}{2}(e^x - e^{-x}) \quad \text{and} \quad Y_2(x) = \sinh x = \tfrac{1}{2}(e^x + e^{-x})$$

being linear combinations of $y_1(x)$ and $y_2(x)$ also represent a fundamental set of solutions to equation (1.3), since each of them is its solution and they are linearly independent on $(-\infty, \infty)$, since their Wronskian is non-zero

$$W(\sinh x, \cosh x) = \begin{vmatrix} \sinh x & \cosh x \\ \cosh x & \sinh x \end{vmatrix}$$
$$= \sinh^2 x - \cosh^2 x = -1 \neq 0 \qquad \blacktriangle$$

Another notion of importance for linear differential equations is the notion of *general (complementary) solution* $y_g(x)$ of a homogeneous equation (1.2). It is introduced as

$$y_g(x) = \sum_{j=1}^{n} C_j y_j(x) \tag{1.4}$$

where the functions $y_j(x)$ represent a fundamental set of solutions to (1.2) while $C_j$ are arbitrary constants.

The *general solution* $Y_g(x)$ of the nonhomogeneous equation (1.1) is defined as a sum of the general solution $y_g(x)$ of the corresponding homogeneous equation with any particular solution $y_p(x)$ of eqn (1.1). That is

$$Y_g(x) = y_p(x) + \sum_{j=1}^{n} C_j y_j(x) \tag{1.5}$$

As to the solution procedure for linear $n$-th order differential equations, we focus first on homogeneous equations. In the case of constant coefficients $p_j$, a solution to eqn (1.2) can be tried in the form of exponential function

$$y(x) = e^{rx} \tag{1.6}$$

containing a parameter $r$. Substituting (1.6) in (1.2) we obtain

$$p_0 r^n e^{rx} + p_1 r^{n-1} e^{rx} + \cdots + p_{n-1} r e^{rx} + p_n e^{rx} = 0$$

or

$$(p_0 r^n + p_1 r^{n-1} + \cdots + p_{n-1} r + p_n) e^{rx} = 0$$

Since the exponential factor in the above equation is never zero, the polynomial factor has to equal zero. That is

$$p_0 r^n + p_1 r^{n-1} + \cdots + p_{n-1} r + p_n = 0 \tag{1.7}$$

This is called the *characteristic* (*auxiliary*) *equation* for eqn (1.2). So, for a solution of eqn (1.2) to exist of the form in eqn (1.6), the parameter $r$ in it has to be a root of the algebraic equation in (1.7). Notice that the latter can formally be obtained by simply replacing the derivatives $d^k y/dx^k$ in (1.2) with $r^k$ for $k = \overline{0, n}$.

As it is known from algebra, roots of eqn (1.7) could be real and complex, distinct and repeated. This results in four different forms of particular solutions of eqn (1.2) that constitute a fundamental set of solutions.

Each distinct real root $r_k$ of eqn (1.7) clearly yields the exponential function $e^{r_k x}$ as a particular solution to eqn (1.2).

Each real root $r_k$ of multiplicity $m$ brings the following $m$ particular solutions

$$e^{r_k x}, \; x e^{r_k x}, \; \ldots, \; x^{m-1} e^{r_k x}$$

to eqn (1.2).

The following two functions

$$e^{\alpha x} \cos \beta x \quad \text{and} \quad e^{\alpha x} \sin \beta x$$

represent particular solutions to eqn (1.2) associated with each pair of distinct complex conjugate roots $r_k = \alpha \pm i\beta$ of the auxiliary equation.

And, lastly, each pair of complex conjugate roots of multiplicity $m$ delivers $2m$ particular solutions to eqn (1.2) of the form

$$e^{\alpha x}\cos\beta x,\ e^{\alpha x}\sin\beta x,\ xe^{\alpha x}\cos\beta x,\ xe^{\alpha x}\sin\beta x,\ \dots,\ x^{m-1}e^{\alpha x}\cos\beta x,\ x^{m-1}e^{\alpha x}\sin\beta x$$

So, a fundamental set of solutions can be routinely obtained for a linear $n$-th order homogeneous equation with constant coefficients. At the same time, in standard undergraduate courses on differential equations (see, for example, [15, 23, 68]), it is shown that linear $n$-th order homogeneous equations with variable coefficients do not allow such a routine approach. There are only a few particular cases of linear differential equations with variable coefficients whose fundamental sets of solutions are expressed in closed form.

The reader is advised to go through Examples 1.7–1.9 below to develop an experience on the construction of fundamental sets of solutions for homogeneous equations of different order.

**Example 1.7:** Find a fundamental set of solutions to the equation

$$\frac{d^2y}{dx^2} - 5\frac{dy}{dx} + 7y = 0 \tag{1.8}$$

The characteristic equation associated with (1.8) has the form

$$r^2 - 5r + 7 = 0$$

It has two complex roots

$$r_1 = \frac{5 + i\sqrt{3}}{2} \quad \text{and} \quad r_2 = \frac{5 - i\sqrt{3}}{2}$$

provided that the functions

$$y_1(x) = \exp\left(\frac{5}{2}x\right)\cos\frac{\sqrt{3}}{2}x \quad \text{and} \quad y_2(x) = \exp\left(\frac{5}{2}x\right)\sin\frac{\sqrt{3}}{2}x$$

represent the fundamental set of solutions to eqn (1.8). ▲

**Example 1.8:** Find the general solution to the equation

$$\frac{d^3y}{dx^3} - 10\frac{dy}{dx} + 9y = 0 \tag{1.9}$$

One of the roots of the auxiliary equation

$$r^3 - 10r + 9 = 0$$

is evident. That is $r_1 = 1$, which suggests that the cubic trinomial in the equation is divisible by $r - 1$. Hence, the synthetic division reduces the auxiliary equation to

$$(r-1)(r^2+r-9)=0$$

and the remaining two roots appear as

$$r_2 = \frac{-1+\sqrt{37}}{2} \quad \text{and} \quad r_3 = \frac{-1-\sqrt{37}}{2}$$

resulting in the general solution of eqn (1.9) as

$$y(x) = C_1 \exp(x) + C_2 \exp\left(\frac{-1+\sqrt{37}}{2}x\right) + C_3 \exp\left(\frac{-1-\sqrt{37}}{2}x\right) \quad \blacktriangle$$

**Example 1.9:** Find the general solution to the equation

$$\frac{d^4y}{dx^4} + \frac{d^2y}{dx^2} - 6y = 0 \tag{1.10}$$

The characteristic equation

$$r^4 + r^2 - 6 = 0$$

is biquadratic and reveals two distinct real roots $r_{1,2} = \pm\sqrt{2}$ and two pure imaginary roots $r_{3,4} = \pm i\sqrt{3}$, based on which the general solution to eqn (1.10) can be written in the form

$$y(x) = C_1 e^{\sqrt{2}x} + C_2 e^{-\sqrt{2}x} + C_3 \cos\sqrt{3}x + C_4 \sin\sqrt{3}x \quad \blacktriangle$$

Let us focus now on nonhomogeneous equations as in (1.1). From what we learned earlier in this section, it follows that once the general solution $y_g(x)$ of the homogeneous equation in (1.2) associated with (1.1) is available, obtaining the general solution $Y_g(x)$ of eqn (1.1) is simply a matter of finding any particular solution $y_p(x)$ to it (see eqn (1.5)).

In simple cases, a particular solution to a nonhomogeneous equation can be found by a guess. To illustrate this point, we bring the following two examples.

**Example 1.10:** Let us find the general solution to the nonhomogeneous equation

$$\frac{d^3y}{dx^3} = \sin x \tag{1.11}$$

Since zero represents a real root of multiplicity three to the characteristic equation associated with (1.11), the three functions

$$y_1(x) = 1, \quad y_2(x) = x \quad \text{and} \quad y_3(x) = x^2$$

represent linearly independent particular solutions of the corresponding homogeneous equation and

$$y_g(x) = C_1 + C_2x + C_3x^2$$

represents its general solution. An apparent guess for a particular solution to eqn (1.11) leads to $y_p(x) = \cos x$. Indeed, this function being substituted in (1.11) makes it true. Hence, the general solution to eqn (1.11) can be written as

$$Y_g(x) = C_1 + C_2x + C_3x^2 + \cos x$$

Note that in such a trivial case as in eqn (1.11), the solution can formally be obtained by three successive integrations. Indeed, the first integration yields

$$\frac{d^2y}{dx^2} = -\cos x + D_1$$

then we have

$$\frac{dy}{dx} = -\sin x + D_1x + D_2$$

and integrating, we finally obtain

$$y(x) = \cos x + \frac{D_1x^2}{2} + D_2x + D_3$$

Since $D_j$ represent arbitrary constants, the above exactly matches $Y_g$. ▲

**Example 1.11:** Since differentiation of an exponential function $e^{kx}$ does not produce functions of a different form, we can reasonably assume that a particular solution $y_p(x)$ to the nonhomogeneous equation

$$\frac{d^2y}{dx^2} + y = e^{2x} \tag{1.12}$$

would be a function $y = Ae^{2x}$, where the constant $A = \frac{1}{5}$ can be readily determined by inspection. So, it turns out, in this case, that $y_p(x) = \frac{1}{5}e^{2x}$. And once $y_p(x)$ is added to the general solution of the corresponding homogeneous equation, the general solution to (1.12) can be written as

$$Y_g(x) = C_1 \cos x + C_2 \sin x + \tfrac{1}{5}e^{2x}$$ ▲

In more complex cases of linear nonhomogeneous equations their particular solutions cannot easily be guessed. But two standard methods are available for this purpose. One of them is the *method of undetermined coefficients* [15, 23, 60, 68]. It is simple to learn and easy in use. But there are two limitations as to the application range of this method. First, the method is limited to equations with constant coefficients and, second, the form of the right-hand side function $f(x)$ in (1.1) cannot be arbitrary. The method works if and only if $f(x)$ is either a constant or a polynomial, or an exponential function $e^{kx}$, or trigonometric functions $\sin kx$ or $\cos kx$, or any finite sums or products of those.

If, however, the right-hand side function $f(x)$ in eqn (1.1) is in neither of the above forms, then the method of undetermined coefficients is not applicable. The *method of variation of parameters* [15, 23, 60, 68] can be recommended in such cases as an option in obtaining $y_p(x)$. The method is associated with a famous

French mathematician J.L. Lagrange (1736–1813) whose huge contribution to the theory of differential equations, in particular, cannot be overestimated.

The method of variation of parameters is potentially applicable to any linear $n$-th order nonhomogeneous equation (1.1) if a fundamental set of solutions to the corresponding homogeneous equation is available. There are no limitations, in this method, as to the form of the right-hand side function $f(x)$. In our text we widely use this method and the reader can find its comprehensive description later in Chapter 2 where Green's functions for linear differential equations are discussed.

### 1.1.2 Statement of problems

Earlier in this section, we have reviewed methods that are usually used for obtaining general solutions of linear ordinary differential equations. Of course, it is fundamentally important to have general solutions of differential equations that simulate phenomena in physics. This is right, in the first place, because in most cases any solution of a differential equation is a member of the family called the general solution.

But in most settings in applied mechanics, we do not, however, look for the general solution itself to an equation that governs the phenomenon under consideration. Instead, a specific particular solution of that equation is a point of our interest. That particular solution ought to satisfy certain conditions which follow from the individual problem setting. We refer to those as the *conditions of uniqueness*.

Three different types of problem settings for ordinary differential equations of high order are usually encountered in applications. Two of them differ by a manner by which conditions of uniqueness are imposed. One of these settings is called the *initial-value problem*, or *Cauchy problem* (named after one of the greatest French mathematicians of all times A.L. Cauchy (1798–1857)).

Giving the reader examples of such settings, we consider a free-falling object of mass $m$, with $g$ representing the standard acceleration of gravity. Assume that the object is released at a height $H_0$ above the ground. If the air resistance is assumed to be, for example, directly proportional to the velocity of the object, with $k$ representing the coefficient of proportionality, then the fall can, according to Newton's Second Law, be governed by the differential equation

$$m\frac{d^2y}{dt^2} = -mg + k\frac{dy}{dt}, \quad t > 0 \tag{1.13}$$

where $y(t)$ represents the displacement of the object (its height at time $t$). In compliance with our problem statement, the equation in (1.13) ought to be subject to conditions of uniqueness formulated in this case as

$$y(0) = H_0, \quad y'(0) = 0 \tag{1.14}$$

which are called the *initial conditions*. The setting in eqns (1.13) and (1.14) represents a typical example of an initial-value problem.

If, in contrast to the statement in (1.13) and (1.14), we assume that the body was not released at $t = 0$, but rather tossed (either upward or downward) and it is known that at time $t = T$ the object was at a height $H_1$, then conditions of uniqueness for the governing equation have to be imposed as

$$y(0) = H_0, \quad y(T) = H_1 \tag{1.15}$$

which are called the *boundary conditions* and the setting in eqns (1.13) and (1.15) is referred to as the boundary-value problem. Hence, a formal distinction between initial and boundary-value problem formulations is in the manner by which conditions of uniqueness are stated. In an initial-value problem, conditions of uniqueness are imposed at a single point, whereas in a boundary-value problem, they are imposed at two distinct points.

Both the initial and the boundary-value problems that have just been stated, can be solved analytically. Since equation (1.13) is linear and has constant coefficients, its general solution can readily be written in the form

$$y(t) = C_1 e^{\alpha t} + C_2 + \alpha g t, \quad \alpha = \frac{k}{m} \tag{1.16}$$

where the constants $C_1$ and $C_2$ can be found by satisfying the conditions of uniqueness. That is, for the initial-value problem (the setting in eqn (1.14)) we obtain the solution in the form

$$y(t) = H_0 + g[(1 - e^{\alpha t}) + \alpha t]$$

while the solution to the boundary-value problem in (1.15) is found as

$$y(t) = \frac{1}{\alpha(1 - e^{\alpha T})}\{\alpha[H_0(e^{\alpha t} - e^{\alpha T}) + H_1(1 - e^{\alpha t})] - gT(1 - e^{\alpha t})\} + \frac{gt}{\alpha}$$

So, if the general solution to the governing equation is obtained in a closed analytic form (like in the case of eqn (1.13)), then to obtain the solution to either initial or boundary-value problems we proceed with a similar technique. Indeed, obtaining the solution is just the matter of satisfying corresponding conditions of uniqueness.

These two types of problem settings require, however, different approaches if the general solution to the governing differential equation ought to be found numerically. We will focus on this issue in Section 1.5.

Both initial and boundary-value problems are potentially applicable to a linear differential equation of high order. Within the scope of this text, we will be overwhelmingly involved with boundary-value problems, although some initial-value problems will also be considered.

The third type of problem settings for ordinary differential equations that this text will be dealing with, is referred to as the *eigenvalue problem*. As an example, consider the homogeneous equation

$$\frac{d^2y(x)}{dx^2} + \lambda^2 y(x) = 0, \quad 0 < x < b \tag{1.17}$$

containing a parameter $\lambda$ and subject to the homogeneous boundary conditions

$$y(0) = 0, \quad y(b) = 0 \tag{1.18}$$

It is evident that $y(x) = 0$ represents a solution to this problem for a fixed value of $\lambda$. This solution is referred to as the *trivial solution*. But it appears that for some specific values of $\lambda$ nontrivial solutions of the problem in (1.17) and (1.18) also exist. And the question that arises in this regard is formulated as follows: "What are the values of $\lambda$ (called the *eigenvalues* of the problem), for which the problem in eqns (1.17) and (1.18) has nontrivial solutions?" Another question related to eigenvalues is: "What are those nontrivial solutions (we call them the *eigenfunction* of the problem)?"

To highlight a specificity of eigenvalue problems, we express the general solution to equation (1.17) as

$$y(x) = C_1 \cos \lambda x + C_2 \sin \lambda x$$

From the first condition in (1.18), it evidently follows that $C_1 = 0$, while the second condition yields

$$C_2 \sin \lambda b = 0$$

The case of $C_2 = 0$ makes no sense because it leads to the trivial solution. The other option for the above equation is $\lambda b = n\pi$. Thus

$$\lambda_n = \frac{n\pi}{b}, \quad n = 1, 2, 3, \ldots \tag{1.19}$$

represent such values of the parameter $\lambda$ that yield the nontrivial solutions

$$y_n(x) = C_2 \sin \frac{n\pi x}{b}, \quad n = 1, 2, 3, \ldots \tag{1.20}$$

to the problem in eqns (1.17) and (1.18). We call $\lambda_n$ *eigenvalues* of the problem while $y_n(x)$ are referred to as its *eigenfunctions*.

Eigenvalue problems for ordinary differential equations are explored in Chapter 4, where transverse natural vibrations and buckling phenomena are examined for elastic beams. In Section 1.5, we review some numerical techniques for approximate solution of eigenvalue problems.

## 1.2 Cramer's Rule

Another mathematical topic that we would like to briefly review in this section is one of the subjects of linear algebra [15, 67, 68]. That is a method called the *Cramer's Rule*, which is frequently employed in analytic developments in this text for solving simultaneous systems of linear algebraic equations. It is named after G. Cramer (1704–1752), an outstanding Swiss mathematician.

This analytic method is not "fast" and cannot, therefore, be recommended for numerical work on systems of high dimension, because it cannot compete against fast convergent numerical methods that are widely used in engineering science for approximate solution of " big" systems. It is, however, convenient for a computer algebra analytic developments.

The purpose in this brief section is not to deliver a complete analysis of the Cramer's Rule and to cover rigorous details of its proof. We aim instead at the description of its algorithm to make sure that the reader comfortably reads the developments where this method is employed.

Let a system of $n$ linear algebraic equations in $n$ unknowns $x_1, x_2, \dots, x_n$ be written in the matrix form $AX = B$ or explicitly

$$\begin{pmatrix} a_{11} & a_{12} & \dots & a_{1n} \\ a_{21} & a_{22} & \dots & a_{2n} \\ \cdot & \cdot & \dots & \cdot \\ a_{n1} & a_{n2} & \dots & a_{nn} \end{pmatrix} \times \begin{pmatrix} x_1 \\ x_2 \\ \cdot \\ x_n \end{pmatrix} = \begin{pmatrix} b_1 \\ b_2 \\ \cdot \\ b_n \end{pmatrix} \tag{1.21}$$

Assume that the determinant of the coefficient matrix $A$ in the above system is non-zero (we say, in such cases, that the matrix is *regular* or *non-singular* and the system itself is said to be *well-posed*). If so then from linear algebra [67] it follows that there exists a unique solution vector $X$ of the system for any right-hand side vector $B$.

The Cramer's Rule represents a straightforward analytic method, which is usually recommended for solving linear systems if the dimension of the system is not high. To briefly review the Cramer's Rule algorithm, let $\Delta$ represent the determinant of the coefficient matrix $A$ of the system in eqn (1.21). That is

$$\Delta = \begin{vmatrix} a_{11} & a_{12} & \dots & a_{1j} & \dots & a_{1n} \\ a_{21} & a_{22} & \dots & a_{2j} & \dots & a_{2n} \\ \cdot & \cdot & \dots & \cdot & \dots & \cdot \\ a_{n1} & a_{n2} & \dots & a_{nj} & \dots & a_{nn} \end{vmatrix}$$

Let $\Delta_j$ represent the determinant of a matrix that is obtained from $A$ by replacing its $j$-th column

$$\begin{pmatrix} a_{1j} \\ a_{2j} \\ \cdot \\ a_{nj} \end{pmatrix}$$

with the right-hand side column $B$. That is

$$\Delta_j = \begin{vmatrix} a_{11} & a_{12} & \dots & b_1 & \dots & a_{1n} \\ a_{21} & a_{22} & \dots & b_2 & \dots & a_{2n} \\ \cdot & \cdot & \dots & \cdot & \dots & \cdot \\ a_{n1} & a_{n2} & \dots & b_n & \dots & a_{nn} \end{vmatrix}$$

The components $x_j$, $(j = \overline{1, n})$ of the vector $X$ can then be obtained in terms of the determinants $\Delta$ and $\Delta_j$ as

$$x_j = \frac{\Delta_j}{\Delta}, \quad \text{for } j = \overline{1, n}$$

In standard texts (see, for example, [67]) on linear algebra, the reader can find justification of the fact that the product of the coefficient matrix $A$ of the system in (1.21) by the vector $X$ whose components $x_j$ are computed with the aid of the above relation is equal to the right-hand side vector $B$ of the system. In other words, this algorithm indeed yields the exact solution to (1.21).

So, the Cramer's Rule procedure requires computation of a total number of $n + 1$ determinants of order $n$. This makes the Rule too expensive computationally and ineffective for numerical solution of systems of a high dimension. That is why for solution of such systems, some other rapidly convergent approximate methods are usually recommended providing high accuracy approximate solutions and being reasonably "fast" at the same time.

Cramer's Rule is, however, especially productive in theoretical developments where systems of relatively low dimension (three or four, at largest) are to be analytically solved. This is exactly the kind of analytic developments that we will repeatedly be involved in within this text. That is why we decided to bring to the reader's attention this brief introduction to the method

## 1.3 Introduction to Fourier Series

To better prepare the reader for a productive voyage through later chapters in this text, we will focus in this section on another topic from mathematics. It is especially important in dealing with such settings of structural mechanics that are simulated with boundary-value problems for partial differential equations. The method for constructing Green's functions for such problems that we apply in this text is based on the *Fourier series* analysis. These series are named after an outstanding French mathematician and physicist J.B. Fourier (1768–1830) who made a decisive contribution to many areas of applied mathematics.

For being consistent with our overall objective of making the presentation as self-descriptive as possible, we will briefly review the principal concepts of the expansion of functions in trigonometric Fourier series. We begin the review with the following definition.

**Definition:** It is said in mathematics that two functions $f(x)$ and $g(x)$ of a single variable are *orthogonal* on an interval $(a, b)$, if the definite integral from $a$ to $b$ of the product of these functions is equal to zero, that is

$$\int_a^b f(x)g(x)\,dx = 0 \tag{1.22}$$

The above relation is called the *condition of orthogonality* of $f(x)$ and $g(x)$ on $(a, b)$. Given two functions on an interval, we can always check out this condition.

**Example 3.1:** Say the functions $f(x) = x^2$ and $g(x) = x^5$. The integral of their product on the interval $(-1, 1)$ is zero. Indeed,

$$\int_{-1}^{1} x^2x^5\,dx = \left.\frac{x^8}{8}\right|_{-1}^{1} = \frac{1}{8} - \frac{1}{8} = 0$$

Hence, these functions are orthogonal on $(-1, 1)$. ▲

**Example 3.2:** The functions $f(x) = \sin 4x$ and $g(x) = \cos 3x$ are orthogonal on $(-\pi, \pi)$, because

$$\begin{aligned}\int_{-\pi}^{\pi} \sin 4x \cos 3x\,dx &= \frac{1}{2}\int_{-\pi}^{\pi} (\sin 7x + \sin x)\,dx \\ &= \frac{1}{2}\left.\left(-\frac{1}{7}\cos 7x - \cos x\right)\right|_{-\pi}^{\pi} = 0\end{aligned}$$

▲

Note that same two functions could be orthogonal on an interval, but non-orthogonal on another interval.

**Example 3.3:** To verify the above point, we consider the two functions from Example 3.2 and check out their orthogonality on the interval $(0, \pi)$, for which

$$\int_0^{\pi} \sin 4x \cos 3x\,dx = \frac{1}{2}\left.\left(-\frac{1}{7}\cos 7x - \cos x\right)\right|_0^{\pi} = \frac{8}{7} \neq 0$$

Hence the functions $\sin 4x$ and $\cos 3x$ are orthogonal on the interval $(-\pi, \pi)$, but are not on $(0, \pi)$. ▲

**Definition:** If every two distinct functions $f_i(x)$ and $f_j(x)$ of a set of functions $f_1(x), f_2(x), \ldots, f_n(x), \ldots$ are orthogonal on an interval, then the set is said to be orthogonal on the interval.

The following theorem can be readily proved.

**Theorem 1.1:** The set of trigonometric functions

$$1, \sin x, \cos x, \sin 2x, \cos 2x, \sin 3x, \cos 3x, \ldots \tag{1.23}$$

is orthogonal on the interval $(-\pi, \pi)$.

**Proof:** The proof is straightforward. We simply check out the condition of orthogonality for each pair of distinct functions from the set in (1.23). That is,

$$\int_{-\pi}^{\pi} 1 \cdot \sin mx\, dx = -\frac{1}{m} \cos mx \Big|_{-\pi}^{\pi} = 0, \quad (m \neq 0)$$

$$\int_{-\pi}^{\pi} 1 \cdot \cos mx\, dx = \frac{1}{m} \sin mx \Big|_{-\pi}^{\pi} = 0, \quad (m \neq 0)$$

$$\begin{aligned}\int_{-\pi}^{\pi} & \cos mx \cdot \cos nx\, dx \\ &= \frac{1}{2} \int_{-\pi}^{\pi} [\cos(m+n)x + \cos(m-n)x]\, dx \\ &= \frac{1}{2}\left[\frac{1}{m+n} \sin(m+n)x + \frac{1}{m-n} \sin(m-n)x\right]\Big|_{-\pi}^{\pi} = 0, \quad (m \neq n)\end{aligned}$$

$$\begin{aligned}\int_{-\pi}^{\pi} & \sin mx \cdot \sin nx\, dx \\ &= \frac{1}{2} \int_{-\pi}^{\pi} [\cos(m-n)x - \cos(m+n)x]\, dx \\ &= \frac{1}{2}\left[\frac{1}{m-n} \sin(m-n)x - \frac{1}{m+n} \sin(m+n)x\right]\Big|_{-\pi}^{\pi} = 0, \quad (m \neq n)\end{aligned}$$

$$\begin{aligned}\int_{-\pi}^{\pi} & \sin mx \cdot \cos nx\, dx \\ &= \frac{1}{2} \int_{-\pi}^{\pi} [\sin(m+n)x + \sin(m-n)x]\, dx \\ &= -\frac{1}{2}\left[\frac{1}{m+n} \cos(m+n)x + \frac{1}{m-n} \cos(m-n)x\right]\Big|_{-\pi}^{\pi} = 0, \quad (m \neq n)\end{aligned}$$

and

$$\int_{-\pi}^{\pi} \sin nx \cdot \cos nx\, dx = \frac{1}{2} \int_{-\pi}^{\pi} \sin 2nx\, dx = -\frac{1}{4n} \cos 2nx \Big|_{-\pi}^{\pi} = 0$$

This completes the proof of Theorem 1.1. ■

So, the set of trigonometric functions in eqn (1.23) is orthogonal on the interval $(-\pi, \pi)$. It is evident that this set is orthogonal on any interval $(\alpha, \alpha + 2\pi)$ of length $2\pi$. This statement can be justified by using the $2\pi$-periodicity of the functions in (1.23).

Theorem 1.1 can be extended to the set of functions

$$1,\ \sin\frac{\pi x}{l},\ \cos\frac{\pi x}{l},\ \sin 2\frac{\pi x}{l},\ \cos 2\frac{\pi x}{l},\ \sin 3\frac{\pi x}{l},\ \cos 3\frac{\pi x}{l},\ \ldots \tag{1.24}$$

proving its orthogonality on a symmetric interval $(-l, l)$ of any length.

The following expression based on the elements of the set of functions in (1.24)

$$\frac{a_0}{2}+\sum_{n=1}^{\infty}\left(a_n\cos n\frac{\pi x}{l}+b_n\sin n\frac{\pi x}{l}\right) \tag{1.25}$$

is called the *trigonometric Fourier series* with period $2l$.

The following theorem plays an important role in the Fourier analysis.

**Theorem 1.2:** If the series in (1.25) converges for all values of $x$ to a certain $2l$-periodic function $f(x)$ and if there exists the integral

$$\int_{-l}^{l}|f(x)|\,dx$$

(either proper or improper), then the coefficients $a_n$ and $b_n$ in (1.25) can be determined in terms of $f(x)$ as

$$a_n=\frac{1}{l}\int_{-l}^{l}f(x)\cos n\frac{\pi x}{l}\,dx,\quad (n=0,1,2,\ldots) \tag{1.26}$$

and

$$b_n=\frac{1}{l}\int_{-l}^{l}f(x)\sin n\frac{\pi x}{l}\,dx,\quad (n=1,2,3,\ldots) \tag{1.27}$$

The relations in eqns (1.26) and (1.27) are referred to as the *Euler–Fourier formulae.*

**Proof:** The assumption in the theorem suggests that

$$f(x)=\frac{a_0}{2}+\sum_{n=1}^{\infty}\left(a_n\cos n\frac{\pi x}{l}+b_n\sin n\frac{\pi x}{l}\right) \tag{1.28}$$

Since we assumed that $f(x)$ is integrable on $(-l, l)$ and the above series converges, the relation in (1.28) can be integrated on the term-by-term basis. This yields

$$\int_{-l}^{l}f(x)\,dx=\frac{a_0}{2}\int_{-l}^{l}dx+a_1\int_{-l}^{l}\cos\frac{\pi x}{l}\,dx+b_1\int_{-l}^{l}\sin\frac{\pi x}{l}\,dx+\cdots$$

Clearly, all the integrals to the right of equal sign, except for the first one, are zero, resulting in

$$a_0=\frac{1}{l}\int_{-l}^{l}f(x)\,dx$$

This verifies the relation from (1.26) for $n=0$. To obtain the rest of the coefficients $a_n$, we multiply (1.28) by $\cos n\frac{\pi x}{l}$ and then integrate it in a term-by-term manner.

This yields

$$\int_{-l}^{l} f(x)\cos n\frac{\pi x}{l}\,dx$$
$$= \frac{a_0}{2}\int_{-l}^{l}\cos n\frac{\pi x}{l}\,dx + a_1\int_{-l}^{l}\cos\frac{\pi x}{l}\cos n\frac{\pi x}{l}\,dx$$
$$+ b_1\int_{-l}^{l}\sin\frac{\pi x}{l}\cos n\frac{\pi x}{l}dx + \cdots + a_n\int_{-l}^{l}\cos^2 n\frac{\pi x}{l}dx + \cdots \quad (1.29)$$

Due to the orthogonality of the set in (1.24), all the integrals to the right of equal sign in (1.29), except for

$$\int_{-l}^{l}\cos^2 n\frac{\pi x}{l}\,dx = l$$

are zero, resulting in (1.26).

To obtain the coefficients $b_n$, we multiply (1.28) by $\sin n\frac{\pi x}{l}$ and then integrate it in a term-by-term manner. This yields

$$\int_{-l}^{l} f(x)\sin n\frac{\pi x}{l}\,dx$$
$$= \frac{a_0}{2}\int_{-l}^{l}\sin n\frac{\pi x}{l}\,dx + a_1\int_{-l}^{l}\cos\frac{\pi x}{l}\sin n\frac{\pi x}{l}\,dx$$
$$+ b_1\int_{-l}^{l}\sin\frac{\pi x}{l}\sin n\frac{\pi x}{l}\,dx + \cdots + b_n\int_{-l}^{l}\sin^2 n\frac{\pi x}{l}\,dx + \cdots \quad (1.30)$$

And again, similarly to the conclusion made for (1.29), all the integrals to the right of equal sign in (1.30), except for

$$\int_{-l}^{l}\sin^2 n\frac{\pi x}{l}\,dx = l$$

are zero, resulting in (1.27). Thus, the theorem is proved. ■

Theorem 1.2 gives the method of obtaining the coefficients of a trigonometric series in terms of the function to which the series converges. In the approximation theory, however, we usually face and solve an inverse problem. That is, given a $2l$-periodic function $f(x)$, find a convergent trigonometric series as of eqn (1.25) (called the Fourier series of $f(x)$), the sum of which is $f(x)$. Such a problem has a solution if the so-called *Dirichlet conditions* are met:

(i) $f(x)$ is either continuous or has a finite number of points of discontinuity of the first kind on $(-l, l)$;

(ii) the interval $(-l, l)$ can be partitioned onto a finite number of subintervals on each of which $f(x)$ is monotone.

The following theorem justifies the solution to the problem.

**Theorem 1.3 (Dirichlet):** If $f(x)$ is defined on $(-l, l)$ and satisfies on that interval the Dirichlet conditions, then the Fourier series of $f(x)$ converges on $(-l, l)$ and its sum is equal to:

(1) $f(x)$ at all points of its continuity;

(2)

$$\tfrac{1}{2}[f(x+0)+f(x-0)]$$

at points of discontinuity;

(3)

$$\tfrac{1}{2}[f(-l+0)+f(l-0)]$$

at the end-points $x=-l$ and $x=l$ of the interval.

Proof of this theorem can be found in specialized texts [60]. The theorem is named after a German mathematician P.G.L. Dirichlet (1805–1859) whose remarkable contribution to contemporary mathematics can hardly be overestimated and makes his name one of the most recognizable in the field. ■

Note that Dirichlet theorem specifies the relation between the Fourier series of $f(x)$ and the function itself on the interval $(-l, l)$ of definition of $f(x)$. As to the exterior of $(-l, l)$, the Fourier series implements the $2l$-periodic extension of $f(x)$. This implies that if $f(x)$ is defined on a wider interval that includes the interval $(-l, l)$ as a portion, then $f(x)$ and its Fourier series might have nothing in common with each other on the exterior of $(-l, l)$.

Illustrative examples that follow are designed to give the reader a firm grasp of the Fourier series analysis of functions.

**Example 3.4:** Expand the function $f(x) \equiv x$ in Fourier series on $(-1, 1)$.

Dirichlet conditions are in this case satisfied ($f(x)$ is continuous and monotone) ensuring, according to Dirichlet theorem, convergence of the Fourier series to $f(x)$. Since the latter is odd, the entire integrand in (1.26) is odd and all the coefficients $a_n$ are zero

$$a_n=\int_{-1}^{1} x\cos n\pi x\,dx=0, \qquad (n=0,1,2,\ldots)$$

whereas the coefficients $b_n$ can be obtained by integration by parts

$$\begin{aligned} b_n=\int_{-1}^{1} x\sin n\pi x\,dx &= -\frac{x}{n\pi}\cos n\pi x\Big|_{-1}^{1}+\frac{1}{n\pi}\int_{-1}^{1}\cos n\pi x\,dx \\ &= 2\frac{(-1)^{n-1}}{n\pi}, \qquad (n=1,2,3,\ldots) \end{aligned}$$

and the Fourier series of $f(x)=x$ is finally obtained as

$$x=\frac{2}{\pi}\sum_{n=1}^{\infty}\frac{(-1)^{n-1}}{n}\sin n\pi x$$ ▲

**Example 3.5:** Expand the function $f(x) \equiv x^2$ in Fourier series on $(-\pi, \pi)$.

Dirichlet conditions are met since $f(x)$ is continuous and monotone on the intervals $(-\pi, 0)$ and $(0, \pi)$. In contrast to the previous example, $f(x)$ is an even function making all the coefficients $b_n$ zero

$$b_n = \frac{1}{\pi}\int_{-\pi}^{\pi} x^2 \sin nx\, dx = 0, \quad (n = 1, 2, 3, \ldots)$$

whereas

$$a_0 = \frac{1}{\pi}\int_{-\pi}^{\pi} x^2\, dx = \frac{1}{\pi}\frac{x^3}{3}\bigg|_{-\pi}^{\pi} = \frac{2\pi^2}{3}$$

and the rest of $a_n$ coefficients is obtained by integration by parts two times in a row

$$\begin{aligned} a_n &= \frac{1}{\pi}\int_{-\pi}^{\pi} x^2 \cos nx\, dx = \frac{1}{\pi}\left\{\frac{x^2 \sin nx}{n}\bigg|_{-\pi}^{\pi} - \frac{2}{n}\int_{-\pi}^{\pi} x \sin nx\, dx\right\} \\ &= \frac{2}{\pi n}\left\{\frac{x \cos nx}{n}\bigg|_{-\pi}^{\pi} - \frac{1}{n}\int_{-\pi}^{\pi} \cos nx\, dx\right\} = (-1)^n \frac{4}{n^2}, \quad (n = 1, 2, 3, \ldots) \end{aligned}$$

Hence, the Fourier series for $x^2$ on the interval $(-\pi, \pi)$ is found as

$$x^2 = \frac{\pi^2}{3} + 4\sum_{n=1}^{\infty} \frac{(-1)^n}{n^2} \cos nx$$ ▲

**Example 3.6:** In this example, we consider a function that fails to be continuous. That is

$$f(x) = \begin{cases} C_1, & \text{if } -\pi < x < \pi/2 \\ C_2, & \text{if } \pi/2 < x < \pi \end{cases}$$

and expand it in Fourier series on $(-\pi, \pi)$.

A feature which makes this case unlike the two in Examples 3.4 and 3.5 is a piecewise definition of $f(x)$. To handle this feature while finding the Fourier coefficients $a_n$ and $b_n$, we break the integrals in eqns (1.26) and (1.27) into two in compliance with the definition of $f(x)$. In doing so for $a_0$, we have

$$a_0 = \frac{1}{\pi}\left(\int_{-\pi}^{\pi/2} C_1\, dx + \int_{\pi/2}^{\pi} C_2\, dx\right) = \frac{1}{2}(3C_1 + C_2)$$

The rest of the coefficients $a_n$ are found as

$$\begin{aligned} a_n &= \frac{1}{\pi}\left(\int_{-\pi}^{\pi/2} C_1 \cos nx\, dx + \int_{\pi/2}^{\pi} C_2 \cos nx\, dx\right) \\ &= \frac{C_1 - C_2}{n\pi} \sin\frac{n\pi}{2}, \quad (n = 1, 2, 3, \ldots) \end{aligned}$$

while for $b_n$ we have

$$b_n = \frac{1}{\pi}\left(\int_{-\pi}^{\pi/2} C_1 \sin nx\, dx + \int_{\pi/2}^{\pi} C_2 \sin nx\, dx\right)$$
$$= \frac{C_1 - C_2}{n\pi}\left((-1)^n - \cos\frac{n\pi}{2}\right), \qquad (n = 1, 2, 3, \ldots)$$

Substituting the coefficients just found in the Fourier series and performing an elementary algebra, we finally obtain the series

$$\frac{1}{4}(3C_1 + C_2) + \frac{C_1 - C_2}{\pi}\sum_{n=1}^{\infty}\frac{1}{n}\left[\sin\left(n\left(\frac{\pi}{2} - x\right)\right) + (-1)^n \sin nx\right]$$

which converges, according to Dirichlet theorem, to the original piecewise constant function at every point of its continuity. As to the point of discontinuity $x = \pi/2$ and the end-points $x = -\pi$ and $x = \pi$, the above series converges to the value of $(C_1 + C_2)/2$. ▲

Note that in many applications we are not required to just obtain the Fourier series of a function $f(x)$ which is defined on an interval $(0, l)$. Instead, the structure of the series is predetermined by the context of a broader problem for which the Fourier series analysis represents just a part. Such a situation occurs when, for example, a boundary-value problem for a partial differential equations is solved by the method of separation of variables. Specific boundary conditions in the problem might require cosine only consisting or sine only consisting series to be obtained.

The *Fourier cosine series* has the form

$$\frac{a_0}{2} + \sum_{n=1}^{\infty} a_n \cos n\frac{\pi x}{l} \tag{1.31}$$

with the coefficients defined as

$$a_n = \frac{2}{l}\int_0^l f(x)\cos n\frac{\pi x}{l}\, dx, \qquad (n = 0, 1, 2, \ldots) \tag{1.32}$$

whereas the *Fourier sine series* is written as

$$\sum_{n=1}^{\infty} b_n \sin n\frac{\pi x}{l} \tag{1.33}$$

and its coefficients are found as

$$a_n = \frac{2}{l}\int_0^l f(x)\sin n\frac{\pi x}{l}\, dx, \qquad (n = 1, 2, 3, \ldots) \tag{1.34}$$

Clearly, if the series in (1.31) is chosen, then we obtain finally an even periodic extension of $f(x)$, whereas with the series of eqn (1.33), the extension is odd.

**Example 3.7:** Consider the exponential function $y = e^x$, with $0 < x < 1$ and obtain its Fourier: (a) cosine series and (b) sine series.

In part (a), the coefficients of the cosine-series are found as

$$a_0 = 2\int_0^1 e^x \, dx = 2(e - 1)$$

and

$$a_n = 2\int_0^1 e^x \cos n\pi x \, dx = 2\frac{(-1)^n e - 1}{1 + n^2\pi^2}, \qquad n = 1, 2, 3, \ldots$$

and an even extension of the exponential function is ultimately obtained in the form

$$e^x = (e - 1) + 2\sum_{n=1}^{\infty} \frac{(-1)^n e - 1}{1 + n^2\pi^2} \cos n\pi x$$

For the coefficients of the sine-series, in part (b) we found

$$b_n = 2\int_0^1 e^x \sin n\pi x \, dx = -2\frac{n\pi[(-1)^n e - 1]}{1 + n^2\pi^2}, \qquad n = 1, 2, 3, \ldots$$

and Fourier series for the exponential function is found as

$$e^x = -2\sum_{n=1}^{\infty} \frac{n\pi[(-1)^n e - 1]}{1 + n^2\pi^2} \sin n\pi x$$

providing its odd periodic extension. ▲

The reader is advised to solve each of a limited number of the End Chapter Exercises related to the Fourier series topic. The topic is especially important for comprehending the material of Chapter 5 where influence functions of a point force are derived and used in solving thin plate and shell problems.

## 1.4 Some Series Summation Formulae

Language of complex variables appears to be constructive in presenting a number of topics that are later covered in this text. The present section is designed to provide the reader with reference material on the basics of complex numbers. These are then used to derive some series summation formulae for a certain type of functional series in Chapter 5 dealing with problems of structural mechanics simulated with partial differential equations.

A complex number $z$ is defined in terms of a pair of real numbers. These are either the real part Re $z$ and the imaginary part Im $z$ that are used to present $z$ in the standard (algebraic) form

$$z = \text{Re}\, z + i\, \text{Im}\, z$$

or the modulus $r = |z|$ and the argument $\varphi = \arg z$ that are used in presenting $z$ in either exponential or trigonometric form

$$z = re^{i\varphi} = r(\cos\varphi + i\sin\varphi)$$

It is convenient to associate the complex number $z = a + ib$ with a point in a plane whose rectangular Cartesian coordinates are $a$ and $b$. The relations between Re $z$, Im $z$ and $r$, $\varphi$ for a complex number $z$ are given as

$$\text{Re}\, z = r\cos\varphi \quad \text{and} \quad \text{Im}\, z = r\sin\varphi$$

or

$$|z| = \sqrt{\text{Re}^2\, z + \text{Im}^2\, z} \quad \text{and} \quad \tan\varphi = \frac{\text{Im}\, z}{\text{Re}\, z}$$

Note that the quadrant containing the point corresponding to $z$ ought to be specified to solve the second of the above relations for $\varphi$, and the latter is defined up to an additive constant which is an integer multiple of $2\pi$.

The possibility of expressing a complex number in three alternative forms is convenient, because certain of these forms always appears to be more convenient, compared to the others, in performing a certain algebraic transformation. Say, in a transformation that requires a separation of real and imaginary parts, either the algebraic or the trigonometric form is obviously preferable, because these forms themselves are "real-imaginary parts separated". If, however, two complex numbers $z_1$ and $z_2$ are, for instance, to multiply or to divide, then the exponential form is preferable to deal with, because such operations are straightforward for this form. Indeed, for the product of $z_1$ and $z_2$ we have

$$z_1 z_2 = r_1 e^{i\varphi_1} r_2 e^{i\varphi_2} = r_1 r_2 e^{i(\varphi_1 + \varphi_2)}$$

while the quotient of $z_1$ and $z_2$ is expressed as

$$\frac{z_1}{z_2} = \frac{r_1 e^{i\varphi_1}}{r_2 e^{i\varphi_2}} = \frac{r_1}{r_2} e^{i(\varphi_1 - \varphi_2)}$$

An integer exponent $z^m$ of a complex number $z$ can be computed with the aid of the so-called DeMoivre's formula [60, 68] which can readily be derived by using the exponential and trigonometric representations. Indeed,

$$z^m = (re^{i\varphi})^m = r^m e^{im\varphi} = r^m(\cos m\varphi + i \sin m\varphi)$$

Taking radicals $\sqrt[n]{z}$ of complex numbers is not that simple because of their multi-valued feature. The DeMoivre's formula takes, in this case [60, 68], the form

$$\begin{aligned}(\sqrt[n]{z})_k &= \sqrt[n]{|z|}\left(\cos\frac{\varphi + 2(k-1)\pi}{n} + i \sin\frac{\varphi + 2(k-1)\pi}{n}\right) \\ &= \sqrt[n]{|z|}\exp\left(i\frac{\varphi + 2(k-1)\pi}{n}\right), \quad k = \overline{1, n} \end{aligned} \tag{1.35}$$

Application of this formula to a certain complex number is not a problem if values of the real and imaginary parts are given numerically. Indeed, all $n$ values of the $\sqrt[n]{z}$ can, in such a case, be obtained, at least, approximately from (1.35) in a straight manner. Whereas, computing values of the $n$-th root of a complex number whose real and imaginary parts are defined symbolically (in terms of parameters) requires some analytic effort. The following examples are designed to illustrate this point.

**Example 4.1:** To find all three values of the cube root of, say, $z = -7 + 3i$, we compute the modulus $r$ and the argument $\varphi$ of $z$

$$r \approx 7.6157731, \quad \varphi \approx 2.7367009$$

convert $z$ itself to the trigonometric form

$$z \approx 7.6157731[\cos(2.7367009) + i \sin(2.7367009)]$$

and present the values of the radical of $z$, as advised by eqn (1.35), as

$$\begin{aligned}(\sqrt[3]{z})_k \approx \sqrt[3]{7.6157731}\Bigg[&\cos\frac{2.7367009 + 2(k-1)\pi}{3} \\ &+ i \sin\frac{2.7367009 + 2(k-1)\pi}{3}\Bigg]\end{aligned}$$

This implies that

$$\begin{aligned}(\sqrt[3]{z})_1 &\approx 1.9674543[\cos(0.9122336) + i \sin(0.9122336)] \\ &= 1.2040442 + 1.5560058i,\end{aligned}$$

$$
\begin{aligned}
(\sqrt[3]{z})_2 &\approx 1.9674543[\cos(3.0066287) + i\,\sin(3.0066287)] \\
&= -1.9495626 + 0.2647300i,
\end{aligned}
$$

and

$$
\begin{aligned}
(\sqrt[3]{z})_3 &\approx 1.99674543[\cos(5.1010238) + i\,\sin(5.1010238)] \\
&= 0.7455184 - 1.8207358i
\end{aligned}
$$

▲

**Example 4.2:** Let us find values of the square root of a complex number $z = a + ib$ analytically. To be certain, let the real and imaginary parts of $z$ be non-negative numbers, $a \geq 0$ and $b \geq 0$, which implies that

$$0 \leq \arg z \leq \pi/2$$

Convert first $z$ itself to the trigonometric form

$$z = \sqrt{a^2 + b^2}\left[\cos\left(\arctan\frac{b}{a}\right) + i\,\sin\left(\arctan\frac{b}{a}\right)\right]$$

and find then the first value of the radical of $z$ as

$$(\sqrt{z})_1 = \sqrt[4]{a^2 + b^2}\left[\cos\left(\frac{1}{2}\arctan\frac{b}{a}\right) + i\,\sin\left(\frac{1}{2}\arctan\frac{b}{a}\right)\right]$$

which follows from eqn (1.35) if $k = 1$. The half-angle identities from trigonometry allow to rewrite the components in the brackets as

$$\cos\left(\frac{1}{2}\arctan\frac{b}{a}\right) = \frac{\sqrt{2}}{2}\sqrt{1 + \cos\left(\arctan\frac{b}{a}\right)}$$

and

$$\sin\left(\frac{1}{2}\arctan\frac{b}{a}\right) = \frac{\sqrt{2}}{2}\sqrt{1 - \cos\left(\arctan\frac{b}{a}\right)}$$

This transforms the above expression for $(\sqrt{z})_1$ to

$$(\sqrt{z})_1 = \sqrt[4]{a^2 + b^2}\frac{\sqrt{2}}{2}\left(\sqrt{1 + \cos\left(\arctan\frac{b}{a}\right)} + i\sqrt{1 - \cos\left(\arctan\frac{b}{a}\right)}\right)$$

Using basic trigonometric identities, we finally obtain

$$
\begin{aligned}
(\sqrt{z})_1 &= \sqrt[4]{a^2 + b^2}\frac{\sqrt{2}}{2}\left(\sqrt{1 + \frac{a}{\sqrt{a^2 + b^2}}} + i\sqrt{1 - \frac{a}{\sqrt{a^2 + b^2}}}\right) \\
&= \frac{\sqrt{2}}{2}\left(\sqrt{\sqrt{a^2 + b^2} + a} + i\sqrt{\sqrt{a^2 + b^2} - a}\right)
\end{aligned}
$$

For $k = 2$ from eqn (1.35) we similarly derive the second value

$$
\begin{aligned}
(\sqrt{z})_2 &= \sqrt[4]{a^2+b^2}\left[\cos\left(\frac{1}{2}\left(2\pi + \arctan\frac{b}{a}\right)\right) + i\,\sin\left(\frac{1}{2}\left(2\pi + \arctan\frac{b}{a}\right)\right)\right] \\
&= \sqrt[4]{a^2+b^2}\left[\cos\left(\pi + \frac{1}{2}\arctan\frac{b}{a}\right) + i\,\sin\left(\pi + \frac{1}{2}\arctan\frac{b}{a}\right)\right] \\
&= -\sqrt[4]{a^2+b^2}\left[\cos\left(\frac{1}{2}\arctan\frac{b}{a}\right) + i\,\sin\left(\frac{1}{2}\arctan\frac{b}{a}\right)\right] \\
&= -\sqrt[4]{a^2+b^2}\frac{\sqrt{2}}{2}\left(\sqrt{1+\cos\left(\arctan\frac{b}{a}\right)} + i\sqrt{1-\cos\left(\arctan\frac{b}{a}\right)}\right) \\
&= -\sqrt[4]{a^2+b^2}\frac{\sqrt{2}}{2}\left(\sqrt{1+\frac{a}{\sqrt{a^2+b^2}}} + i\sqrt{1-\frac{a}{\sqrt{a^2+b^2}}}\right) \\
&= -\frac{\sqrt{2}}{2}\left(\sqrt{\sqrt{a^2+b^2}+a} + i\sqrt{\sqrt{a^2+b^2}-a}\right)
\end{aligned}
$$

of the square root of a complex number $z = a + ib$ with non-negative real and imaginary parts. ▲

Later in this text, a special approach is developed and successfully used for the construction of influence functions of a point force for thin plates and shells. One of the key parts in the approach is either complete or partial summation of series expressing singular components of influence functions. In the present brief section, we derive some standard [1, 30] summation formulae that appeared to be useful in accomplishing the required summation.

The derivation is conducted by means of complex variables. This represents a typical approach widely used in mathematics, where often times the theory of complex variables is productively serving the real analysis.

To derive summation formulae that is required in later sections of this text, we recall the relation

$$\sum_{m=0}^{\infty} z^m = \frac{1}{1-z}, \quad \text{where } |z| < 1 \tag{1.36}$$

which represents a summation formula for the infinite geometric series whose first term equals 1 and the common ratio is $z$. This formula is valid not only for a real but for a complex variable $z$ as well.

Using the trigonometric form for $z$

$$z = r(\cos\varphi + i\,\sin\varphi) \tag{1.37}$$

and the DeMoivre's formula where $m$ is an integer

$$z^m = r^m(\cos m\varphi + i\,\sin m\varphi)$$

we rewrite the relation in (1.36) as

$$\sum_{m=0}^{\infty} r^m(\cos m\varphi + i \sin m\varphi) = \frac{1}{1 - r(\cos\varphi - i \sin\varphi)}$$

Eliminating the irrationality at the denominator in the right-hand side, we multiply the numerator and the denominator by the conjugate of the denominator. This yields

$$\sum_{m=0}^{\infty} r^m(\cos m\varphi + i \sin m\varphi) = \sum_{m=0}^{\infty} r^m \cos m\varphi + i \sum_{m=0}^{\infty} r^m \sin m\varphi$$
$$= \frac{(1 - r\cos\varphi) + ir\sin\varphi}{1 - 2r\cos\varphi + r^2}$$

When we equate the real and imaginary parts in the above relation, we obtain the following summation formulae

$$\sum_{m=0}^{\infty} r^m \cos m\varphi = \frac{1 - r\cos\varphi}{1 - 2r\cos\varphi + r^2} \tag{1.38}$$

and

$$\sum_{m=1}^{\infty} r^m \sin m\varphi = \frac{r\sin\varphi}{1 - 2r\cos\varphi + r^2} \tag{1.39}$$

that are valid for real variables $r$ and $\varphi$ satisfying the limitations $r^2 < 1$ and $0 \le \varphi < 2\pi$. Note that the summation in the series of eqn (1.39) should formally begin with $m = 0$, but, since the zero term in the series vanishes anyway, we can begin the summation with $m = 1$.

It is worth noting that the summation formulae that are just obtained in eqns (1.38) and (1.39) touch upon real-valued functions of real variables.

Some other summation formulae that are of great importance in our further developments will be derived in this section. In doing so, recall the summation formula in eqn (1.36). By integrating the latter, with the integral in the left-hand side being taken in the term-by-term fashion, one obtains

$$\sum_{m=0}^{\infty} \frac{z^{m+1}}{m+1} = -\ln(1 - z), \quad |z| < 1$$

When the summation index in this relation is changed as $m + 1 = n$, the above transforms to

$$\sum_{n=1}^{\infty} \frac{z^n}{n} = -\ln(1 - z), \quad |z| < 1 \tag{1.40}$$

Recall that, taking advantage of the exponential form of a complex variable $z$, a logarithmic function of $z$ can be written as

$$\ln z = \ln(re^{i\varphi}) = \ln|z| + i \arg z$$

The right-hand side in (1.40) can therefore be written as

$$-\ln(1-z) = -\ln|1-z| - i\arg(1-z) \tag{1.41}$$

where the modulus and the argument of $(1-z)$ are defined in terms of the modulus $r$ and the argument $\varphi$ of $z$ as

$$|1-z| = \sqrt{1-2r\cos\varphi + r^2}$$

and

$$\arg(1-z) = -\arctan\frac{r\sin\varphi}{1-r\cos\varphi}$$

allowing (1.41) to be rewritten as

$$-\ln(1-z) = -\ln\sqrt{1-2r\cos\varphi+r^2} + i\arctan\frac{r\sin\varphi}{1-r\cos\varphi}$$

So, the relation in (1.40) can be rewritten in the form

$$\sum_{n=1}^{\infty}\frac{r^n}{n}(\cos n\varphi + i\sin n\varphi) = -\ln\sqrt{1-2r\cos\varphi+r^2} + i\arctan\frac{r\sin\varphi}{1-r\cos\varphi}$$

Equating the real and the imaginary parts in the above, we arrive at another summation formulae

$$\sum_{n=1}^{\infty}\frac{r^n}{n}\cos n\varphi = -\ln\sqrt{1-2r\cos\varphi+r^2} \tag{1.42}$$

and

$$\sum_{n=1}^{\infty}\frac{r^n}{n}\sin n\varphi = \arctan\frac{r\sin\varphi}{1-r\cos\varphi} \tag{1.43}$$

Both of the above are valid for two real variables $r$ and $\varphi$ which must not necessarily be interpreted as polar coordinates of a point anymore. What is important however is that the domain for these real variables is preconditioned by the derivation procedure and is therefore defined as: $r^2 < 1$ and $0 \le \varphi < 2\pi$.

## 1.5 Numerical Solution of ODE

While exploring, in this text, the behavior of such structural elements as beams, plates and shells, we will be involved with three types of problem statements for ordinary differential equations. These are the *initial-value problems*, the *boundary-value problem* and the *eigenvalue problems*. In Section 1.1 the reader can find a brief description of each of these types of statements.

We will develop, in this text, a semi-analytical influence function method-based approach to these types of problems. The objective in this section is to focus on some of the simplest pure numerical methods (different of the influence function) that are traditionally [5, 6, 19, 60, 68] recommended. The reader will later be able to analyze the relative effectiveness of our semi-analytic approach by comparing its outcome against that of pure numerical schemes.

### 1.5.1 Initial-value problems

Let an initial-value problem for a first-order equation be stated as

$$\frac{dy(x)}{dx} = f(x, y(x)), \quad x_0 < x < X \tag{1.44}$$

$$y(x_0) = y_0 \tag{1.45}$$

Divide the interval $[x_0, X]$ into subintervals according to the partition

$$\{x_0 < x_1 < x_2 < \cdots < x_n = X\}$$

which is not necessarily uniform, but in most practical cases the users do prefer uniform partition, with the partition points $x_k$ determined as

$$x_k = x_{k-1} + h, \quad k = \overline{1, n} \tag{1.46}$$

where the constant $h = (X - x_0)/n$ is referred to herein as the *step of partition.*

To find a set of approximate values $y_k$ of the function $y(x)$ at the partition points $x_k$

$$y_k \approx y(x_k), \quad k = \overline{1, n} \tag{1.47}$$

we do not approach equation (1.44) directly but integrate it rather on each subinterval $[x_k, x_{k+1}]$ of the partition (1.46)

$$y(x) = \int_{x_k}^{x} f(x, y(x))\, dx + C_k, \quad k = \overline{0, n-1}$$

The constant $C_k$ can be found by fixing the upper limit of the integral at $x_k$. This yields

$$y(x) = y(x_k) + \int_{x_k}^{x} f(x, y(x))\, dx, \quad k = \overline{0, n-1}$$

from which it follows that

$$y(x_{k+1}) = y(x_k) + \int_{x_k}^{x_{k+1}} f(x, y(x))\, dx, \quad k = \overline{0, n-1} \tag{1.48}$$

Hence, (1.48) represents an implicit integral form which is equivalent to the initial-value problem of (1.44) and (1.45) on each subinterval $[x_k, x_{k+1}]$. Note that the values $y(x_{k+1})$ and $y(x_k)$ of the unknown function $y(x)$ are expressed in (1.48) in terms of the definite integral of the right-hand side function of equation (1.44) on the interval $[x_k, x_{k+1}]$.

At this stage, we approximate equation (1.48) by replacing $y(x_{k+1})$ and $y(x_k)$ with their approximate values $y_{k+1}$ and $y_k$, respectively, and by applying the rectangle rule to the integral term. This yields the recurrence relation

$$y_{k+1} = y_k + hf(x_k, y_k), \quad k = \overline{0, n-1} \tag{1.49}$$

which allows one to successively determine approximate values $y_1, y_2, \ldots, y_n$ of $y(x)$ starting with $x_0$ and $y_0$. This scheme is called the *Euler method* for the initial-value problem in (1.44) and (1.45). It is directly applicable to the initial-value problem

$$\frac{dy_j(x)}{dx} = f(x, y_1(x), y_2(x), \ldots, y_m(x)), \quad j = \overline{1, m}, \quad x_0 < x < X \quad (1.50)$$

$$y_j(x_0) = y_{j0}, \quad j = \overline{1, m}$$

for a normal system of $m$ first-order differential equations. The recurrence relation (1.49) applies to each equation of the system in (1.50) and reads in this case as

$$y_{j,k+1} = y_{j,k} + hf(x_k, y_{1,k}, y_{2,k}, \ldots, y_{m,k}), \quad j = \overline{1, m}, \quad k = \overline{0, n-1}$$

A notable drawback of the Euler method is its slow convergence. That is, one is forced to choose a really small step of partition $h$ in (1.46) to attain a required accuracy level. Another group of numerical methods, which converges at a much higher rate, can be recommended for initial-value problems. It is called the group of *Runge–Kutta methods* [5, 6, 19]. We present just the algorithm of one of the versions of this method (called the *fourth order* Runge–Kutta scheme), which suggests the following recurrence relation

$$y_{k+1} = y_k + \tfrac{1}{6}[r_1^{(k)} + 2r_2^{(k)} + 2r_3^{(k)} + r_4^{(k)}] \quad (1.51)$$

for the approximate values (1.47) of the solution to the initial-value problem in eqns (1.44) and (1.45) at the partition points (1.46). The parameters $r_i^{(k)}$, $(i = \overline{1, 4})$ in (1.51) are determined as

$$r_1^{(k)} = hf(x_k, y_k)$$

$$r_2^{(k)} = hf\left(x_k + \frac{h}{2}, y_k + \frac{r_1^{(k)}}{2}\right)$$

$$r_3^{(k)} = hf\left(x_k + \frac{h}{2}, y_k + \frac{r_2^{(k)}}{2}\right)$$

and

$$r_4^{(k)} = hf(x_k + h, y_k + r_3^{(k)})$$

Note that the fourth order Runge–Kutta scheme is very popular in engineering applications, because it combines a relatively simple algorithm with a high accuracy level attainable.

### 1.5.2 Boundary-value problems

A boundary-value problem for a high-order ordinary differential equation can potentially be solved by the so-called *shooting method* [2, 15, 23, 68]. That is when a well-established step-by-step method, developed originally for initial-value problems, is employed to solve a boundary-value problem. We can think of such as either the Euler, or the Runge–Kutta, or some other step-by-step method.

To briefly review this technique, we take as a sample the second-order boundary-value problem

$$F\left(x, y, \frac{dy}{dx}, \frac{d^2y}{dx^2}\right) = 0, \quad a < x < b \tag{1.52}$$

$$y(a) = y_a, \quad y(b) = y_b \tag{1.53}$$

Instead of directly tackling this problem, we state, for the governing equation in (1.52), conditions of uniqueness in the form

$$y(a) = y_a, \quad \frac{dy(a)}{dx} = Y_a \tag{1.54}$$

which are initial conditions, since both of them are imposed at $x = a$. The parameter $Y_a$ in (1.54) is chosen by guess. By carrying out a step-by-step numerical procedure for the solution of the initial-value problem in (1.52) and (1.54) as far as $x = b$, we are able to compute an approximate value $y(b)$ which, unlikely, "hits the target" of $y_b$ (see the second condition in (1.53)). Repeating, however, this procedure, with different values of $Y_a$, until the target is hit with a required accuracy level, we do approximately solve the boundary-value problem in (1.52) and (1.53).

The shooting technique is especially productive in the case of a linear differential equation, where the information from just a few trials of the described procedure can instruct on a choice of $Y_a$ that allows us "to hit the target". For nonlinear equations the procedure could be too expensive computationally.

We turn now to an approach to a two-point boundary-value problem, which does not utilize the initial-value problem-targeted techniques. The *finite difference method* [2, 5, 6, 19, 32, 60, 68] procedure will be reviewed, which is driven by a completely different strategy.

To be specific, we consider a problem from structural mechanics. Let an elastic beam of length $a$ (with $EI = const$ representing the flexural rigidity) be loaded with a tensile force $S$ and a transverse distributed force $q(x)$. Let also the edge $x = 0$ be clamped while the edge $x = a$ simply-supported. The deflection $w(x)$ of such a beam is, in compliance with the classical Kirchhoff theory [12, 27, 63], modelled with the boundary-value problem

$$\frac{d^4w(x)}{dx^4} - \frac{S}{EI}\frac{d^2w(x)}{dx^2} = q(x) \tag{1.55}$$

$$w(0) = \frac{dw(0)}{dx} = 0, \quad w(a) = \frac{d^2w(a)}{dx^2} = 0 \tag{1.56}$$

Use a uniform partition

$$\{0 = x_0 < x_1 < x_2 < \cdots < x_n = a\}$$

of the interval $[0, a]$, where the partition points $x_k$ are determined as

$$x_k = x_{k-1} + h, \quad k = \overline{1, n} \tag{1.57}$$

and approximate the problem in (1.55) and (1.56) by finite differences. In doing so, recall two Taylor series expansions of $w(x)$ about a partition point $x_k$

$$w(x_{k+1}) = w(x_k) + w'(x_k)\frac{h}{1!} + w''(x_k)\frac{h^2}{2!} + w'''(x_k)\frac{h^3}{3!} + w^{IV}(x_k)\frac{h^4}{4!} + \cdots \tag{1.58}$$

and

$$w(x_{k-1}) = w(x_k) - w'(x_k)\frac{h}{1!} + w''(x_k)\frac{h^2}{2!} - w'''(x_k)\frac{h^3}{3!} + w^{IV}(x_k)\frac{h^4}{4!} - \cdots \tag{1.59}$$

If all the terms involving $h^2$ and higher are ignored in the expansions of (1.58) and (1.59), then the first-order derivative of $w(x)$ can be approximated at a partition (1.57) point $x_k$ from the expansion of (1.58) as

$$\frac{dw(x_k)}{dx} \approx \frac{1}{h}[w(x_{k+1}) - w(x_k)] \tag{1.60}$$

referred to as the *forward approximation*, or from the expansion of (1.59)

$$\frac{dw(x_k)}{dx} \approx \frac{1}{h}[w(x_k) - w(x_{k-1})] \tag{1.61}$$

called the *backward approximation*, or from both (1.58) and (1.59)

$$\frac{dw(x_k)}{dx} \approx \frac{1}{2h}[w(x_{k+1}) - w(x_{k-1})] \tag{1.62}$$

which is referred to as the *central approximation*. At the same time, if all the terms involving $h^3$ and higher are ignored in (1.58) and (1.59), then an approximate expression for the second-order derivative of $w(x)$ at $x_k$

$$\frac{d^2w(x_k)}{dx^2} \approx \frac{1}{h^2}[w(x_{k+1}) - 2w(x_k) + w(x_{k-1})] \tag{1.63}$$

can be obtained by adding the expansions of (1.58) and (1.59).

To approximate the higher-order derivatives in the problem of eqns (1.55) and (1.56), we recall the two-step Taylor expansions

$$\begin{aligned} w(x_{k+2}) = w(x_k) &+ w'(x_k)\frac{2h}{1!} + w''(x_k)\frac{(2h)^2}{2!} \\ &+ w'''(x_k)\frac{(2h)^3}{3!} + w^{IV}(x_k)\frac{(2h)^4}{4!} + \cdots \end{aligned} \tag{1.64}$$

and

$$\begin{aligned} w(x_{k-2}) = w(x_k) &- w'(x_k)\frac{2h}{1!} + w''(x_k)\frac{(2h)^2}{2!} \\ &- w'''(x_k)\frac{(2h)^3}{3!} + w^{IV}(x_k)\frac{(2h)^4}{4!} - \cdots \end{aligned} \tag{1.65}$$

A finite difference approximation of the fourth-order derivative of $w(x)$ at $x_k$ can be written if we ignore all the terms involving $h^5$ and higher in (1.58), (1.59),

(1.64) and (1.65), multiply (1.58) and (1.59) by a factor of "−4 ", and add then all four of them together. This yields

$$\frac{d^4w(x_k)}{dx^4} \approx \frac{1}{h^4}[w(x_{k+2}) - 4w(x_{k+1}) + 6w(x_k) - 4w(x_{k-1}) + w(x_{k-2})] \tag{1.66}$$

Replacing now the derivatives of the governing equation in (1.55) with their approximate expressions from eqns (1.63) and (1.66) at a partition point $x_k$, for $k = \overline{2, n-2}$, and combining the like terms, we come up with $n-3$ linear algebraic equations

$$\begin{aligned} w(x_{k+2}) + Aw(x_{k+1}) + Bw(x_k) + Aw(x_{k-1}) \\ + w(x_{k-2}) = h^4q(x_k), \quad k = \overline{2, n-2} \end{aligned} \tag{1.67}$$

in $n+1$ unknowns $w(x_k)$, for $k = \overline{0, n}$, where the parameters $A$ and $B$ are defined as

$$A = -\left(4 + \frac{Sh^2}{EI}\right), \quad B = 2\left(3 + \frac{Sh^2}{EI}\right)$$

It is evident that the system in (1.67) is ill-posed because the number of equations $(n-3)$ in it is four fewer than the number of unknowns which is $(n+1)$. This deficiency can, however, be evaded by adding to the system another four finite difference equations when the boundary conditions in (1.56) are accordingly approximated. Indeed, the conditions at $x = 0$ in (1.56) imply

$$w(x_0) = 0 \quad \text{and} \quad w(x_1) = 0$$

while the conditions at $x = a$ yield

$$w(x_n) = 0 \quad \text{and} \quad w(x_{n-2}) - 2w(x_{n-1}) = 0$$

where the second relation follows from (1.60) with $k = n-1$. Hence, a well-posed system of linear algebraic equations is eventually obtained, from which we compute approximate values $w(x_k)$ of the solution to the boundary-value problem of eqns (1.55) and (1.56) at the partition points.

Note that the finite difference method can potentially be applied to a nonlinear boundary-value problem as well. This yields, however, a system of nonlinear algebraic equations in $w(x_k)$, whose numerical solution is seldom easy.

### 1.5.3 Eigenvalue problems

In Section 1.1 we found out that eigenvalue problems represent a special class of boundary-value problems for differential equations. In structural mechanics they arise when either natural vibrations are examined of a structural element or a buckling phenomenon is analyzed. These problems also emerge as a part of the *separation of variables* method [15, 20, 32, 53, 60, 61, 68] in solving partial differential equations simulating behavior of structural elements.

As an example, consider the sample eigenvalue problem

$$\frac{d^2y(x)}{dx^2} + \lambda^2 y(x) = 0, \quad 0 < x < a \tag{1.68}$$

$$y(0) = 0, \quad y(a) = 0 \tag{1.69}$$

whose eigenvalues $\lambda_n$ and corresponding eigenfunctions $y_n(x)$ were presented in Section 1.1 as

$$\lambda_n = \frac{n\pi}{a}, \qquad y_n(x) = C \sin\frac{n\pi x}{a}, \qquad n = 1, 2, 3, \ldots$$

where $C$ is an arbitrary constant.

We are going to use the problem in eqns (1.68) and (1.69) to describe the finite difference method-based procedure that might be applied to a variety of other eigenvalue problems arising in structural mechanics. For the sake of simplicity, use a uniform partition

$$\{0 = x_0 < x_1 < x_2 < \cdots < x_n = a\}$$

of the interval $[0, a]$, where the partition points $x_k$ are determined as

$$x_k = x_{k-1} + h, \quad k = \overline{1, n} \tag{1.70}$$

and approximate the problem in (1.68) and (1.69) with finite differences. In doing so, recall the approximate expression for the second-order derivative of $y(x)$ from eqn (1.63) and approximate the governing equation in (1.68) at the interior partition points $x_k$, for $k = \overline{1, n-1}$, with the homogeneous system of linear algebraic equations

$$\frac{1}{h^2}[y(x_{k+1}) - 2y(x_k) + y(x_{k-1})] + \lambda^2 y(x_k) = 0, \quad k = \overline{1, n-1}$$

or

$$-y(x_{k+1}) + (2 - \Lambda)y(x_k) - y(x_{k-1}) = 0, \quad \Lambda = (\lambda h)^2 \tag{1.71}$$

in $y(x_k)$, $(k = \overline{0, n+1})$, where $\Lambda = (\lambda h)^2$. It looks like this system is ill-posed, because the number of equations $(n - 1)$ is fewer than the number of unknowns $(n + 1)$. However, in compliance with the boundary conditions of eqn (1.69), two of those unknowns are prescribed. Indeed, we have

$$y(x_0) = y(0) = 0 \quad \text{and} \quad y(x_n) = y(a) = 0$$

which makes the system in (1.71) well-posed (the total of $n - 1$ equations in the same number of unknowns). The system can be written in a matrix form as

$$\begin{pmatrix} 2-\Lambda & -1 & 0 & \ldots & 0 & 0 \\ -1 & 2-\Lambda & -1 & \ldots & 0 & 0 \\ \ldots & \ldots & \ldots & \ldots & \ldots & \ldots \\ 0 & 0 & 0 & \ldots & 2-\Lambda & -1 \\ 0 & 0 & 0 & \ldots & -1 & 2-\Lambda \end{pmatrix} \times \begin{pmatrix} y(x_1) \\ y(x_2) \\ \ldots \\ y(x_{n-2}) \\ y(x_{n-1}) \end{pmatrix} = \begin{pmatrix} 0 \\ 0 \\ \ldots \\ 0 \\ 0 \end{pmatrix}$$

representing, in linear algebra [60, 67], a standard eigenvalue problem for the matrix

$$\begin{pmatrix} 2 & -1 & 0 & \dots & 0 & 0 \\ -1 & 2 & -1 & \dots & 0 & 0 \\ \dots & \dots & \dots & \dots & \dots & \dots \\ 0 & 0 & 0 & \dots & 2 & -1 \\ 0 & 0 & 0 & \dots & -1 & 2 \end{pmatrix}$$

Hence, the finite difference method converts the problem in (1.68) and (1.69) to the eigenvalue problem of linear algebra, which can readily be attacked by standard computer routines.

The finite difference method-based approach can be applied to eigenvalue problems for equations of higher order. Consider, as an example, the homogeneous boundary-value problem

$$\frac{d^4 w(x)}{dx^4} - \lambda^4 w(x) = 0 \tag{1.72}$$

$$w(0) = \frac{dw(0)}{dx} = 0, \quad w(a) = \frac{dw(a)}{dx} = 0 \tag{1.73}$$

which arises when transverse natural vibrations are explored for a clamped at both edges Kirchhoff elastic beam of a uniform flexural rigidity.

The statement in eqns (1.72) and (1.73) is really an eigenvalue problem. Upon utilizing the uniform partition of the interval $[0, a]$ as of eqn (1.70) and replacing the fourth-order derivative in (1.72) with the approximate expression from eqn (1.66) at a partition point $x_k$, for $k = \overline{2, n-2}$, we obtain a system of $n - 3$ linear algebraic equations

$$\begin{aligned} & w(x_{k+2}) - 4w(x_{k+1}) + [6 - (\lambda h)^4] w(x_k) \\ & - 4w(x_{k-1}) + w(x_{k-2}) = 0, \quad k = \overline{2, n-2} \end{aligned}$$

in $n + 1$ unknowns $w(x_k)$, for $k = \overline{0, n}$.

The deficiency in the above system can be eliminated with the aid of the boundary conditions of eqn (1.73). This reduces the system to a well-posed form, with the coefficient matrix expressed as

$$\begin{pmatrix} 6-\Lambda & -4 & 1 & 0 & 0 & \dots & 0 & 0 & 0 & 0 & 0 \\ -4 & 6-\Lambda & -4 & 1 & 0 & \dots & 0 & 0 & 0 & 0 & 0 \\ 1 & -4 & 6-\Lambda & -4 & 1 & \dots & 0 & 0 & 0 & 0 & 0 \\ \dots & \dots & \dots & \dots & \dots & \dots & \dots & \dots & \dots & \dots & \dots \\ 0 & 0 & 0 & 0 & 0 & \dots & 1 & -4 & 6-\Lambda & -4 & 1 \\ 0 & 0 & 0 & 0 & 0 & \dots & 0 & 1 & -4 & 6-\Lambda & -4 \\ 0 & 0 & 0 & 0 & 0 & \dots & 0 & 0 & 1 & -4 & 6-\Lambda \end{pmatrix}$$

So, the finite difference method-based approach converts the problem in eqns (1.72) and (1.73) to a standard eigenvalue problem of linear algebra.

## 1.6 Introduction to Integral Equations

The reader will find, in this text, some other types of functional equations that are different of *differential*. We will briefly review one of those types called integral equations. A functional equation containing an unknown function under the integral sign is said to be the *integral equation*. As an example consider

$$y(x) + \lambda \int_a^b K(x, s) y(s)\, ds = f(x), \quad a \leq x \leq b \tag{1.74}$$

where $K(x, s)$ and $f(x)$ are given functions, $\lambda$ is a given parameter, while $y(x)$ represents the unknown function. Other examples of integral equations are

$$y(x) + \lambda \int_a^b K(x, s) y(s)\, ds = 0, \quad a \leq x \leq b \tag{1.75}$$

or

$$y(x) + \lambda \int_a^x K(x, s) y(s)\, ds = f(x), \quad a \leq x \leq b \tag{1.76}$$

or

$$y(x) + \lambda \int_a^x K(x, s) y(s)\, ds = 0, \quad a \leq x \leq b \tag{1.77}$$

or

$$\int_a^b K(x, s) y(s)\, ds = f(x), \quad a \leq x \leq b \tag{1.78}$$

Notice that all the above integral equations are *linear*. The function $K(x, s)$ is called the *kernel*, $f(x)$ is called the *right-hand side*.

Nonlinear integral equations like, for example,

$$y(x) + \int_a^x K(x, s, y(s))\, ds = f(x), \quad a \leq x \leq b$$

also arise in applications. In most cases of this text, we deal with linear equations. However, nonlinear ones are also considered in Chapter 4, where one of the influence function approaches is applied to a beam buckling problem.

Linear integral equations are often classified [60]. If the kernel $K(x, s)$ is bounded, then the equation is said to be *regular*. A linear regular integral equation with fixed limits of integration (as in (1.74), or (1.75), or (1.78)) belongs to the *Fredholm* type, named after Swedish mathematician E.I. Fredholm (1866–1927). If the limits of integration are variable (as in (1.76) or (1.77)), then we say the equation is of the *Volterra* type (Italian mathematician V. Volterra (1860–1940)). Equations (1.74), (1.76) and (1.78) are said to be *nonhomogeneous*, whereas equations (1.75) and (1.77), where the right-hand side function is zero, are called *homogeneous*. If the unknown function is contained only under the integral sign (like in (1.78)), then the equation is of the *first kind*, otherwise it belongs to the *second kind* (as in (1.74)–(1.77)).

Hence, classifying the equation in (1.74), for example, we call it a *nonhomogeneous Fredholm integral equation of the second kind.*

Often in applications, the kernel of integral equation appears to be unbounded. Such equations are called *singular*. Singular integral equations play an essential role in the boundary element method [7, 16] which has recently risen to the level of one of the most applicable numerical techniques in engineering and science.

Volterra equation in (1.76) can formally be considered as a particular case of Fredholm equation in (1.74). Indeed, the integral with variable upper limit in (1.76) can be replaced with a definite integral from $a$ to $b$ if the kernel $K(x, s)$ is formally replaced with the function

$$\widetilde{K}(x, s) = \begin{cases} K(x, s), & \text{for } s \leq x \\ 0, & \text{for } s > x \end{cases}$$

which is defined in two pieces. Notice, however, that physical phenomena that yield Fredholm and Volterra equations, as well as properties of solutions of these equations, are very different of each other. That is why Fredholm and Volterra equations are usually [60] considered separately.

It is worth noting that equation (1.75) has an evident trivial solution. But since the parameter $\lambda$ in it is not fixed, this equation represents an eigenvalue problem, in which we look for such values of $\lambda$ (we call them *eigenvalues* of $K(x, s)$) that deliver nontrivial solutions (*eigenfunctions* of $K(x, s)$) to (1.75).

Note that integral equations represent an independent topic in mathematics. These equations may directly arise as mathematical models for some phenomena or processes in physics. They may also emerge from differential equation-based problem statements in mathematics. As an illustration to the latter point, consider, for example, the initial-value problem

$$\frac{d^n z(x)}{dx^n} + p_1(x)\frac{d^{n-1}z(x)}{dx^{n-1}} + p_2(x)\frac{d^{n-2}z(x)}{dx^{n-2}} + \cdots + p_n(x)z(x) = f(x) \tag{1.79}$$

$$z(a) = 0, \quad \frac{dz(a)}{dx} = 0, \ldots, \frac{d^{n-1}z(a)}{dx^{n-1}} = 0 \tag{1.80}$$

where $z(x)$ is a function to find. Introducing a new unknown function $y(x)$ expressed in terms of $z(x)$ by the integral relation

$$z(x) = \frac{1}{(n-1)!}\int_a^x y(s)(x-s)^{n-1}\,ds \tag{1.81}$$

and successively differentiating this relation $n - 1$ times with respect to $x$, we obtain

$$\frac{d^k z(x)}{dx^k} = \frac{1}{(n-1-k)!}\int_a^x y(s)(x-s)^{n-1-k}\,ds, \quad 1 \leq k \leq n-1 \tag{1.82}$$

Clearly, for $k = n - 1$, the above relation reads

$$\frac{d^{n-1}z(x)}{dx^{n-1}} = \int_a^x y(s)\,ds$$

implying that

$$\frac{d^n z(x)}{dx^n} = y(x) \tag{1.83}$$

The relations in (1.81) and (1.82) evidently suggest that the initial conditions in (1.80) are satisfied. Substituting the relations from (1.81)–(1.83) in (1.79) and combining all the integral terms we finally obtain

$$y(x) + \int_a^x K(x, s)y(s)\,ds = f(x) \tag{1.84}$$

where the kernel function is defined as

$$K(x, s) = p_1(x) + p_2(x)\frac{x - s}{1!} + \cdots + p_n(x)\frac{(x - s)^{n-1}}{(n - 1)!}$$

Thus, the initial-value problem stated by eqns (1.79) and (1.80) reduces to the nonhomogeneous Volterra integral equation of the second kind (1.84). This text brings a number of differential equation-stated problems in structural mechanics that reduce to integral equations. In Chapter 4, for example, the reader can find integral equations that arise from the use of the influence function approach to some linear elastic beam problems.

Analytic solution of integral equations is rarely possible. This means that, in many cases, numerical approaches represent the only choice with which we are left. One of the most widely used numerical methods (which is, by the way, applied to integral equations that are treated in this text) is based on an approximation of the definite integral with a finite sum [60]. Consider, for example, equation (1.74) and partition the interval of integration $[a, b]$ in, say, uniform manner as

$$s_k = s_0 + k\frac{b - a}{n}, \qquad k = \overline{0, n}$$

with $s_0 = a$ and $s_n = b$. This allows the integral term in (1.74) to be approximated with the aid of a quadrature formula in the form

$$\int_a^b K(x, s)y(s)\,ds \approx \sum_{i=0}^{n} A_i K(x, s_i)y_i$$

where $A_i$ are coefficients of a quadrature formula applied, while $y_i$ represent approximate values of $y(x)$ at $x = s_i$. Fixing now the variable $x$ in eqn (1.74) at $x = s_j$, with $j = \overline{0, n}$, we arrive at the system of linear algebraic equations

$$y_j + \lambda\sum_{i=0}^{n} A_i K(x_j, s_i)y_i = f(x_j), \qquad j = \overline{0, n}$$

in approximate values $y_i$ of the solution to (1.74).

The choice of the quadrature formula and location of the quadrature (partition) points $s_i$ depend upon a variety of circumstances and is usually made by taking into account the accuracy level required for a particular integral equation. In most engineering applications, such elementary quadrature formulae as the trapezoidal rule or Simpson's rule are quite satisfactory. If the accuracy requirements are, however, tougher, then the highly accurate Gaussian quadrature formulae can be applied.

## 1.7 End Chapter Exercises

**1.1** Determine whether the following functions are linearly independent on the indicated interval:

**a)** $3x + 1$, $2 - x$, and 5 for $x \in (-1, 1)$;
**b)** 1, $\ln x$, $\ln 2x$, and $\ln 3x$ for $x \in (0, \infty)$;
**c)** $e^x$, $e^{2x}$, $e^{3x}$ and $e^{4x}$ for $x \in (-\infty, \infty)$;
**d)** $xe^x$, $x^2e^x$, and $x^3e^x$ for $x \in (-\infty, \infty)$;
**e)** $e^{-3x}\sin 2x$, and $e^{-3x}\cos 2x$ for $x \in (-\infty, \infty)$;
**f)** $|x + 5|$, $x - 1$, and $5 - 4x$ for $x \in (-5, 5)$;
**g)** 1, $x^{-1}$, $x^{-2}$ and $x^{-3}$ for $x \in (0, \infty)$;
**h)** $2x^2 + x - 1$, $x^2 + 1$, and $3x - 2$ for $x \in (-1, 1)$.

**1.2** Find a fundamental set of solutions for the following linear homogeneous differential equations:

**a)** $y''(x) + 4y'(x) = 0$;
**b)** $y''(x) + 2y'(x) - 8y(x) = 0$;
**c)** $y''(x) + 3y'(x) + 5y(x) = 0$;
**d)** $y'''(x) - 9y'(x) = 0$;
**e)** $y'''(x) - 2y''(x) - 15y'(x) = 0$;
**f)** $y'''(x) - 3y''(x) + 2y(x) = 0$;
**g)** $y^{IV}(x) - y(x) = 0$;
**h)** $y^{IV}(x) + y''(x) - 12y(x) = 0$.

**1.3** Decide the method and find a particular solution for the following linear nonhomogeneous equations:

**a)** $y''(x) = 4x^{-2}$;
**b)** $y''(x) - 4y(x) = 3\cos 2x$;
**c)** $y''(x) + 16y(x) = 2e^{-3x}$;
**d)** $y'''(x) - 3y''(x) + 2y(x) = -\sin^3 x$;

**e)** $y'''(x) - 2y''(x) - 5y'(x) = 2x^2 \sin x + 3e^{-x} \cos x$;
**f)** $y^{IV}(x) - 4y''(x) = 2x^2 - 3x + 7$;
**g)** $y^{IV}(x) + y''(x) = x^2 - 4x$.

**1.4** Determine whether the following boundary-value problems have only the trivial solution:

**a)** $y''(x) = 0$, with $y'(0) = 0$ and $y'(a) + my(a) = 0$;
**b)** $y''(x) - k^2 y(x) = 0$, with $y(0) = 0$ and $|y(\infty)| < \infty$;
**c)** $y''(x) - k^2 y(x) = 0$, with $y(0) = y(a)$ and $y'(0) = y'(a)$;
**d)** $((mx + p)y'(x))' = 0$, with $y'(0) = 0$ and $y(a) = 0$;
**e)** $(xy'(x))' = 0$, with $|y(0)| < \infty$ and $y'(a) + hy(a) = 0$;
**f)** $y''(x) + y'(x) - 2y(x) = 0$, with $y(0) = 0$ and $|y(\infty)| < \infty$;
**g)** $y''(x) + y'(x) = 0$, with $y'(0) = 0$ and $y'(a) = 0$;
**h)** $y^{IV}(x) = 0$, with $y(0) = y''(0) = y''(a) = y'''(a) = 0$.

**1.5** For the following functions, obtain the Fourier series:

**a)** $f(x) = e^{2x}$, with $-1 < x < 1$;
**b)** $f(x) = \cos^8 x$, with $-\pi < x < \pi$;
**c)** $f(x) = \begin{cases} 1, & \text{if } -\pi < x < 0 \\ x, & \text{if } 0 < x < \pi. \end{cases}$

**1.6** For the following functions, obtain Fourier cosine and sine series:

**a)** $f(x) = e^x - 1$, with $0 < x < \pi$;
**b)** $f(x) = x - x^2$, with $0 < x < 2$.

**1.7** Find all values of the radicals as indicated:

**a)** $\sqrt[3]{8 - 6i}$; **b)** $\sqrt{-5 - 4i}$; **c)** $\sqrt[4]{-4 + 5i}$.

# Chapter 2

# Green's Functions

The concept of Green's function is overwhelmingly the most important mathematical issue in this volume. It supports the keystone notion of influence function of a point force which is actually what all this text is about. The author thinks that presenting the Green's function topic within a separate chapter is right from the methodology point of view. It helps the reader in comprehending the significance of this topic for the entire text.

Two alternative approaches are drawn and described in detail in this text, as they are commonly used in the existing literature [15, 25, 43, 45, 60] for the construction of Green's functions for ordinary differential equations. One of these approaches, which we focus on in Section 2.1, flows down from the proof of existence and uniqueness theorem for the Green's function. It is constructive by nature and uses defining properties of this function. Another approach is under discussion in Section 2.3. It uses the fact that a solution of a nonhomogeneous equation can be written in terms of Green's function and utilizes the classical method of variation of parameters. Practicality of the two approaches for various types of problem settings is illustrated through numerous examples.

A special feature of Green's functions goes with an issue that is closely examined in Section 2.2. That is the relation between the symmetry of these functions with respect to the observation point and the source point and the so-called self-adjointness of the differential equation involved. This feature of Green's functions is of a great theoretical and practical importance.

Section 2.4 extends the notion of Green's function to the so-called *multi-point posed* boundary-contact value problems for specific sets of linear ordinary differential equations. These are not, however, systems of equations in a common sense, because each of the involved equations governs a single unknown function, each of which is defined over an interval that represents a subinterval of the basic interval in the problem. The system is formed by imposing contact conditions at the end-points of the subintervals. Such an extension gives birth to a new notion that is refereed to in this text as the *matrix of Green's type*.

A generalization of the material of Section 2.4 is provided in Section 2.5, where sets of linear ordinary differential equations with individual domains are also considered. But, in contrast to the settings in Section 2.4, every governing equation is posed on a single edge of a finite weighted graph. Continuity conditions and boundary conditions imposed at the vertices and the end-points of the graph, respectively, arrange the setting into a specific system of differential equations. Matrices of Green's type for the latter are constructed with the method proposed in Section 2.4.

## 2.1 Construction Based on Defining Properties

The notion of Green's function, named after the brilliant British mathematician George Green (1793–1841), will be introduced in this section to a linear boundary-value problem stated for an ordinary linear differential equation of the $n$-th order. We then provide a detailed description of a traditional method for the construction of Green's functions, which is based on their defining properties. A number of examples presented here highlight different aspects of the derivation procedure and are related to various problems of continuum mechanics.

The discussion that follows is concerned with a linear homogeneous boundary-value problem for the ordinary differential equation

$$L[y(x)] \equiv p_0(x)\frac{d^n y}{dx^n} + p_1(x)\frac{d^{n-1} y}{dx^{n-1}} + \cdots + p_n(x)y = 0 \qquad (2.1)$$

posed on an interval $(a, b)$. In this setting, the coefficients $p_i(x)$, $(i = \overline{0, n})$ are continuous functions on $(a, b)$, where the leading coefficient is supposed to be non-zero, $p_0(x) \neq 0$.

The governing differential equation is subject to uniqueness (boundary) conditions imposed as

$$M_i(y(a), y(b)) \equiv \sum_{k=0}^{n-1}\left[\alpha_k^i \frac{d^k y(a)}{dx^k} + \beta_k^i \frac{d^k y(b)}{dx^k}\right] = 0, \quad (i = \overline{1, n}) \qquad (2.2)$$

A brief form of the boundary conditions in the above equation is informative but, probably, requires a special explanation. The total number of conditions in (2.2) is $n$, and it is important to note that the superscript $i$ on $\alpha_k^i$ and $\beta_k^i$ is not an exponent but rather represents just a superscript. The relations in (2.2) are written in a *two-point form* where each of them involves both end-points $a$ and $b$ of the interval. This generalization allows such specific uniqueness conditions as conditions of symmetry or periodicity to be also included in this form. At the same time, if a certain uniqueness condition in the setting is in a *single-point* form, or, in other words, is imposed at, say, $a$ only, then all the coefficients $\beta_k^i$ in (2.2) are zero, while at least one of the coefficients $\alpha_k^i \neq 0$. Similarly, if a certain condition is imposed at $b$ only, then all the coefficients $\alpha_k^i$ are zero, while at least one of the coefficients $\beta_k^i$ is non-zero.

Note that most of boundary-value problem settings that are usually considered in structural mechanics are single-point. Although, two-point settings are also not strangers in reality (see, for example, the conditions in eqn (2.23) that are imposed in Example 1.5 of this Section).

The forms $M_i$, $(i = \overline{1, n})$ represent in (2.2) linearly independent forms with constant coefficients $\alpha_k^i$ and $\beta_k^i$. Note that in each of $n$ boundary conditions in eqn (2.2), at least one of the coefficients $\alpha_k^i$ and $\beta_k^i$ is not zero. This implies that none of the boundary conditions degenerates or, in other words, we have exactly $n$ conditions imposed ensuring that the formulation in eqns (2.1) and (2.2) is not ill-posed.

The boundary conditions in eqn (2.2) are written in a general form, which implies that a certain physically feasible formulation can be obtained from this form as a particular case. If, for example, the displacement formulation of a problem from the theory of elasticity is considered (beam, plate, shell problems, etc.), then the condition in eqn (2.2) may model either clamped, or simply supported, or free, or even an elastically supported edge.

In this text, we will be using conventional and customary interval notations, where $(a, b)$, $[a, b]$, and $(a, b]$ or $[a, b)$ specify open, closed, and half-open intervals, respectively.

### 2.1.1 Existence and uniqueness

At this point we turn the reader's attention to one of the key issues of this text. Let us define the *Green's function* for the homogeneous boundary-value problem that appears in eqns (2.1) and (2.2).

**Definition:** The function $g(x, s)$ is said to be the *Green's function* for the boundary-value problem in eqns (2.1) and (2.2), if as a function of its first variable $x$, it meets the following defining properties, for any $s \in (a, b)$:

1. On both of the intervals $[a, s)$ and $(s, b]$, $g(x, s)$ is a continuous function having continuous derivatives of up to the $n$-th order included, and satisfies the governing equation (2.1) on $(a, s)$ and $(s, b)$, i.e.:

$$L[g(x, s)] = 0, \;\; x \in (a, s); \quad L[g(x, s)] = 0, \;\; x \in (s, b)$$

2. For $x = s$, $g(x, s)$ is continuous along with all its derivatives of up to $(n-2)$ order included

$$\lim_{x \to s^+} \frac{\partial^k g(x, s)}{\partial x^k} - \lim_{x \to s^-} \frac{\partial^k g(x, s)}{\partial x^k} = 0, \quad (k = \overline{0, n-2})$$

3. The $(n-1)$-th derivative of $g(x, s)$ is discontinuous when $x = s$, providing

$$\lim_{x \to s^+} \frac{\partial^{n-1} g(x, s)}{\partial x^{n-1}} - \lim_{x \to s^-} \frac{\partial^{n-1} g(x, s)}{\partial x^{n-1}} = -\frac{1}{p_0(s)}$$

   where $p_0(s)$ represents the leading coefficient of eqn (2.1);
4. $g(x, s)$ satisfies the boundary conditions in eqn (2.2), i.e.:

$$M_i(g(a, s), g(b, s)) = 0, \quad (i = \overline{1, n})$$

The arguments $x$ and $s$ in the Green's function are conventionally referred to as the *observation (field) point* and the *source point*, respectively. The following theorem is valid specifying the existence and uniqueness conditions for the Green's function.

**Theorem 2.1 (of existence and uniqueness):** If the homogeneous boundary-value problem in eqns (2.1) and (2.2) has only the trivial (zero) solution, then there exists its unique Green's function $g(x, s)$.

We suggest the reader carefully read through the proof below, because it provides a straightforward algorithm for the actual construction of Green's functions. Throughout the present text, we will be frequently using this algorithm for a variety of settings in applied mechanics.

**Proof:** Let functions $y_j(x)$, $(j = \overline{1, n})$ constitute a fundamental set of solutions for eqn (2.1). That is, $y_j(x)$ are linearly independent on $(a, b)$ particular solutions of eqn (2.1).

In numerous practical situations, one can find an analytic form for $y_j(x)$. This can, in particular, be easily done for equations with constant coefficients. If, however, the governing differential equation does not allow an analytical solution, then appropriate numerical procedures may be employed for obtaining approximate ones. Later in this book we will discuss this point in more detail.

In compliance with property 1 of the definition, for any arbitrarily fixed value of $s \in (a, b)$, the Green's function $g(x, s)$ ought to be a solution of eqn (2.1) in $(a, s)$ (on the left of $s$), as well as in $(s, b)$ (on the right of $s$). Since any solution of eqn (2.1) can be expressed as a linear combination of the components $y_j(x)$ of the fundamental set of solutions, one may write $g(x, s)$ in the following form

$$g(x, s) = \sum_{j=1}^{n} \begin{cases} y_j(x)A_j(s), & \text{for } a \le x \le s \\ y_j(x)B_j(s), & \text{for } s \le x \le b \end{cases} \tag{2.3}$$

where $A_j(s)$ and $B_j(s)$ represent functions to be determined. Clearly, the number of these functions is $2n$ and the number of the relations, which can be derived from properties 2, 3, and 4 of the definition, is also $2n$. Thus, the situation is promising so far. Indeed, we are going to derive a system of $2n$ equations in $2n$ unknowns ($(n-1)$ equations can be obtained from property 2, one equation comes out from property 3, and $n$ equations follow from property 4).

Hence, the key issue to be highlighted in the remaining part of this proof is a clarification of two facts. These are, whether that system is going to be consistent and whether it has a unique solution.

By virtue of property 2, which stipulates the continuity of $g(x, s)$ itself and its partial derivatives with respect to $x$ of up to $(n-2)$ order, as $x = s$, one derives the following system of $(n-1)$ linear algebraic equations

$$\sum_{i=1}^{n} C_j(s)\frac{d^k y_j(s)}{dx^k} = 0, \quad (k = \overline{0, n-2}) \tag{2.4}$$

in $n$ unknown functions

$$C_j(s) = B_j(s) - A_j(s), \quad (j = \overline{1, n}) \tag{2.5}$$

The system in eqn (2.4) is under-determined, because the number of equations in it $(n-1)$ is fewer than the number of unknowns $(n)$ involved. This shortage can be eluded, however, by applying property 3 to the expression in eqn (2.3). This yields a single linear algebraic equation

$$\sum_{i=1}^{n} C_i(s)\frac{d^{n-1}y_i(s)}{dx^{n-1}} = -\frac{1}{p_0(s)} \tag{2.6}$$

in the same set $\{C_i(s) \mid i = \overline{1, n}\}$ of unknowns. Hence, the relations in eqn (2.4) along with that in eqn (2.6) form a system of $n$ simultaneous linear algebraic equations in $n$ unknowns. The determinant of the coefficient matrix in this system is not zero, because it represents the Wronskian for the fundamental set of solutions $\{y_j(x) \mid j = \overline{1, n}\}$. Thus, the system has a unique solution. In other words, one can readily obtain the explicit expressions for $C_j(s)$.

In order to obtain the values of $A_j(s)$ and $B_j(s)$, we take advantage of property 4. In doing so, let us first break down the forms $M_i(y(a), y(b))$ in eqn (2.2) into two additive parts as

$$M_i(y(a), y(b)) = P_i(y(a)) + Q_i(y(b)), \quad (i = \overline{1, n})$$

with $P_i(a)$ and $Q_i(b)$ being defined as

$$P_i(y(a)) = \sum_{k=0}^{n-1} \alpha_k^i y^{(k)}(a), \quad Q_i(y(b)) = \sum_{k=0}^{n-1} \beta_k^i y^{(k)}(b)$$

In compliance with property 4, we now substitute the expression for $g(x, s)$ from eqn (2.3) into eqn (2.2)

$$M_i(g(a, s), g(b, s)) \equiv P_i(g(a, s)) + Q_i(g(b, s)) = 0, \quad (i = \overline{1, n}) \tag{2.7}$$

Since $P_i$ in eqn (2.7) governs the values of $g(a, s)$ at the left-end point $x = a$ of the interval $[a, b]$, while $Q_i$ governs the values of $g(b, s)$ at the right-end point $x = b$, the upper branch

$$\sum_{j=1}^{n} y_j(x)A_j(s)$$

of $g(x, s)$ from eqn (2.3) goes to $P_i(g(a, s))$, while the lower branch

$$\sum_{j=1}^{n} y_j(x)B_j(s)$$

of $g(x, s)$ ought to be substituted in $Q_i(g(b, s))$, resulting in

$$M_i(g(a, s), g(b, s))$$
$$\equiv \sum_{j=1}^{n} [P_i(g(a, s))A_j(s) + Q_i(g(b, s))B_j(s)] = 0, \quad (i = \overline{1, n})$$

Replacing the values of $A_j(s)$ in the above equation with $B_j(s) - C_j(s)$ in accordance with eqn (2.5), one rewrites it in the form

$$\sum_{j=1}^{n}[P_i(g(a,s))(B_j(s) - C_j(s)) + Q_j(g(b,s))B_j(s)] = 0, \quad (i = \overline{1,n})$$

Combining then the terms with $B_j(s)$ and taking the term with $C_j(s)$ to the right-hand side, one obtains

$$\sum_{j=1}^{n}[P_i(g(a,s)) + Q_i(g(b,s))]B_j(s) = \sum_{j=1}^{n} P_i(g(a,s))C_j(s), \quad (i = \overline{1,n})$$

Upon recalling eqn (2.7), the above relations can finally be rewritten in the form

$$\sum_{j=1}^{n} M_i(g(a,s), g(b,s))B_j(s) = \sum_{j=1}^{n} P_i(g(a,s))C_j(s), \quad (i = \overline{1,n}) \tag{2.8}$$

These relations constitute a system of $n$ linear algebraic equations in $n$ unknowns $B_j(s)$. The coefficient matrix of this system is not singular, since the forms $M_i$ are linearly independent. The right-hand side vector in eqn (2.8) is defined in terms of the known values of $C_j(s)$. This system has, consequently, a unique solution for $B_j(s)$. Based on this, the values of $A_j(s)$ can readily be obtained from eqn (2.5). Hence, this final step completes the proof of Theorem 2.1, because upon substituting the values of $A_j(s)$ and $B_j(s)$ into eqn (2.3), we finally obtain an explicit expression for $g(x,s)$. ■

### 2.1.2 Illustrative examples

As we have already mentioned, the proof just completed suggests a consistent way to practically construct the Green's function. This point is supported below with a series of particular examples, in each of which we present and analyze different peculiarities in statements of boundary-value problems, which may occur while considering practical situations in computational mechanics.

**Example 1.1:** Consider the following differential equation

$$\frac{d^2y(x)}{dx^2} = 0, \quad x \in (0,a) \tag{2.9}$$

subject to boundary conditions written as

$$y(0) = y(a) = 0 \tag{2.10}$$

This boundary-value problem can be related to many phenomena in continuum mechanics (it associates, in particular, with deflection of elastic cable whose end-points are fixed).

The most elementary set of functions constituting a fundamental set of solutions for eqn (2.9) is represented by

$$y_1(x) \equiv 1, \quad y_2(x) \equiv x$$

Therefore, the general solution $y_g(x)$ for this equation can be written as a linear combination of the above functions

$$y_g(x) = D_1 + D_2 x$$

where $D_1$ and $D_2$ represent arbitrary constants.

A substitution of this function in the boundary conditions of eqn (2.10) yields the homogeneous system of linear algebraic equations in $D_1$ and $D_2$, with a well-posed coefficient matrix. Hence, the problem in eqns (2.9) and (2.10) has only the trivial solution. Thus, there exists a unique Green's function for this problem. According to the procedure described in the proof of Theorem 2.1, it can be sought in the form

$$g(x,s) = \begin{cases} A_1(s) + xA_2(s), & \text{for } 0 \le x \le s \\ B_1(s) + xB_2(s), & \text{for } s \le x \le a \end{cases} \tag{2.11}$$

Introducing then, as it is suggested in eqn (2.5), $C_1(s) = B_1(s) - A_1(s)$ and $C_2(s) = B_2(s) - A_2(s)$, we form a system of linear algebraic equations in these unknowns (see the system in eqns (2.4) and (2.6)) written as

$$\begin{cases} C_1(s) + sC_2(s) = 0 \\ \qquad\quad C_2(s) = -1 \end{cases} \tag{2.12}$$

Its obvious solution is $C_1(s) = s$ and $C_2(s) = -1$.

The first boundary condition $y(0) = 0$ in eqn (2.10), being satisfied with the upper branch of $g(x, s)$, results in $A_1(s) = 0$. The upper branch is chosen because $x = 0$ belongs to its domain $0 \le x \le s$. Since $B_1(s) = C_1(s) + A_1(s)$, we conclude that $B_1(s) = s$.

The second condition $y(a) = 0$ in eqn (2.10), being treated with the lower branch of $g(x, s)$, yields $B_1(s) + aB_2(s) = 0$. Hence, $B_2(s) = -s/a$, and finally, since $A_2(s) = B_2(s) - C_2(s)$, we find that $A_2(s) = 1 - s/a$. Substituting these into eqn (2.11), we ultimately obtain the Green's function in the form

$$g(x,s) = \begin{cases} a^{-1}x(a-s), & \text{for } 0 \le x \le s \\ a^{-1}s(a-x), & \text{for } s \le x \le a \end{cases} \tag{2.13}$$

▲

**Example 1.2:** We formulate now another boundary-value problem

$$\frac{dy(0)}{dx} = 0, \quad \frac{dy(a)}{dx} = 0$$

for eqn (2.9) over the same interval $(0, a)$.

This problem is not uniquely solvable. Indeed, one can clearly see that any constant function represents the solution to it. Hence, the condition of existence and uniqueness for Green's function does not hold for the above statement. Therefore, a Green's function does not exist. ▲

**Example 1.3:** Consider another boundary-value problem

$$\frac{dy(0)}{dx}=0, \qquad \frac{dy(a)}{dx}+hy(a)=0 \tag{2.14}$$

for equation (2.9) over $(0, a)$, where $h$ is thought to be a non-zero constant.

It can easily be shown (see Exercise 2.1(b) in this End Chapter Exercises) that the problem in eqns (2.9) and (2.14) has only the trivial solution. Consequently, there exists a unique Green's function for this problem.

The first part of the construction procedure precisely resembles that from the problem stated in Example 1.2. The Green's function is again expressed by eqn (2.11), the coefficients $C_1(s)$ and $C_2(s)$ again satisfy the system in eqn (2.12), resulting in $C_1(s)=s$ and $C_2(s)=-1$.

The first boundary condition in eqn (2.27), being treated by the upper branch in eqn (2.11), yields $A_2(s)=0$. This immediately results in $B_2(s)=-1$. The second condition in (2.14), being treated by the lower branch in eqn (2.11), yields the following equation

$$B_2(s)+h[B_1(s)+aB_2(s)]=0$$

in $B_1(s)$ and $B_2(s)$. Based on the known value of $B_2(s)$, one obtains $B_1(s)=(1+ha)/h$. This in turn yields $A_1(s)=[1+h(a-s)]/h$.

Substituting the values of $A_j(s)$ and $B_j(s)$ just found, into eqn (2.11), we finally obtain the Green's function to the boundary-value problem posed by eqns (2.9) and (2.14) in the form

$$g(x, s)=\begin{cases}(a-s)+h^{-1}, & \text{for } 0\leq x\leq s\\ (a-x)+h^{-1}, & \text{for } s\leq x\leq a\end{cases} \tag{2.15}$$

Notice that as $h$ is taken to infinity, the second term $h^{-1}$ in (2.15) vanishes yielding the Green's function

$$g(x, s)=\begin{cases}a-s, & \text{for } 0\leq x\leq s\\ a-x, & \text{for } s\leq x\leq a\end{cases}$$

for equation (2.9) subject to the following boundary conditions

$$\frac{dy(0)}{dx}=0, \qquad y(a)=0$$ ▲

In applied mechanics, it is frequently required to work out research projects for phenomena occurring in infinite media. The influence (Green's) function formalism can successfully be applied to the associated boundary-value problems formulated over infinite intervals. As our next example, we construct the Green's function for such a problem.

**Example 1.4:** Consider the following differential equation

$$\frac{d^2y(x)}{dx^2} - k^2y(x) = 0, \quad x \in (0, \infty) \tag{2.16}$$

subject to boundary conditions imposed as

$$y(0) = 0, \quad |y(\infty)| < \infty \tag{2.17}$$

It can readily be shown that the conditions of existence and uniqueness for the Green's function are met in this case. This assures a unique Green's function of the above formulation.

Since roots of the characteristic equation are, in this case, $k$ and $-k$, the following two exponential functions

$$y_1(x) \equiv \exp(kx), \quad y_2(x) \equiv \exp(-kx)$$

represent a fundamental set of solutions for eqn (2.16), one can express the Green's function for the boundary-value problem in eqns (2.16) and (2.17) in the following form

$$g(x, s) = \begin{cases} A_1(s)\exp(kx) + A_2(s)\exp(-kx), & \text{for } x \le s \\ B_1(s)\exp(kx) + B_2(s)\exp(-kx), & \text{for } s \le x \end{cases} \tag{2.18}$$

Denoting $C_i(s) = B_i(s) - A_i(s)$, $(i = 1, 2)$, one obtains the following system of linear algebraic equations

$$\begin{cases} \exp(ks)C_1(s) + \exp(-ks)C_2(s) = 0 \\ k\exp(ks)C_1(s) - k\exp(-ks)C_2(s) = -1 \end{cases}$$

in $C_1(s)$ and $C_2(s)$. Its solution is expressed as

$$C_1(s) = -\frac{1}{2k}\exp(-ks), \quad C_2(s) = \frac{1}{2k}\exp(ks) \tag{2.19}$$

The first condition in eqn (2.17) implies

$$A_1(s) + A_2(s) = 0 \tag{2.20}$$

while the second condition results in $B_1(s) = 0$, because the exponential function $\exp(kx)$ is unbounded as $x$ approaches infinity. And the only way to satisfy the second condition in eqn (2.17) is to set $B_1(s)$ equals zero. This immediately yields

$$A_1(s) = \frac{1}{2k} \exp(-ks)$$

and the relation in eqn (2.20) consequently provides

$$A_2(s) = -\frac{1}{2k} \exp(-ks)$$

Hence, based on the known values of $C_2(s)$ and $A_2(s)$, one obtains

$$B_2(s) = \frac{1}{2k}[\exp(ks) - \exp(-ks)]$$

Upon substituting the values of the coefficients $A_j(s)$ and $B_j(s)$ just found into eqn (2.18), one finally obtains the Green's function to the problem posed by eqns (2.16) and (2.17) in the form

$$g(x, s) = \frac{1}{2k} \begin{cases} \exp(k(x-s)) - \exp(-k(x+s)), & \text{for } x \le s \\ \exp(k(s-x)) - \exp(-k(s+x)), & \text{for } s \le x \end{cases} \tag{2.21}$$

▲

**Example 1.5:** Consider a boundary-value problem for the same equation as in Example 1.4 but formulated over a different domain

$$\frac{d^2y(x)}{dx^2} - k^2 y(x) = 0, \quad x \in (0, a) \tag{2.22}$$

and subject to boundary conditions written as

$$y(0) = y(a), \quad \frac{dy(0)}{dx} = \frac{dy(a)}{dx} \tag{2.23}$$

This boundary-value problem represents an important type of formulations in applied mechanics. The relations in eqn (2.23) specify conditions of the $a$-periodicity of the solution.

Using the experience gained at the moment, the reader can easily show that this boundary-value problem has only the trivial solution, providing existence of a unique Green's function for it.

Since the formulation in eqns (2.22) and (2.23) again entails the same differential equation which was considered in Example 1.4, the beginning stage of the construction procedure for the Green's function resembles that from the previous problem. We again express the Green's function by eqn (2.18), and the coefficients $C_1(s)$ and $C_2(s)$ are again given with eqn (2.19).

Satisfying the first condition in eqn (2.23), we utilize the upper branch in eqn (2.18) in order to compute the value of $y(0)$, while its lower branch is used for computing the value of $y(a)$. This results in

$$A_1(s) + A_2(s) = B_1(s) \exp(ka) + B_2(s) \exp(-ka) \tag{2.24}$$

Satisfying the second condition in eqn (2.23), we compute the derivative of $y(x)$ at $x = 0$ by using the upper branch in eqn (2.18), while the value of the derivative of $y(x)$ at $x = a$ is computed by using the lower branch of eqn (2.18). This yields

$$A_1(s) - A_2(s) = B_1(s) \exp(ka) - B_2(s) \exp(-ka) \tag{2.25}$$

So the relations in eqns (2.24) and (2.25) along with those in eqn (2.19) form a system of four linear algebraic equations in $A_1(s)$, $A_2(s)$, $B_1(s)$, and $B_2(s)$. To find the values of $A_1(s)$ and $B_1(s)$, we add eqns (2.24) and (2.25) to each other. This provides

$$A_1(s) - B_1(s) \exp(ka) = 0 \tag{2.26}$$

And the first relation in eqn (2.19) can be rewritten in the form

$$-A_1(s) + B_1(s) = -\frac{1}{2k} \exp(-ks) \tag{2.27}$$

Solving eqns (2.26) and (2.27) simultaneously, one obtains

$$A_1(s) = \frac{\exp(k(a-s))}{2k[\exp(ka) - 1]}, \quad B_1(s) = \frac{\exp(-ks)}{2k[\exp(ka) - 1]}$$

To find the values of $A_2(s)$ and $B_2(s)$, we subtract eqn (2.25) from eqn (2.24). This results in

$$A_2(s) - B_2(s) \exp(-ka) = 0 \tag{2.28}$$

Rewriting then the second relation from eqn (2.19) in the form

$$-A_2(s) + B_2(s) = \frac{1}{2k} \exp(ks) \tag{2.29}$$

we then solve eqns (2.28) and (2.29) simultaneously. This yields

$$A_2(s) = \frac{\exp(ks)}{2k[\exp(ka) - 1]}, \quad B_2(s) = \frac{\exp(k(s+a))}{2k[\exp(ka) - 1]}$$

Substituting the values of $A_1(s)$, $A_2(s)$, $B_1(s)$, and $B_2(s)$ just found into eqn (2.18), we finally obtain

$$g(x, s) = K_0 \begin{cases} \exp(k(x - s + a)) + \exp(k(s - x)), & \text{for } x \leq s \\ \exp(k(s - x + a)) + \exp(k(x - s)), & \text{for } s \leq x \end{cases} \tag{2.30}$$

where $K_0 = \{2k[\exp(ka) - 1]\}^{-1}$. ▲

So far we have dealt with differential equations with constant coefficients. Notice that variable coefficients do not bring limitations to the described algorithm, if a fundamental set of solutions to the governing equation is obtained in terms of elementary functions. In other words, if the governing equation allows exact solution, one can readily construct a Green's function by means of this algorithm. The following three examples are designed to address this point.

**Example 1.6:** Consider the equation

$$\frac{d}{dx}\left((mx+b)\frac{dy}{dx}\right)=0, \quad x\in(0,a) \tag{2.31}$$

with boundary conditions imposed as

$$\frac{dy(0)}{dx}=0, \quad y(a)=0 \tag{2.32}$$

where we assume $m>0$ and $b>0$, which implies that $mx+b\neq 0$ on $x\in[0,a]$.

The fundamental set of solutions

$$y_1(x)\equiv 1, \quad y_2(x)\equiv \ln(mx+b)$$

required for the construction of the Green's function for the problem in eqns (2.31) and (2.32) can be obtained by two successive integrations of eqn (2.31). Indeed, integration yields

$$(mx+b)\frac{dy}{dx}=C_1$$

dividing then the above equation through by $mx+b$ and multiplying by $dx$, we separate variables

$$dy=C_1\frac{dx}{mx+b}$$

and finally we have

$$y(x)=\frac{C_1}{m}\ln(mx+b)+C_2$$

At this point, as usual, we state that the boundary-value problem in eqns (2.31) and (2.32) has only the trivial solution. Hence, there exists a unique Green's function which can be presented in the form

$$g(x,s)=\begin{cases}A_1(s)+\ln(mx+b)A_2(s), & \text{for } 0\le x\le s\\ B_1(s)+\ln(mx+b)B_2(s), & \text{for } s\le x\le a\end{cases} \tag{2.33}$$

Tracing out then our customary procedure, one obtains the system of linear algebraic equations

$$\begin{cases}C_1(s)+\ln(ms+b)C_2(s)=0\\ \qquad m(ms+b)^{-1}C_2(s)=-(ms+b)^{-1}\end{cases}$$

in $C_j(s) = B_j(s) - A_j(s)$, $(j = 1, 2)$. Its solution is

$$C_1(s) = \frac{1}{m} \ln (ms + b), \quad C_2(s) = -\frac{1}{m} \tag{2.34}$$

The first boundary condition in eqn (2.19) yields $A_2(s) = 0$. Consequently, $B_2(s) = -1/m$. The second condition in eqn (2.19) gives

$$B_1(s) + \ln (ma + p) B_2(s) = 0$$

resulting in $B_1(s) = [\ln(ma + b)]/m$, which provides

$$A_1(s) = \frac{1}{m} \ln \frac{ma + b}{ms + b}$$

Substituting the values of $A_j(s)$ and $B_j(s)$ just found into eqn (2.33), one obtains the Green's function that we are looking for in the form

$$g(x, s) = \frac{1}{m} \begin{cases} \ln[(ma + b)(ms + b)^{-1}], & \text{for } 0 \le x \le s \\ \ln[(ma + b)(mx + b)^{-1}], & \text{for } s \le x \le a \end{cases} \tag{2.35}$$

▲

Sometimes in applied mechanics, we consider boundary-value problems set up on finite intervals, where one of the end-points is singular for the governing differential equation. The algorithm described in this section can also be used to construct Green's functions for such problems. As an illustration of this point, we offer the following example.

**Example 1.7:** Consider a boundary-value problem for the following differential equation

$$\frac{d}{dx}\left(x \frac{dy(x)}{dx}\right) = 0, \quad x \in (0, a) \tag{2.36}$$

subject to boundary conditions written as

$$|y(0)| < \infty, \quad \frac{dy(a)}{dx} + hy(a) = 0 \tag{2.37}$$

Clearly, the left end-point $x = 0$ of the domain is a point of singularity for eqn (2.36). Therefore, instead of formulating a traditional boundary condition at this point, we require in eqn (2.37) for $y(0)$ to be bounded.

Integrating eqn (2.36) successively two times, one obtains, similarly to the case in Example 1.6, its fundamental set of solutions that can be written as

$$y_1(x) \equiv 1, \quad y_2(x) \equiv \ln x \tag{2.38}$$

The problem in eqns (2.36) and (2.37) has only the trivial solution, allowing a unique Green's function in the form

$$g(x, s) = \begin{cases} A_1(s) + \ln x A_2(s), & \text{for } 0 \le x \le s \\ B_1(s) + \ln x B_2(s), & \text{for } s \le x \le a \end{cases} \tag{2.39}$$

As our customary procedure, we form a system of linear algebraic equations

$$\begin{cases} C_1(s) + \ln s\, C_2(s) = 0 \\ \qquad\qquad s^{-1}\, C_2(s) = -s^{-1} \end{cases}$$

whose solution is $C_1(s) = \ln s$ and $C_2(s) = -1$.

The boundedness of the Green's function at $x = 0$ implies $A_2(s) = 0$. Consequently, $B_2(s) = -1$. The second condition in eqn (2.37) yields

$$B_2(s)/a + h[B_1(s) + \ln a B_2(s)] = 0$$

Hence, $B_1(s) = 1/ah + \ln a$, and ultimately, $A_1(s) = 1/ah - \ln s/a$. Thus, we finally obtain

$$g(x, s) = \begin{cases} (ah)^{-1} - \ln[(a)^{-1}s], & \text{for } 0 \le x \le s \\ (ah)^{-1} - \ln[(a)^{-1}x], & \text{for } s \le x \le a \end{cases} \tag{2.40}$$

Notice that as the value of $h$ is taken to infinity, the first term $(ah)^{-1}$ in eqn (2.40) vanishes, yielding the Green's function

$$g(x, s) = \begin{cases} -\ln[(a)^{-1}s], & \text{for } 0 \le x \le s \\ -\ln[(a)^{-1}x], & \text{for } s \le x \le a \end{cases}$$

for eqn (2.36) subject to the boundary conditions $|y(0)| < \infty$ and $y(a) = 0$. ▲

**Example 1.8:** We state a boundary-value problem for the simplest fourth order differential equation on the unit interval

$$\frac{d^4y(x)}{dx^4} = 0, \quad x \in (0, 1) \tag{2.41}$$

with boundary conditions written as

$$y(0) = \frac{dy(0)}{dx} = 0, \quad y(1) = \frac{d^2y(1)}{dx^2} = 0 \tag{2.42}$$

As it is known from applied mechanics, this setting relates to the bending of a beam of unit length, if its left edge is clamped while the right edge is simply supported. Later in this text, we consider a number of other problems from the beam theory.

The following set of functions

$$y_1(x) \equiv 1, \; y_2(x) \equiv x, \; y_3(x) \equiv x^2, \; y_4(x) \equiv x^3 \tag{2.43}$$

constitutes the simplest fundamental set of solutions for eqn (2.41). Hence, its general solution is

$$y_g(x) = D_1 + D_2 x + D_3 x^2 + D_4 x^3$$

Applying the first two boundary conditions from eqn (2.42), we obtain $D_1 = D_2 = 0$. The conditions at $x = 1$ yield $D_3 = D_4 = 0$. Hence, the boundary-value problem in eqns (2.41) and (2.42) has only the trivial solution. There exists, consequently, a unique Green's function for the problem posed by eqns (2.41) and (2.42).

Based on the fundamental set of solutions presented in eqn (2.43), the Green's function can be written in the form

$$g(x,s) = \begin{cases} A_1(s) + A_2(s)x + A_3(s)x^2 + A_4(s)x^3, & \text{for } x \le s \\ B_1(s) + B_2(s)x + B_3(s)x^2 + B_4(s)x^3, & \text{for } s \le x \end{cases} \tag{2.44}$$

From properties 2 and 3 of the definition of the Green's function, one derives the following system of linear equations in $C_i(s) = B_i(s) - A_i(s)$, written in a matrix form

$$\begin{pmatrix} 1 & s & s^2 & s^3 \\ 0 & 1 & 2s & 3s^2 \\ 0 & 0 & 2 & 6s \\ 0 & 0 & 0 & 6 \end{pmatrix} \times \begin{pmatrix} C_1(s) \\ C_2(s) \\ C_3(s) \\ C_4(s) \end{pmatrix} = \begin{pmatrix} 0 \\ 0 \\ 0 \\ -1 \end{pmatrix}$$

whose solution

$$C_1 = \tfrac{1}{6}s^3, \quad C_2 = -\tfrac{1}{2}s^2, \quad C_3 = \tfrac{1}{2}s, \quad C_4 = -\tfrac{1}{6} \tag{2.45}$$

is easily obtained because of the triangular form of its coefficient matrix.

By virtue of property 4 in the definition, the boundary conditions in eqn (2.42) provide

$$A_1 = 0, \quad A_2 = 0, \quad B_1 + B_2 + B_3 + B_4 = 0, \quad 2B_3 + 6B_4 = 0$$

while the rest of the coefficients for $g(x, s)$

$$A_3 = -\tfrac{1}{4}s^3 + \tfrac{3}{4}s^2 - \tfrac{1}{2}s, \quad A_4 = \tfrac{1}{12}s^3 - \tfrac{1}{4}s^2 + \tfrac{1}{6}$$

$$B_1 = \tfrac{1}{6}s^3, \quad B_2 = -\tfrac{1}{2}s^2, \quad B_3 = -\tfrac{1}{4}s^3 + \tfrac{3}{4}s^2, \quad B_4 = \tfrac{1}{12}s^3 - \tfrac{1}{4}s^2$$

are computed through the values of $C_j(s)$ presented in eqn (2.45).

Substituting all the coefficients $A_j(s)$ and $B_j(s)$ just obtained into eqn (2.44), we obtain the Green's function $g(x, s)$ for the boundary-value problem posed by eqns (2.41) and (2.42). For $x \leq s$, it is found in the form

$$g(x, s) = -(\tfrac{1}{4}s^3 - \tfrac{3}{4}s^2 + \tfrac{1}{2}s)x^2 + (\tfrac{1}{12}s^3 - \tfrac{1}{4}s^2 + \tfrac{1}{6})x^3 \tag{2.46}$$

while for $x \geq s$, its expression is

$$g(x, s) = -(\tfrac{1}{4}x^3 - \tfrac{3}{4}x^2 + \tfrac{1}{2}x)s^2 + (\tfrac{1}{12}x^3 - \tfrac{1}{4}x^2 + \tfrac{1}{6})s^3$$

This example shows that even for equations of higher order, the procedure for the construction of Green's functions utilized in this section is compact enough and results in a quite reasonable amount of computation. ▲

Observing the form of all of the Green's functions constructed so far in this section, one may notice their common property. Indeed, they are symmetric in a certain sense. That is, the interchange of $x$ with $s$ in the expression of Green's function that is valid for $x \leq s$ yields that one is valid for $x \geq s$ and vice versa. In the next section, we will discuss this point in more detail. The conditions will be discovered, under which the symmetry takes place.

**Example 1.9:** To close the discussion in this section, we consider a problem whose Green's function, in contrast to all previous settings, appears to be in a non-symmetric form. Namely, a boundary-value problem is set up for the following equation

$$\frac{d^2y(x)}{dx^2} + \frac{dy(x)}{dx} - 2y(x) = 0, \quad x \in (0, \infty) \tag{2.47}$$

subject to the boundary conditions

$$y(0) = 0, \quad |y(\infty)| < \infty \tag{2.48}$$

It is evident that this problem has only the trivial solution, allowing, subsequently, a unique Green's function. Since $y_1(x) = \exp(x)$ and $y_2(x) = \exp(-2x)$ represent a fundamental set of solutions to eqn (2.47), one can express the Green's function to this problem in the form

$$g(x, s) = \begin{cases} A_1(s)\exp(x) + A_2(s)\exp(-2x), & \text{for } x \leq s \\ B_1(s)\exp(x) + B_2(s)\exp(-2x), & \text{for } s \leq x \end{cases} \tag{2.49}$$

This results in the system of linear equations in $C_j(s) = B_j(s) - A_j(s)$

$$\begin{pmatrix} \exp(s) & \exp(-2s) \\ \exp(s) & -2\exp(-2s) \end{pmatrix} \times \begin{pmatrix} C_1(s) \\ C_2(s) \end{pmatrix} = \begin{pmatrix} 0 \\ -1 \end{pmatrix}$$

whose solution is found as

$$C_1(s) = -\tfrac{1}{3}\exp(-s), \quad C_2(s) = \tfrac{1}{3}\exp(2s)$$

The first condition in eqn (2.48) provides $A_1(s) + A_2(s) = 0$, while the second condition implies $B_1(s) = 0$. Therefore $A_1(s) = [\exp(-s)]/3$, resulting in

$A_2(s) = -[\exp(-s)]/3$, and, finally, $B_2(s) = [\exp(2s) - \exp(-s)]/3$. Substituting these values in eqn (2.49), one obtains the Green's function to the problem in eqns (2.47) and (2.48) as

$$g(x, s) = \frac{1}{3}\begin{cases} \exp(-s)[\exp(x) - \exp(-2x)], & \text{for } x \leq s \\ \exp(-2x)[\exp(2s) - \exp(-s)], & \text{for } s \leq x \end{cases} \tag{2.50}$$

It is absolutely evident that $g(x, s)$ fails, in this case, to be symmetric. The question that naturally arises with regard to this fact is, why? What makes the statement of the boundary-value problem posed by eqns (2.47) and (2.48) different from all the others considered earlier in this section? The reader will find the reasoning for this occurrence in the next section. ▲

## 2.2 Symmetry of Green's Functions

In order to address the issue of symmetry of Green's function with respect to the observation and the source point, a certain preparatory work ought to be accomplished in this section.

### 2.2.1 Self-adjoint equations

Let us write down the linear $n$-th order homogeneous differential equation

$$L[y(x)] \equiv p_0(x)\frac{d^n y(x)}{dx^n} + p_1(x)\frac{d^{n-1}y(x)}{dx^{n-1}} + \cdots + p_n(x)y(x) = 0$$

From the qualitative theory of linear equations (see, for example, [15, 60, 61]), it is known that the equation

$$L_a[y(x)] \equiv (-1)^n\frac{d^n[p_0(x)y(x)]}{dx^n} + (-1)^{n-1}\frac{d^{n-1}[p_1(x)y(x)]}{dx^{n-1}} + \cdots + p_n(x)y(x) = 0$$

is said to be *adjoint* to $L[y(x)] = 0$. The operator $L_a$ is called *adjoint* to $L$, and if $L \equiv L_a$, then $L$ is said to be a *self-adjoint* operator and the equation $L[y(x)] = 0$ is said to be a *self-adjoint* equation.

For the sake of simplicity, the discussion in this section is limited to equations of the second order

$$L[y(x)] \equiv p_0(x)\frac{d^2y(x)}{dx^2} + p_1(x)\frac{dy(x)}{dx} + p_2(x)y(x) = 0 \tag{2.51}$$

This limitation does not radically affect the generality of the presentation but notably condense it and makes it easier to comprehend.

The leading coefficient $p_0(x)$ is not supposed to equal zero at any single point in $(a, b)$ except, maybe, for one of its end-points. In addition, we require the coefficient $p_0(x)$ to be two times differentiable and $p_1(x)$ just differentiable on $(a, b)$.

According to the definition that was just introduced, the following equation

$$L_a[y(x)] \equiv \frac{d^2[p_0(x)y(x)]}{dx^2} - \frac{d[p_1(x)y(x)]}{dx} + p_2(x)y(x) = 0 \tag{2.52}$$

is adjoint to that in (2.51).

We will briefly review here the self-adjointness of differential equations and other relevant issues that are important in the analysis of symmetry of Green's functions. A more complete discussion on this subject can be found in standard graduate texts on differential equations.

Using the product rule of differentiation, the operator $L_a$ in eqn (2.52) can be rewritten in the form

$$L_a[y(x)] \equiv \frac{d}{dx}\left(y\frac{dp_0}{dx} + p_0\frac{dy}{dx}\right) - \left(y\frac{dp_1}{dx} + p_1\frac{dy}{dx}\right) + p_2 y$$

Differentiating further and combining the like terms, one obtains

$$L_a[y(x)] \equiv p_0\frac{d^2y}{dx^2} + \left(2\frac{dp_0}{dx} - p_1\right)\frac{dy}{dx} + \left(\frac{d^2p_0}{dx^2} - \frac{dp_1}{dx} + p_2\right)y \tag{2.53}$$

Suppose eqn (2.51) is self-adjoint, that is $L[y(x)] \equiv L_a[y(x)]$. If so, then by comparison of the coefficients of $dy/dx$ in $L[y(x)]$ and $L_a[y(x)]$ in eqns (2.51) and (2.53), one obtains the following relation for the coefficients $p_0(x)$ and $p_1(x)$

$$2\frac{dp_0(x)}{dx} - p_1(x) = p_1(x)$$

which ought to hold for the self-adjointness of eqn (2.51). This implies

$$p_1(x) = \frac{dp_0(x)}{dx} \tag{2.54}$$

Differentiating the above relation, we realize that the sum of the first two terms in the coefficient

$$\frac{d^2p_0(x)}{dx^2} - \frac{dp_1(x)}{dx} + p_2(x)$$

of $y(x)$ in eqn (2.53) equals zero. This means that self-adjointness of eqn (2.51) implies the relation between the coefficients $p_0(x)$ and $p_1(x)$ in eqn (2.54) and puts no constraints on the coefficient $p_2(x)$ in eqn (2.51). In other words, if eqn (2.51) is self-adjoint, then it can be written as

$$p_0(x)\frac{d^2y(x)}{dx^2} + \frac{dp_0(x)}{dx}\frac{dy(x)}{dx} + p_2(x)y(x) = 0$$

which reads in a short-hand form as

$$\frac{d}{dx}\left(p_0(x)\frac{dy(x)}{dx}\right) + p_2(x)y(x) = 0 \tag{2.55}$$

The above is usually referred to as the *standard form of a self-adjoint equation of the second order.*

Thus, if the coefficients $p_0(x)$ and $p_1(x)$ in eqn (2.51) satisfy the relation in eqn (2.54), then eqn (2.51) is in a self-adjoint form, if, however, the condition in eqn (2.54) is not met, then eqn (2.51) is not in a self-adjoint form. The fact that eqn (2.54) does not involve the coefficient $p_2(x)$ prompts a simple idea of how a linear second order differential equation can be reduced to a self-adjoint form.

Indeed, multiplying eqn (2.51) through by a certain non-zero function (we call it the *integrating factor*) and applying then the relation in eqn (2.54) to the coefficients of $d^2y/dx^2$ and $dy/dx$ of the resultant equation, one can readily formulate a relation from which the integrating factor can afterwards be found. The procedure for finding the integrating factor is quite straightforward. In Example 2.1 below, we consider a particular equation and go through that procedure in detail.

**Example 2.1:** Find out if the equation

$$e^x \frac{d^2y(x)}{dx^2} + (1 - \cos 2x)y(x) = 0 \tag{2.56}$$

is in a self-adjoint form and if not, then reduce it to such.

It is clearly seen that this equation is not in a self-adjoint form, since $p_0(x)$ is $e^x$ while $p_1(x)$ equals zero and the condition in eqn (2.54) is not met. The integrating factor $e^{-x}$ is also evident, in this case, because if equation (2.56) is multiplied by $e^{-x}$, then it reduces to the self-adjoint equation

$$\frac{d^2y(x)}{dx^2} + e^{-x}(1 - \cos 2x)y(x) = 0$$

▲

**Example 2.2:** It is evident that the equation

$$x^3 \frac{d^2y(x)}{dx^2} + 3x^2 \frac{dy(x)}{dx} - y(x) = 0$$

is in a self-adjoint form. Indeed, the condition in (2.54) is met in this case. ▲

**Example 2.3:** The condition in eqn (2.54) is not met for the equation

$$\frac{d^2y(x)}{dx^2} + 4x \frac{dy(x)}{dx} - 2y(x) = 0 \tag{2.57}$$

so, it is not in a self-adjoint form, and a guess of the integrating factor is not easy in this case. However, in compliance with the procedure sketched earlier, we multiply this equation by an integrating factor $\mu(x)$

$$\mu(x) \frac{d^2y(x)}{dx^2} + 4x\mu(x) \frac{dy(x)}{dx} - 2\mu(x)y(x) = 0 \tag{2.58}$$

The coefficient $p_0(x)$ of this equation is $\mu(x)$, while the coefficient $p_1(x)$ equals $4x\mu(x)$. Thus, the equation in (2.58) would be self-adjoint if (according to the condition in eqn (2.54))

$$\frac{d\mu(x)}{dx} = 4x\mu(x) \tag{2.59}$$

which is a separable first order differential equation in $\mu(x)$. Multiplying it by $dx$ and dividing by $\mu(x)$, we separate variables

$$\frac{d\mu(x)}{\mu(x)} = 4x\, dx$$

and then integrate both sides

$$\ln|\mu(x)| = 2x^2 + C$$

Solving this equation for $\mu(x)$, we obtain general solution to (2.59) as

$$\mu(x) = e^{2x^2+C}$$

Any function from this family can be considered as the integrating factor for equation (2.57). In other words, constant $C$ can be arbitrarily fixed and we assume, say, $C = 0$, which yields

$$\mu(x) = e^{2x^2} \tag{2.60}$$

Substituting now (2.60) in (2.58), we reduce finally eqn (2.57) to the self-adjoint form

$$e^{2x^2}\frac{d^2y(x)}{dx^2} + 4xe^{2x^2}\frac{dy(x)}{dx} - 2e^{2x^2}y(x) = 0$$ ▲

At this point in our development we assume that $L$ represents a self-adjoint operator of the second order. That is

$$L \equiv \frac{d}{dx}\left(p_0(x)\frac{d}{dx}\right) + p_2(x)$$

Consider two functions $u(x)$ and $v(x)$ both being two times continuously differentiable on $(a, b)$, and form the following bilinear combination of them

$$u(x)\, L[v(x)] - v(x)\, L[u(x)] \tag{2.61}$$

which can be rewritten explicitly as

$$u\left(\frac{d}{dx}\left(p_0(x)\frac{dv}{dx}\right) + p_2(x)v\right) - v\left(\frac{d}{dx}\left(p_0(x)\frac{du}{dx}\right) + p_2(x)u\right)$$

Removing the outer parentheses in both the components above and cancelling the terms $p_2(x)uv$, we have

$$uL(v) - vL(u) = u\frac{d}{dx}\left(p_0(x)\frac{dv}{dx}\right) - v\frac{d}{dx}\left(p_0(x)\frac{du}{dx}\right)$$

When the product rule is applied and some regrouping accomplished, the above expression transforms as

$$
\begin{aligned}
& u\frac{d}{dx}\left(p_0(x)\frac{dv}{dx}\right) - v\frac{d}{dx}\left(p_0(x)\frac{du}{dx}\right) \\
&= u\left(\frac{dp_0(x)}{dx}\frac{dv}{dx} + p_0(x)\frac{d^2v}{dx^2}\right) - v\left(\frac{dp_0(x)}{dx}\frac{du}{dx} + p_0(x)\frac{d^2u}{dx^2}\right) \\
&= u\frac{dp_0(x)}{dx}\frac{dv}{dx} - v\frac{dp_0(x)}{dx}\frac{du}{dx} + p_0(x)u\frac{d^2v}{dx^2} - p_0(x)v\frac{d^2u}{dx^2} \\
&= \frac{dp_0(x)}{dx}\left(u\frac{dv}{dx} - v\frac{du}{dx}\right) + p_0(x)\frac{d}{dx}\left(u\frac{dv}{dx} - v\frac{du}{dx}\right) \\
&= \frac{d}{dx}\left(p_0(x)\left(u\frac{dv}{dx} - v\frac{du}{dx}\right)\right)
\end{aligned}
$$

Hence, the bilinear combination in eqn (2.61) reduces to

$$u\,L(v) - v\,L(u) = \frac{d}{dx}\left(p_0(x)\left(u\frac{dv}{dx} - v\frac{du}{dx}\right)\right) \tag{2.62}$$

Integrating both sides of eqn (2.62) from $a$ to $b$, one obtains the following relation

$$\int_a^b [u\,L(v) - v\,L(u)]\,dx = p_0(x)\left(u\frac{dv}{dx} - v\frac{du}{dx}\right)\Big|_a^b \tag{2.63}$$

which is usually referred to as the *Green's formula* for a self-adjoint operator. From the recent development, it follows that the Green's formula holds for a self-adjoint operator $L$ and continuously differentiable on $(a, b)$ functions $u(x)$ and $v(x)$.

If in addition to being two times continuously differentiable on $(a, b)$, $u(x)$ and $v(x)$ are functions, for which the right-hand side in eqn (2.63) vanishes, then the Green's formula reduces to a compact form. That is, if

$$p_0(x)\left(u\frac{dv}{dx} - v\frac{du}{dx}\right)\Big|_a^b = 0 \tag{2.64}$$

then we have

$$\int_a^b [u\,L(v) - v\,L(u)]\,dx = 0 \tag{2.65}$$

So, the Green's formula in eqn (2.65) is valid for a self-adjoint operator $L$, with $u(x)$ and $v(x)$ being two times continuously differentiable on $(a, b)$ and satisfying the relation in eqn (2.64). This relation is, however, implicit in nature, which makes it too cumbersome to deal with over and over again in actual computations. Therefore, it is important to find some of its explicit equivalents which are more convenient for practical use.

In doing so, we rewrite the relation in eqn (2.64) in the extended form

$$p_0(b)\left(u(b)\frac{dv(b)}{dx}-v(b)\frac{du(b)}{dx}\right)-p_0(a)\left(u(a)\frac{dv(a)}{dx}-v(a)\frac{du(a)}{dx}\right)=0 \tag{2.66}$$

Since this relation contains the values of $u(x)$, $v(x)$, and their derivatives at the end-points of the interval $[a, b]$, it is directly seen that the equation in eqn (2.66) holds, if both $u(x)$ and $v(x)$ satisfy one of the following types of boundary conditions at $x = a$ and $x = b$:

**(1)** $y(a) = 0,\ y(b) = 0$
**(2)** $y(a) = 0,\ y'(b) = 0$
**(3)** $y'(a) = 0,\ y'(b) = 0$

It is also directly seen that the condition in eqn (2.66) is valid in the so-called *singular* case, when the leading coefficient $p_0(x)$ in eqn (2.55) equals zero at one of the end-points of $[a, b]$. In such a case we usually require $y(x)$ to be bounded at that end-point, with a value of either $y(x)$ or $y'(x)$ being zero at the other end-point, that is:

**(4)** $|y(a)| < \infty,\ y(b) = 0$
**(5)** $|y(a)| < \infty,\ y'(b) = 0$

In addition, from those exercises in the End Chapter Exercises that relate to this section, it follows that the condition in eqn (2.66) holds also for both $u(x)$ and $v(x)$ satisfying one of the following sets of boundary conditions:

**(6)** $y(a) = 0,\ y'(b) + hy(b) = 0$
**(7)** $y'(a) = 0,\ y'(b) + hy(b) = 0$
**(8)** $y'(a) + h_1 y(a) = 0,\ y'(b) + h_2 y(b) = 0$
**(9)** $y(a) = y(b),\ p_0(a)y'(a) = p_0(b)y'(b)$
**(10)** $|y(a)| < \infty,\ y'(b) + hy(b) = 0$

The last set of conditions presumes (similarly to cases **(4)** and **(5)**) that the leading coefficient $p_0(x)$ of eqn (2.55) equals zero at $x = a$.

Note that the end-points $a$ and $b$, in all the types of boundary conditions **(1)**–**(10)**, are interchangeable. Namely, the set of conditions

$$y(b) = 0, \quad y'(a) + hy(a) = 0$$

falls into type **(6)**. This is also true for the boundary conditions of types **(4)**, **(5)**, **(7)** and **(10)**.

The recent development allows us to introduce another important terminological issue. A boundary-value problem formulated for eqn (2.55) subject to either one of the types of boundary conditions listed above, belongs to the class of the so-called *self-adjoint boundary-value problems*.

### 2.2.2 Property of symmetry

We now turn the reader's attention to the basic question in this section. That is, what makes a Green's function symmetric in the sense mentioned in Section 2.1. The following theorem specifies conditions that the boundary-value problem ought to meet for its Green's function to be symmetric.

**Theorem 2.2:** If the boundary-value problem

$$M_1[y(a), y(b)] = 0, \quad M_2[y(a), y(b)] = 0 \tag{2.67}$$

stated for eqn (2.55) is self-adjoint and has only the trivial solution, then its Green's function $g(x, s)$ is symmetric, provided that its expression $g^-(x, s)$ (valid for $x \geq s$) can be obtained from $g^+(x, s)$ (valid for $x \leq s$) by interchanging of $x$ with $s$ in the latter one.

**Proof:** This proof is based on a slight modification of that procedure which has been used in the proof of Theorem 2.1. Here we also choose two linearly independent particular solutions $y_1(x)$ and $y_2(x)$ of the governing equation (2.55). But contrary to Theorem 2.1, we put some additional constraints on $y_1(x)$ and $y_2(x)$, choosing them in a special manner.

First, let $y_1(x)$ and $y_2(x)$ be two non-zero linearly independent particular solutions to eqn (2.55) and let $y_1(x)$ satisfy the first boundary condition in eqn (2.67) while $y_2(x)$ satisfies the second condition in eqn (2.67). Clearly, neither $y_1(x)$ nor $y_2(x)$ can satisfies both boundary conditions in eqn (2.67), because assuming otherwise we come to a conflict with the statement that the trivial solution is the only solution to the problem in eqns (2.55) and (2.67).

Second, let us form the bilinear combination based on $y_1(x)$ and $y_2(x)$

$$y_1(x)\, L[y_2(x)] - y_2(x)\, L[y_1(x)]$$

which identically equals zero on $(a, b)$, since $L[y_1(x)] \equiv 0$ and $L[y_2(x)] \equiv 0$ for $x \in (a, b)$.

Recalling the relation in eqn (2.62), derived earlier in this section, and rewriting it in terms of $y_1(x)$ and $y_2(x)$ yields

$$y_1\, L(y_2) - y_2\, L(y_1) = \frac{d}{dx}\left(p_0(x)\left(y_1\frac{dy_2}{dx} - y_2\frac{dy_1}{dx}\right)\right)$$

Since the left-hand side of the relation is identically zero, so is the right-hand side. That is

$$\frac{d}{dx}\left(p_0(x)\left(y_1\frac{dy_2}{dx} - y_2\frac{dy_1}{dx}\right)\right) = 0$$

which implies

$$p_0(x)\left(y_1\frac{dy_2}{dx} - y_2\frac{dy_1}{dx}\right) = C \tag{2.68}$$

where $C$ is a constant.

Notice that $y_1(x)$ and $y_2(x)$ are determined up to a constant multiple. Indeed, if $y_1(x)$, for example, satisfies both the governing equation in eqn (2.55) and the first boundary condition in eqn (2.67), then, for any non-zero constant $\alpha$, the product $\alpha\, y_1(x)$ also satisfies both of these relations. This is equally true for $y_2(x)$, which allows us to arbitrarily fix the constant $C$ in eqn (2.68). We choose $C = -1$ and rewrite (2.68) in the form

$$p_0(x)\left(y_1\frac{dy_2}{dx} - y_2\frac{dy_1}{dx}\right) = -1 \tag{2.69}$$

Hence, without losing generality, we can assume that $y_1(x)$ and $y_2(x)$ meet the condition in eqn (2.69) throughout $(a, b)$. Hence, for any location of point $s \in (a, b)$ we express the Green's function $g(x, s)$ to the problem in eqns (2.55) and (2.67) in the form

$$g(x, s) = \begin{cases} c_1(s)y_1(x), & \text{for } a \le x \le s \\ c_2(s)y_2(x), & \text{for } s \le x \le b \end{cases} \tag{2.70}$$

This function satisfies the boundary conditions in eqn (2.67) regardless of the values of $c_1(s)$ and $c_2(s)$. This occurs because $y_1(x)$ and $y_2(x)$ satisfy the first and the second of those boundary conditions, respectively. Hence, $g(x, s)$ in the form of eqn (2.70) already meets properties 1 and 4 of the definition of Green's function.

By virtue of properties 2 and 3 of the definition, we obtain the following system of linear algebraic equations

$$\begin{pmatrix} y_2(s) & -y_1(s) \\ y_2'(s) & -y_1'(s) \end{pmatrix} \times \begin{pmatrix} c_2(s) \\ c_1(s) \end{pmatrix} = \begin{pmatrix} 0 \\ -p_0^{-1}(s) \end{pmatrix}$$

in $c_1(s)$ and $c_2(s)$. The coefficient matrix of this system is not singular, because its determinant $y_1(s)y_2'(s) - y_2(s)y_1'(s)$ is the Wronskian for the two linearly independent functions $y_1(s)$ and $y_2(s)$. Hence, the above system has a unique solution which appears in the form

$$c_1(s) = -\frac{y_2(s)}{p_0(s)W(s)}, \quad c_2(s) = -\frac{y_1(s)}{p_0(s)W(s)}$$

Upon substituting these values of $c_1(s)$ and $c_2(s)$ in eqn (2.70), one obtains, for the upper branch of the Green's function

$$g^+(x, s) = -\frac{y_1(x)y_2(s)}{p_0(s)W(s)}, \quad x \le s \tag{2.71}$$

while for the lower branch, we have

$$g^-(x, s) = -\frac{y_2(x)y_1(s)}{p_0(s)W(s)}, \quad s \le x \tag{2.72}$$

According to the relation in eqn (2.69), the denominator in eqns (2.71) and (2.72) meets the condition

$$p_0(s)W(s) \equiv p_0(s)\left(y_1(s)\frac{dy_2(s)}{dx} - y_2(s)\frac{dy_1(s)}{dx}\right) \equiv -1$$

This allows us to finally write the Green's function $g(x, s)$ for the boundary-value problem posed by eqns (2.55) and (2.67) in the following "symmetric" form

$$g(x, s) = \begin{cases} y_2(s)y_1(x), & \text{for } a \le x \le s \\ y_1(s)y_2(x), & \text{for } s \le x \le b \end{cases}$$

Thus, the theorem has been proven. Indeed, from the above representation, it follows that the Green's function $g(x, s)$ of a self-adjoint boundary-value problem is invariant to the interchange of the observation point $x$ with the source point $s$. In other words, the Green's function is symmetric in the sense that whenever $x$ variable is interchanged with the $s$ variable in one of the branches (either $g^+(x, s)$ or $g^-(x, s)$) of $g(x, s)$, we obtain the other branch. ■

The analysis of symmetry of Green's functions, completed in this section, has direct implementations in mechanics, which will be discussed in later sections of this text. In the next section, we will revisit the basic issue of this chapter, which is the construction of Green's functions. Another construction procedure that is recommended for this purpose in the existing literature [45, 60] will be presented below in detail.

## 2.3 Alternative Construction of Green's Functions

From our presentation in preceding sections, one learns that the notion of Green's function is introduced for a boundary-value problem where both the governing differential equation and the boundary conditions are homogeneous. Such settings are referred to as the *homogeneous boundary-value problems*. In this section, we will turn the reader's attention to nonhomogeneous linear differential equations subject to homogeneous boundary conditions.

Later, we will state and prove an important theorem that builds up a theoretical background for the utilization of Green's functions in solving boundary-value problems for nonhomogeneous equations. Then we will review the classical procedure for the construction of Green's functions, which is based on that theorem and the Lagrange method of variation of parameters that is traditionally used in ODE to analytically solve nonhomogeneous linear differential equations if the fundamental set of solutions is available for the corresponding homogeneous equation.

Consider a boundary-value problem for a linear nonhomogeneous equation

$$L[y(x)] \equiv p_0(x)\frac{d^n y}{dx^n} + p_1(x)\frac{d^{n-1} y}{dx^{n-1}} + \cdots + p_n(x)y = -f(x) \qquad (2.73)$$

subject to the homogeneous boundary conditions

$$M_i(y(a), y(b)) \equiv \sum_{k=0}^{n-1}\left[\alpha_k^i \frac{d^k y(a)}{dx^k} + \beta_k^i \frac{d^k y(b)}{dx^k}\right] = 0, \quad (i = \overline{1, n}) \qquad (2.74)$$

where the coefficients $p_j(x)$ in the governing equation are continuous functions, with $p_0(x) \neq 0$ on $(a, b)$, and $M_i$ represent linearly independent forms with constant coefficients.

The following theorem establishes a connection between the uniqueness of the solution of the setting in eqns (2.73), (2.74) and the corresponding homogeneous problem. It also prepares a background for the use of Green's function, constructed for a homogeneous problem, in solving nonhomogeneous equations.

**Theorem 2.3:** If the homogeneous boundary-value problem corresponding to that in eqns (2.73) and (2.74) has only the trivial solution, then the setting in eqns (2.73) and (2.74) has a unique solution.

**Proof:** The statement of this theorem follows from the linearity of the setting in eqns (2.73) and (2.74). Indeed, let $Y_1(x)$ and $Y_2(x)$ represent two distinct solutions to (2.73) and (2.74). This means that each of these solutions is supposed to make eqn (2.73) true. That is

$$p_0(x)\frac{d^n Y_1}{dx^n} + p_1(x)\frac{d^{n-1} Y_1}{dx^{n-1}} + \cdots + p_n(x)Y_1 = -f(x)$$

and

$$p_0(x)\frac{d^n Y_2}{dx^n} + p_1(x)\frac{d^{n-1} Y_2}{dx^{n-1}} + \cdots + p_n(x)Y_2 = -f(x)$$

Subtracting these in a term-by-term manner, we have

$$p_0(x)\frac{d^n (Y_1 - Y_2)}{dx^n} + p_1(x)\frac{d^{n-1} (Y_1 - Y_2)}{dx^{n-1}} + \cdots + p_n(x)(Y_1 - Y_2) = 0$$

Thus, if $Y_1(x)$ and $Y_2(x)$ represent two distinct solutions to eqn (2.73), then their difference $Y_{12}(x) = Y_1(x) - Y_2(x)$ is a solution to the corresponding homogeneous equation. In the same fashion, taking advantage of the linearity of the forms $M_i$, we can show that $Y_{12}(x)$ ought to satisfy the homogeneous boundary conditions in eqn (2.74). In other words, $Y_{12}(x)$ represents a solution to the homogeneous boundary-value problem corresponding to (2.73) and (2.74). But, according to the statement in this theorem, the corresponding homogeneous problem has only the trivial solution, which means that the difference $Y_1(x) - Y_2(x)$ ought to be identically zero.

So, our assumption about the existence of two distinct solutions of the original setting in eqns (2.73) and (2.74) is wrong and there exists, therefore, a unique solution of that problem, if the corresponding homogeneous problem has only the trivial solution. ■

### 2.3.1 Method of variation of parameters

Recall now from Section 2.1 that if the homogeneous boundary-value problem corresponding to that in eqns (2.73) and (2.74) has only the trivial solution, then there exists its unique Green's function. The theorem below establishes a direct way for expressing the solution to the problem in eqns (2.73) and (2.74) in terms of the Green's function constructed for the corresponding homogeneous boundary-value problem.

**Theorem 2.4:** If $g(x, s)$ represents the Green's function of the homogeneous boundary-value problem corresponding to that posed by eqns (2.73) and (2.74), then the unique solution of that problem itself can be expressed by the integral

$$y(x) = \int_a^b g(x, s) f(s)\, ds \tag{2.75}$$

**Proof:** It is clear that two independent points require to be proven. First, that the integral in eqn (2.75) satisfies the equation (2.73), and second, that it satisfies the boundary conditions in eqn (2.74).

Since the Green's function $g(x, s)$ is defined in two pieces, we break down the integral in eqn (2.75) into two integrals as shown

$$y(x) = \int_a^x g^-(x, s) f(s)\, ds + \int_x^b g^+(x, s) f(s)\, ds \tag{2.76}$$

where, as it was accustomed earlier, by $g^+(x, s)$ and $g^-(x, s)$, we denote the upper (valid for $x \leq s$) and the lower (valid for $x \geq s$), respectively, branches of $g(x, s)$.

To find out if the nonhomogeneous equation in (2.73) is satisfied by $y(x)$ as defined in eqn (2.76), we need to differentiate it. In doing so, we ought to take into account a specific occurrence of $y(x)$ in eqn (2.76). The point is that it is defined in terms of definite integrals (with respect to $s$), which contain a parameter $x$ and have variable limits depending on $x$. Therefore, one has to recall from the first part of the fundamental theorem of integral calculus [60] that if a function $\Phi(x)$ is defined in the integral form

$$\Phi(x) = \int_{\alpha(x)}^{\beta(x)} F(x, s)\, ds$$

then its derivative (with respect to $x$) is written as

$$\frac{d\Phi(x)}{dx} = F(x, \beta(x))\beta'(x) - F(x, \alpha(x))\alpha'(x) + \int_{\alpha(x)}^{\beta(x)} \frac{\partial F(x, s)}{\partial x}\, ds \tag{2.77}$$

Hence, since both of the integrals in eqn (2.76) contain $x$ as a parameter and their limits depend on $x$, we obtain

$$\frac{dy(x)}{dx} = \int_a^x \frac{\partial g^-(x, s)}{\partial x} f(s)\, ds + g^-(x, x) f(x)$$
$$+ \int_x^b \frac{\partial g^+(x, s)}{\partial x} f(s)\, ds - g^+(x, x) f(x)$$

The above integrals can be combined and non-integral terms are eliminated due to the continuity of the Green's function as $x = s$. This yields

$$\frac{dy(x)}{dx} = \int_a^b \frac{\partial g(x, s)}{\partial x} f(s)\, ds \tag{2.78}$$

Recalling the continuity of the derivatives of the Green's function of up to the $(n-2)$-nd order included as $x = s$ (see property 2 of the definition), the higher order derivatives of the integral in eqn (2.76) of up to the $(n-1)$-st order included can be computed analogously to the first derivative in eqn (2.78) as

$$\frac{d^k y(x)}{dx^k} = \int_a^b \frac{\partial^k g(x, s)}{\partial x^k} f(s)\, ds, \quad (k = \overline{1, n-1}) \tag{2.79}$$

Thus, the boundary conditions in eqn (2.74) are satisfied with $y(x)$ expressed by eqn (2.76), since all the derivatives of $y(x)$ of order up to $n-1$ in $M_i(y(a), y(b))$ can be taken under the integral sign. Indeed, substituting the derivatives of $y(x)$ from (2.79) in (2.74) and interchanging the order of the integration and the summation, we obtain

$$M_i(y(a), y(b)) \equiv \sum_{k=0}^{n-1}\left[\alpha_k^i \int_a^b \frac{\partial^k g(a, s)}{\partial x^k} f(s)\, ds + \beta_k^i \int_a^b \frac{\partial^k g(b, s)}{\partial x^k} f(s)\, ds\right]$$

$$= \int_a^b \sum_{k=0}^{n-1}\left[\alpha_k^i \frac{\partial^k g(a, s)}{\partial x^k} + \beta_k^i \frac{\partial^k g(b, s)}{\partial x^k}\right] f(s)\, ds = 0, \quad i = \overline{1, n}$$

because the expressions in the brackets equal zero due to the defining property of the Green's function.

In order to substitute $y(x)$ from (2.76) into eqn (2.73), we compute the $n$-th order derivative of $y(x)$ by differentiating the relation in eqn (2.79), where $k$ is fixed as $n-1$. This yields

$$\frac{d^n y(x)}{dx^n} = \int_a^b \frac{\partial^n g(x, s)}{\partial x^n} f(s)\, ds + \left[\frac{\partial^{n-1} g^-(x, x)}{\partial x^{n-1}} - \frac{\partial^{n-1} g^+(x, x)}{\partial x^{n-1}}\right] f(x)$$

which, in compliance with property 3 of the definition of Green's function, transforms into

$$\frac{d^n y(x)}{dx^n} = \int_a^b \frac{\partial^n g(x, s)}{\partial x^n} f(s)\, ds - f(x) p_0^{-1}(x)$$

Upon substituting $y(x)$ and its derivatives found above into eqn (2.73) and combining all the integral terms into a single term, one finally obtains

$$\int_a^b L[g(x, s)]\, f(s)\, ds - f(x) = -f(x)$$

The above equality is an identity, since $L[g(x, s)] = 0$ on $(a, b)$. Thus, the theorem has been proven. ∎

In Section 2.1, we showed how Green's functions can be constructed for $n$-th order linear ordinary differential equations based on the defining properties. Utilizing the above theorem, we will present another approach which can be used for the construction of Green's functions. The idea behind this approach is to employ Lagrange's method of variation of parameters which is traditionally used in solving nonhomogeneous linear differential equations.

For the sake of simplicity, we consider an equation of the second order

$$L[y(x)] \equiv p_0(x)\frac{d^2y(x)}{dx^2} + p_1(x)\frac{dy(x)}{dx} + p_2(x)y(x) = -f(x) \tag{2.80}$$

subject to the simplest set of boundary conditions

$$y(a) = 0, \quad y(b) = 0 \tag{2.81}$$

Assume that the above boundary-value problem has a unique solution, which implies, as we know, that the corresponding homogeneous problem has only the trivial solution. Let $y_1(x)$ and $y_2(x)$ represent two linearly independent particular solutions of the homogeneous equation associated with that in (2.80). Express then the general solution of eqn (2.80), in compliance with the method of variation of parameters, in the form

$$y(x) = C_1(x)y_1(x) + C_2(x)y_2(x) \tag{2.82}$$

where $C_1(x)$ and $C_2(x)$ are differentiable functions to be found in what follows.

The expression in eqn (2.82) does not look well-posed, since eqn (2.80) proposes the only relation on $(a, b)$ available at this point of derivation for determining $C_1(x)$ and $C_2(x)$. This presumes a certain degree of freedom in choosing a second relation, which would allow us to uniquely define $C_1(x)$ and $C_2(x)$. Lagrange's method provides an effective and elegant choice of such a relation.

The direct substitution of $y(x)$ from eqn (2.82) into eqn (2.80) would result in a cumbersome single differential equation of the second order in two unknown functions $C_1(x)$ and $C_2(x)$. In order to avoid such an unfortunate complication, the procedure in Lagrange's method suggests as follows. First, differentiate $y(x)$ in eqn (2.82) using the product rule

$$y'(x) = C_1'(x)y_1(x) + C_1(x)y_1'(x) + C_2'(x)y_2(x) + C_2(x)y_2'(x)$$

and then, keeping in mind the degree of freedom mentioned above, we make a simplifying assumption as

$$C_1'(x)y_1(x) + C_2'(x)y_2(x) = 0 \tag{2.83}$$

resulting in

$$y'(x) = C_1(x)y_1'(x) + C_2(x)y_2'(x) \tag{2.84}$$

Hence, the second derivative of $y(x)$ is now expressed as follows

$$y''(x) = C_1'(x)y_1'(x) + C_1(x)y_1''(x) + C_2'(x)y_2'(x) + C_2(x)y_2''(x) \tag{2.85}$$

Substitute $y(x)$, $y'(x)$, and $y''(x)$ from eqns (2.82), (2.84), and (2.85) into eqn (2.80). This yields

$$\begin{aligned} p_0(C_1'y_1' + C_1y_1'' + C_2'y_2' + C_2y_2'') + p_1(C_1y_1' + C_2y_2') \\ + p_2(C_1y_1 + C_2y_2) = -f(x) \end{aligned}$$

Rearranging the order of terms, we rewrite this as

$$\begin{aligned} C_1(p_0y_1'' + p_1y_1' + p_2y_1) + C_2(p_0y_2'' + p_1y_2' + p_2y_2) \\ + p_0(C_1'y_1' + C_2'y_2') = -f(x) \end{aligned}$$

Since $y_1(x)$ and $y_2(x)$ represent particular solutions of the homogeneous equation associated with eqn (2.80), the coefficients of $C_1(x)$ and $C_2(x)$ in the above equation are zero. This yields

$$C_1'(x)y_1'(x) + C_2'(x)y_2'(x) = -f(x)p_0^{-1}(x) \tag{2.86}$$

Solving eqns (2.83) and (2.86) simultaneously, we obtain

$$C_1'(x) = -\frac{y_2(x)f(x)}{p_0(x)W(x)}, \quad C_2'(x) = \frac{y_1(x)f(x)}{p_0(x)W(x)}$$

where $W(x) = y_1(x)y_2'(x) - y_2(x)y_1'(x)$ is the Wronskian of the fundamental set of solutions $y_1(x)$ and $y_2(x)$. Straightforward integration of the derivatives $C_1'(x)$ and $C_2'(x)$ yields

$$C_1(x) = -\int_a^x \frac{y_2(s)f(s)}{p_0(s)W(s)}\,ds + H_1, \quad C_2(x) = \int_a^x \frac{y_1(s)f(s)}{p_0(s)W(s)}\,ds + H_2$$

Substituting these values of $C_1(x)$ and $C_2(x)$ into eqn (2.82), we notice that, since $s$ represents the integration variable, the factors $y_1(x)$ and $y_2(x)$ can be formally taken inside of the integrals. And after the two integrals are combined, we obtain

$$y(x) = \int_a^x \frac{y_1(s)y_2(x) - y_1(x)y_2(s)}{p_0(s)W(s)} f(s)\,ds + H_1y_1(x) + H_2y_2(x) \tag{2.87}$$

Let us satisfy now the boundary conditions in eqn (2.81) with $y(x)$ as expressed above. This yields the following system of linear algebraic equations

$$\begin{pmatrix} y_1(a) & y_2(a) \\ y_1(b) & y_2(b) \end{pmatrix} \times \begin{pmatrix} H_1 \\ H_2 \end{pmatrix} = \begin{pmatrix} 0 \\ P(a,b) \end{pmatrix} \tag{2.88}$$

in the coefficients $H_1$ and $H_2$, where $P(a, b)$ is defined as

$$P(a, b) = \int_a^b \frac{R(b, s)}{p_0(s)W(s)} f(s)\, ds$$

where $R(b, s) = y_1(b)y_2(s) - y_1(s)y_2(b)$. This brings the solution to the system in eqn (2.88) in the form

$$H_1 = -\int_a^b \frac{y_2(a)R(b, s)f(s)}{p_0(s)R(a, b)W(s)}\, ds, \quad H_2 = \int_a^b \frac{y_1(a)R(b, s)f(s)}{p_0(s)R(a, b)W(s)}\, ds$$

Upon substituting these in eqn (2.87), we obtain the solution of the boundary-value problem in eqns (2.80) and (2.81) as

$$y(x) = -\int_a^x \frac{R(x, s)f(s)}{p_0(s)W(s)}\, ds + \int_a^b \frac{R(a, x)R(b, s)f(s)}{p_0(s)R(a, b)W(s)}\, ds$$

This representation can be rewritten (later in Example 3.3 we will clarify this transformation) as a single integral

$$y(x) = \int_a^b g(x, s)f(s)\, ds \tag{2.89}$$

whose kernel function $g(x, s)$ is expressed in two pieces. For the range $x \leq s$, it is defined as

$$g(x, s) = \frac{R(a, x)R(b, s)}{p_0(s)R(a, b)W(s)}, \quad x \leq s \tag{2.90}$$

while for the range $x \geq s$, one readily obtains

$$g(x, s) = \frac{R(a, x)R(b, s) - R(x, s)R(a, b)}{p_0(s)R(a, b)W(s)}, \quad x \geq s$$

After a trivial but quite cumbersome transformation, the above expression can be simplified to

$$g(x, s) = \frac{R(a, s)R(b, x)}{p_0(s)R(a, b)W(s)}, \quad x \geq s \tag{2.91}$$

Thus, since the solution to the problem posed by eqns (2.80) and (2.81) is found as the integral in eqn (2.89), by virtue of Theorem 2.4 we conclude that the kernel function $g(x, s)$ does in fact represent the Green's function to the corresponding homogeneous boundary-value problem.

As it was shown earlier, if the setting in eqns (2.80) and (2.81) is self-adjoint, then the product $p_0(s)W(s)$ is equal to a constant (see eqn (2.69)), which obviously makes the expressions in eqns (2.90) and (2.91) symmetric in the sense discussed in Section 2.1.

### 2.3.2 Examples of the construction

So, the approach based on the method of variation of parameters can successfully be used for actual construction of Green's functions as an alternative to the method described in Section 2.1. We present below a number of examples illustrating some peculiarities of this approach that emerge in practical situations.

**Example 3.1:** Find the derivative of the following function

$$\Phi(x) = \int_{x^2}^{x} (x+s)^2\, ds \tag{2.92}$$

expressed in integral form, containing a parameter $x$ and having variable limits. To find the derivative $d\Phi/dx$, we apply the formula from eqn (2.77). This yields

$$\frac{d\Phi(x)}{dx} = (x+x)^2 - 2x(x+x^2)^2 + \int_{x^2}^{x} 2(x+s)\, ds$$

Integrating and carrying out a trivial transformation, one finally obtains

$$\frac{d\Phi(x)}{dx} = 7x^2 - 4x^3 - 5x^4 - 2x^5$$

The same expression for $d\Phi/dx$ is obtained by directly evaluating the integral in eqn (2.92) and differentiating the result afterwards. ▲

**Example 3.2:** Differentiate the function

$$\Phi(x) = \int_{0}^{x} (x-s)^2\, ds$$

Since the integrand $(x-s)^2$ equals zero if $x=s$, the non-integral terms in eqn (2.69) vanish and the derivative $d\Phi/dx$ can, in this case, be obtained by formal differentiation under the integral sign. That is

$$\frac{d\Phi}{dx} = \int_{0}^{x} 2(x-s)\, ds = x^2$$

Note that such an occurrence of the integrand usually takes place when the Lagrange's method is used for the construction of Green's functions. ▲

**Example 3.3:** Apply the procedure based on the method of variation of parameters to the construction of the Green's function for the nonhomogeneous equation

$$\frac{d^2y(x)}{dx^2} + k^2y(x) = -f(x), \quad x \in (0, a) \tag{2.93}$$

subject to homogeneous boundary conditions imposed as

$$y'(0) = 0, \quad y'(a) = 0 \tag{2.94}$$

We assume that the right-hand side function $f(x)$ in eqn (2.93) is continuous on $(0, a)$.

It can easily be shown that the homogeneous problem associated with that in eqns (2.93) and (2.94) has only the trivial solution. This implies that the conditions of existence and uniqueness of the Green's function are met and the latter can be constructed.

Since the functions $y_1(x) \equiv \sin kx$ and $y_2(x) \equiv \cos kx$ represent a fundamental set of solutions for the corresponding homogeneous equation, the general solution to (2.93) can be expressed as

$$y(x) = C_1(x) \sin kx + C_2(x) \cos kx \tag{2.95}$$

The system of linear algebraic equations in $C_1'(x)$ and $C_2'(x)$, which has been derived in eqns (2.83) and (2.86) appears, in this case, as

$$\begin{pmatrix} \sin kx & \cos kx \\ k \cos kx & -k \sin kx \end{pmatrix} \times \begin{pmatrix} C_1'(x) \\ C_2'(x) \end{pmatrix} = \begin{pmatrix} 0 \\ -f(x) \end{pmatrix}$$

providing us with the following solution

$$C_1'(x) = \frac{1}{k} \cos kx f(x), \quad C_2'(x) = -\frac{1}{k} \sin kx f(x)$$

Integrating, one obtains

$$C_1(x) = \int_0^x \frac{1}{k} \cos ks f(s)\, ds + H_1, \quad C_2(x) = -\int_0^x \frac{1}{k} \sin ks f(s)\, ds + H_2$$

Upon substituting these into eqn (2.95) and carrying out an obvious transformation, we obtain

$$y(x) = \int_0^x \frac{1}{k} \sin k(x-s) f(s)\, ds + H_1 \sin kx + H_2 \cos kx \tag{2.96}$$

To determine the values of $H_1$ and $H_2$, we differentiate $y(x)$

$$y'(x) = \int_0^x \cos k(x-s) f(s)\, ds + H_1 k \cos kx - H_2 k \sin kx$$

Note that in performing the above differentiation, it appears that the non-integral terms (which are present in eqn (2.77)) do not, in this case, show up. It happens because of the specific form of the integrand $\sin k(x-s)$ in (2.96) which vanishes as $x = s$ (see the comment provided in Example 3.2).

From the first condition in eqn (2.94), it follows that $H_1 = 0$, while the second condition yields

$$\int_0^a \cos k(a-s) f(s)\, ds - H_2 k \sin ka = 0$$

from which we immediately obtain

$$H_2 = \int_0^a \frac{\cos k(a-s)}{k \sin ka} f(s)\, ds$$

Upon substituting the values of $H_1$ and $H_2$ just found into eqn (2.96) and correspondingly regrouping the integrals, one obtains

$$y(x) = \int_0^x \frac{\sin k(x-s)}{k} f(s)\, ds + \int_0^a \cos(kx) \frac{\cos k(a-s)}{k \sin ka} f(s)\, ds \tag{2.97}$$

Both of the above integrals can be combined and written in the form of a compact single integral. In helping the reader to easier proceed through this transformation, we add formally the term

$$\int_x^a 0 \cdot f(s)\, ds$$

to the first of the two integrals in eqn (2.97) and break down the second as

$$\int_0^a \cos kx \frac{\cos k(a-s)}{k \sin ka} f(s)\, ds = \int_0^x \cos kx \frac{\cos k(a-s)}{k \sin ka} f(s)\, ds$$
$$+ \int_x^a \cos kx \frac{\cos k(a-s)}{k \sin ka} f(s)\, ds.$$

If so, then $y(x)$ is presented as a sum of four definite integrals, in two of which we integrate from 0 to $x$, in the other two – from $x$ to $a$. That is

$$y(x) = \int_0^x \frac{\sin k(x-s)}{k} f(s)\, ds + \int_0^x \cos kx \frac{\cos k(a-s)}{k \sin ka} f(s)\, ds$$
$$+ \int_x^a 0 \cdot f(s)\, ds + \int_x^a \cos kx \frac{\cos k(a-s)}{k \sin ka} f(s)\, ds$$

Combining the first two integrals and the other two, we have

$$y(x) = \int_0^x \left[ \frac{\sin k(x-s)}{k} + \cos kx \frac{\cos k(a-s)}{k \sin ka} \right] f(s)\, ds$$
$$+ \int_x^a \cos kx \frac{\cos k(a-s)}{k \sin ka} f(s)\, ds$$
$$= \int_0^x \cos ks \frac{\cos k(a-x)}{k \sin ka} f(s)\, ds + \int_x^a \cos kx \frac{\cos k(a-s)}{k \sin ka} f(s)\, ds$$

Note that in the first integral, the variables $x$ and $s$ satisfy the inequality $x \geq s$, since $x$ represents the upper limit of integration, whereas in the second integral $x$ is the lower limit, so $x \leq s$.

Hence, the above representation for $y(x)$ can be viewed as a single integral

$$y(x) = \int_0^a g(x, s) f(s)\, ds \tag{2.98}$$

whose kernel function $g(x, s)$ is defined in two pieces as

$$g(x, s) = \frac{1}{k \sin ka} \begin{cases} \cos kx \cos k(a - s), & \text{for } x \le s \\ \cos ks \cos k(a - x), & \text{for } s \le x \end{cases} \tag{2.99}$$

Thus, since the solution of the boundary-value problem stated with eqns (2.93) and (2.94) is expressed as the integral in eqn (2.98), $g(x, s)$ represents, in compliance with Theorem 2.4, the Green's function to the homogeneous boundary-value problem associated with that in eqns (2.93) and (2.94). ▲

**Example 3.4:** Consider the nonhomogeneous equation

$$\frac{d^2y(x)}{dx^2} - k^2 y(x) = -f(x) \tag{2.100}$$

subject to the homogeneous boundary conditions

$$y'(0) = 0, \quad y(a) = 0 \tag{2.101}$$

Evidently, the homogeneous boundary-value problem corresponding to that of eqns (2.100) and (2.101) has only the trivial solution. This justifies the existence and uniqueness of the Green's function.

It is evident that the following set of functions

$$y_1(x) \equiv \exp(kx), \quad y_2(x) \equiv \exp(-kx)$$

represents a fundamental set of solutions for the homogeneous equation corresponding to that in (2.100). Hence, the general solution of eqn (2.100) can be represented by

$$y(x) = C_1(x) \exp(kx) + C_2(x) \exp(-kx) \tag{2.102}$$

In compliance with the procedure of Lagrange's method, one obtains the values of $C_1(x)$ and $C_2(x)$ in the form

$$C_1(x) = -\int_0^x \frac{1}{2k} \exp(-ks) f(s)\, ds + H_1,$$

$$C_2(x) = \int_0^x \frac{1}{2k} \exp(ks) f(s)\, ds + H_2$$

The substitution of these into eqn (2.102) yields

$$y(x) = -\int_0^x \frac{1}{k} \sinh k(x - s) f(s)\, ds + H_1 \exp(kx) + H_2 \exp(-kx) \tag{2.103}$$

The first boundary condition $y'(0) = 0$ in eqn (2.101) implies that $H_1 = H_2$, while the second condition $y(a) = 0$ yields

$$H_1 = H_2 = \int_0^a \frac{\sinh k(a-s)}{2k \cosh ka} f(s)\, ds$$

Substituting these in (2.103), one obtains

$$y(x) = \int_0^a \frac{\cosh kx \sinh k(a-s)}{k \cosh ka} f(s)\, ds - \int_0^x \frac{1}{k} \sinh k(x-s) f(s)\, ds$$

Hence, following again the transformation of the above integrals as in Example 3.3, we obtain the Green's function $g(x, s)$ to the homogeneous boundary-value problem associated with that in eqns (2.100) and (2.101) as

$$g(x, s) = \frac{1}{k \cosh ka} \begin{cases} \cosh kx \sinh k(a-s), & \text{for } x \leq s \\ \cosh ks \sinh k(a-x), & \text{for } s \leq x \end{cases} \tag{2.104}$$

▲

**Example 3.5:** Let us consider eqn (2.100) again, but we subject it to a different set of boundary conditions. Namely, we consider the case of

$$y'(0) - hy(0) = 0, \qquad |y(\infty)| < \infty \tag{2.105}$$

This example is designed to show how Lagrange's method manages to treat the boundedness conditions of the type that occurs in eqn (2.105).

It can be easily checked out that there exists a unique Green's function for the homogeneous boundary-value problem corresponding to that posed by eqns (2.100) and (2.105).

The general solution of eqn (2.100) was derived in eqn (2.103). In this case, however, we prefer to express it completely in terms of exponential functions

$$y(x) = \int_0^x \frac{1}{2k} [e^{k(s-x)} - e^{k(x-s)}] f(s)\, ds + H_1 e^{kx} + H_2 e^{-kx} \tag{2.106}$$

in contrast to the mixed hyperbolic-exponential form in eqn (2.103). The point is that the form in eqn (2.106) will be more practical in view of the necessity to treat the boundedness condition $|y(\infty)| < \infty$ in the discussion that follows. Indeed, splitting off both of the exponential terms under the integral sign and grouping then both of the terms containing $\exp(kx)$ and both of the terms containing $\exp(-kx)$, we rewrite eqn (2.106) as

$$y(x) = \left( H_1 - \int_0^x \frac{e^{-ks}}{2k} f(s)\, ds \right) e^{kx} + \left( H_2 + \int_0^x \frac{e^{ks}}{2k} f(s)\, ds \right) e^{-kx} \tag{2.107}$$

It is clearly seen that the condition of boundedness $|y(\infty)| < \infty$ implies that the coefficient of the positive exponential term $\exp(kx)$ in eqn (2.107) ought to equal zero as $x$ approaches infinity. This yields

$$H_1 = \int_0^\infty \frac{1}{2k} \exp(-ks) f(s)\, ds$$

while the first condition in eqn (2.105) subsequently yields

$$H_2 = \frac{k-h}{k+h} H_1 = \int_0^\infty \frac{k-h}{2k(k+h)} \exp(-ks) f(s)\, ds$$

Upon substituting the values of $H_1$ and $H_2$ just found in eqn (2.106) and rewriting then its first term again in a more compact hyperbolic form, we obtain

$$y(x) = -\int_0^x \frac{1}{k} \sinh k(x-s) f(s)\, ds$$
$$+ \int_0^\infty \frac{1}{2k} \exp(-ks)\Big(\exp(kx) + h^* \exp(-kx)\Big) f(s)\, ds$$

where $h^* = (k-h)/(k+h)$. From this representation it follows that the Green's function $g(x, s)$ to the problem in eqns (2.100) and (2.105) is written as

$$g(x, s) = \frac{1}{2k} \begin{cases} \exp(-ks)(\exp(kx) + h^* \exp(-kx)), & \text{for } x \le s \\ \exp(-kx)(\exp(ks) + h^* \exp(-ks)), & \text{for } s \le x \end{cases} \tag{2.108}$$

▲

**Example 3.6:** We consider a boundary-value problem stated for the equation with variable coefficients

$$\frac{d}{dx}\left((\beta^2 x^2 + 1)\frac{dy(x)}{dx}\right) = f(x), \quad x \in (0, a)$$

with boundary conditions imposed as

$$y(0) = 0, \quad y(a) = 0$$

and briefly describe the construction procedure for the Green's function of the corresponding homogeneous problem.

A fundamental set of solutions can, in this case, be formed with the functions $y_1(x) \equiv 1$ and $y_2(x) \equiv \arctan \beta x$. This yields the general solution to the governing equation in the form

$$y(x) = \int_0^x \frac{1}{\beta} \arctan\left(\frac{\beta(s-x)}{1+\beta^2 xs}\right) f(s)\, ds + D_1 + D_2 \arctan \beta x$$

By satisfying the boundary conditions, the values of $D_1$ and $D_2$ are found as

$$D_1 = 0, \quad D_2 = \int_0^a \frac{\arctan \beta a - \arctan \beta s}{\beta \arctan \beta a} f(s)\, ds$$

Substituting these into the above expression for the general solution and rearranging the integral terms, we obtain the solution to the original boundary-value problem as

$$y(x) = \int_0^a g(x, s) f(s) \, ds$$

where the kernel $g(x, s)$ represents the Green's function that we are looking for as defined in two pieces

$$g(x, s) = \frac{1}{A} \begin{cases} \arctan \beta x (A - \arctan \beta s), & \text{for } 0 \leq x \leq s \\ \arctan \beta s (A - \arctan \beta x), & \text{for } x \leq s \leq a \end{cases}$$

where $A = \beta \arctan \beta a$. ▲

**Example 3.7:** We construct here the Green's function for the homogeneous boundary-value problem associated with the following equation

$$\frac{d^4 y(x)}{dx^4} - 2k^2 \frac{d^2 y(x)}{dx^2} + k^4 y(x) = -f(x) \tag{2.109}$$

subject to the boundary conditions

$$y(0) = 0, \quad y'(0) = 0, \quad |y(\infty)| < \infty, \quad |y'(\infty)| < \infty \tag{2.110}$$

This setting simulates, in structural mechanics [12, 21, 27, 55, 58, 64], a special case of a semi-infinite elastic beam resting on an elastic foundation, with clamped edge $x = 0$.

Existence and uniqueness of the Green's function for the above problem can be routinely justified. And since the characteristic equation

$$m^4 - 2k^2 m^2 + k^4 = 0$$

associated with eqn (2.109) has two pairs of repeated roots: $m_{1,2} = k$ and $m_{3,4} = -k$, the general solution for eqn (2.109) can be expressed as

$$y(x) = C_1(x) e^{kx} + C_2(x) e^{-kx} + C_3(x) x e^{kx} + C_4(x) x e^{-kx} \tag{2.111}$$

The coefficient matrix for the system of linear algebraic equations in $C_i'(x)$, $(i = 1, 2, 3, 4)$ is obtained in this case as

$$\begin{pmatrix} e^{kx} & e^{-kx} & xe^{kx} & xe^{-kx} \\ ke^{kx} & -ke^{-kx} & (1+kx)e^{kx} & (1-kx)e^{-kx} \\ k^2 e^{kx} & k^2 e^{-kx} & k(2+kx)e^{kx} & -k(2-kx)e^{-kx} \\ k^3 e^{kx} & -k^3 e^{-kx} & k^2(3+kx)e^{kx} & k^2(3-kx)e^{-kx} \end{pmatrix}$$

while the right-hand side vector is $(0, 0, 0, -f(x))^T$. Clearly, the above matrix is non-singular, because its determinant represents the Wronskian of the fundamental set of solutions used in eqn (2.111). Therefore, one can readily obtain

$$C_1'(x) = \frac{1+kx}{4k^3} e^{-kx} f(x), \quad C_2'(x) = -\frac{1-kx}{4k^3} e^{kx} f(x)$$
$$C_3'(x) = -\frac{1}{4k^2} e^{-kx} f(x), \quad C_4'(x) = -\frac{1}{4k^2} e^{kx} f(x)$$

Upon integrating these, the functions $C_i(x)$ are found as

$$C_1(x) = \int_0^x \frac{1+ks}{4k^3} e^{-ks} f(s)\, ds + H_1$$
$$C_2(x) = -\int_0^x \frac{1-ks}{4k^3} e^{ks} f(s)\, ds + H_2$$
$$C_3(x) = -\int_0^x \frac{1}{4k^2} e^{-ks} f(s)\, ds + H_3$$
$$C_4(x) = -\int_0^x \frac{1}{4k^2} e^{ks} f(s)\, ds + H_4$$

Hence, $y(x)$ in eqn (2.111), with $C_i(x)$ just obtained, provides the general solution to eqn (2.109). The constants $H_i$ are to be computed when satisfying the boundary conditions in eqn (2.110). Before going any further with this, for the better clarity in the development that follows, we first differentiate $y(x)$ in eqn (2.111) by using the product rule

$$y'(x) = ke^{kx} C_1(x) + \underbrace{e^{kx}\left(\frac{1+kx}{4k^3} e^{-kx} f(x)\right)}$$
$$- ke^{-kx} C_2(x) + \underbrace{e^{-kx}\left(-\frac{1-kx}{4k^3} e^{kx} f(x)\right)}$$
$$+ (1+kx)e^{kx} C_3(x) + \underbrace{xe^{kx}\left(-\frac{1}{4k^2} e^{-kx} f(x)\right)}$$
$$+ (1-kx)e^{-kx} C_4(x) + \underbrace{xe^{-kx}\left(-\frac{1}{4k^2} e^{kx} f(x)\right)}$$

By observation, we conclude that the sum of the underlined terms vanishes. Substituting then the values of $C_i(x)$ into the remaining part of $y'(x)$, one obtains

$$y'(x) = ke^{kx}\left(\int_0^x \frac{1+ks}{4k^3} e^{-ks} f(s)\, ds + H_1\right)$$
$$+ ke^{-kx}\left(-\int_0^x \frac{1-ks}{4k^3} e^{ks} f(s)\, ds + H_2\right)$$

$$+ (1+kx)e^{kx}\left(-\int_0^x \frac{1}{4k^2}e^{-ks} f(s)\, ds + H_3\right)$$
$$+ (1-kx)e^{-kx}\left(-\int_0^x \frac{1}{4k^2}e^{ks} f(s)\, ds + H_4\right)$$

Let us now return to the boundary conditions imposed by eqn (2.110). The first of them $y(0) = 0$ yields

$$H_1 + H_2 = 0 \tag{2.112}$$

while the second $y'(0) = 0$ results in

$$kH_1 + H_3 - kH_2 + H_4 = 0 \tag{2.113}$$

The boundedness conditions $|y(\infty)| < \infty$ and $|y'(\infty)| < \infty$ formulated in eqn (2.110) provide

$$H_1 = -\int_0^\infty \frac{1+ks}{4k^3} e^{-ks} f(s)\, ds, \quad H_3 = \int_0^\infty \frac{1}{4k^2} e^{-ks} f(s)\, ds$$

Substituting these into eqns (2.112) and (2.113), one obtains

$$H_2 = \int_0^\infty \frac{1+ks}{4k^3} e^{-ks} f(s)\, ds, \quad H_4 = \int_0^\infty \frac{1+2ks}{4k^2} e^{-ks} f(s)\, ds$$

Upon substituting the values of $H_1$, $H_2$, $H_3$, and $H_4$ into eqn (2.111) and combining then the like integrals, one finally obtains

$$y(x) = \int_0^x \left(\frac{1-k(x-s)}{4k^3} e^{k(x-s)} - \frac{1+k(x-s)}{4k^3} e^{-k(x-s)}\right) f(s)\, ds$$
$$+ \int_0^\infty \left(\frac{1+k(x+s)+2k^2xs}{4k^3} e^{-k(x+s)} - \frac{1-k(x-s)}{4k^3} e^{k(x-s)}\right) f(s)\, ds.$$

This representation for $y(x)$ can be rewritten as a single integral

$$y(x) = \int_0^\infty g(x,s) f(s)\, ds$$

whose kernel-function $g(x, s)$ is found in the form

$$g(x,s) = \frac{1}{4k^3}\begin{cases}(1+k(x+s)+2k^2xs)e^{-k(x+s)} - (1-k(x-s))e^{k(x-s)}\\ (1+k(x+s)+2k^2xs)e^{-k(x+s)} - (1-k(s-x))e^{k(s-x)}\end{cases} \tag{2.114}$$

where the upper branch is defined for $x \le s$, while the lower branch is defined for $s \le x$. Hence, in view of Theorem 2.4, the kernel-function $g(x, s)$ represents the Green's function for the boundary-value problem posed by eqns (2.109) and (2.110). ▲

**Example 3.8:** Using a corresponding Green's function and following Theorem 2.4, find a solution of the nonhomogeneous equation

$$\frac{d^4y(x)}{dx^4} = P_0 \sin \pi x, \qquad P_0 = \text{const} \tag{2.115}$$

subject to the homogeneous boundary conditions

$$y(0) = \frac{dy(0)}{dx} = 0, \quad y(1) = \frac{d^2y(1)}{dx^2} = 0 \tag{2.116}$$

This setting models [55, 58] the deflection $y(x)$ of an elastic beam of unit length whose left edge is clamped while the right edge is simply supported. The beam is subject to a transverse load given as $f(x) = P_0 \sin \pi x$.

The Green's function for the homogeneous problem associated with that in eqns (2.115) and (2.116) was earlier derived in Section 2.1 (see eqn (2.45)). Its branch valid for $x \leq s$ was found as

$$g^+(x, s) = -\left(\frac{s^3}{4} - \frac{3s^2}{4} + \frac{s}{2}\right)x^2 + \left(\frac{s^3}{12} - \frac{s^2}{4} + \frac{1}{6}\right)x^3$$

Due to the self-adjointness of the original statement, the second branch $g^-(x, s)$ of the Green's function that is valid for $x \geq s$, can be obtained from that above by interchanging $x$ with $s$.

Hence, in compliance with Theorem 2.4, the solution of the problem in eqns (2.115) and (2.116) can be found by the straightforward integration

$$y(x) = -P_0 \int_0^1 g(x, s)(\sin \pi s)\, ds \tag{2.117}$$

In order to evaluate this integral, we recall that $g(x, s)$ is defined in two pieces and break accordingly the integral as

$$\begin{aligned} y(x) &= -P_0\left[\int_0^x g^-(x, s)(\sin \pi s)\, ds + \int_x^1 g^+(x, s)(\sin \pi s)\, ds\right] \\ &= -P_0\left[\int_0^x \left(\left(\frac{x^3}{12} - \frac{x^2}{4} + \frac{1}{6}\right)s^3 - \left(\frac{x^3}{4} - \frac{3x^2}{4} + \frac{x}{2}\right)s^2\right)(\sin \pi s)\, ds\right. \\ &\quad \left. + \int_x^1 \left(\left(\frac{s^3}{12} - \frac{s^2}{4} + \frac{1}{6}\right)x^3 - \left(\frac{s^3}{4} - \frac{3s^2}{4} + \frac{s}{2}\right)x^2\right)(\sin \pi s)\, ds\right] \end{aligned} \tag{2.118}$$

The actual computation of the integrals in eqn (2.118) is a routine procedure, by which we finally obtain

$$y(x) = \frac{P_0}{2\pi^4}[2 \sin \pi x - \pi x(x-1)(x-2)]$$

Note that the Green's function-based approach to boundary-value problems of the type in eqns (2.115) and (2.116) is especially effective if we are required to compute a number of solutions of same boundary-value problem with a variety of different right-hand side functions. In such a situation, the direct use of Theorem 2.4 provides a notable computational convenience.

To illustrate the last point, let us assume that we are required to find a solution to the setting in eqns (2.115) and (2.116), where the right-hand side in eqn (2.115) is given as a linear function $f(x) = mx + b$. Substituting it in eqn (2.117), after an elementary computation, we come up with

$$y(x) = \frac{x^2(1-x)}{240}[2mx^2 + 2(5b+m)x - (15b+7m)] \qquad \blacktriangle$$

The material in this chapter so far touched upon classical boundary-value problems for linear ordinary differential equations, for which the Green's function method is well developed and its use is a quite straightforward procedure. In the next section, we will present an extension of the Green's function formalism to a new sphere of possible applications where this method has not been used until recently [44, 45]. A specific class of problems is considered, which often occur in various areas of applied mechanics. The so-called *multi-point posed boundary-value problems* for some systems of linear ordinary differential equations will be treated by means of the Green's function method.

## 2.4 Boundary-contact Value Problems

In this section, we will be concerned with some nontraditional implementations of the Green's function method. We deal here with such statements that reduce to the so-called multi-point posed boundary-contact value problems for systems of linear ordinary differential equations. These are not, however, systems of equations in a common sense, where several unknown functions have a common domain, and at least one of the equations in the system involves more than one unknown function.

Each equation in the systems that are discussed in this section, governs a single unknown function and is formulated over an individual domain. The system is formed by letting the domains contact each other at their end-points (we call these the contact points). Subsequently, the single differential equations are put in a system format by some contact conditions imposed at the contact points.

### 2.4.1 Matrix of Green's type

The notion of a *matrix of Green's type* [44, 45] is introduced for a piecewise homogeneous media of a sandwich type. Later, in Section 2.6, we extend the notion of a matrix of Green's type to problems stated on more complex assemblies of one-dimensional elements.

To present a typical formulation of a multi-point posed boundary-value problem of the kind to be considered, let the interval $[a_0, a_k]$ be partitioned with a set of internal points $a_i$, $(i = \overline{1, k-1})$ into $k$ arbitrary subintervals $(a_{i-1}, a_i)$. Consider a set of nonhomogeneous linear differential equations

$$L_i[y_i(x)] \equiv \sum_{j=0}^{n} p_{ij}(x)\frac{dy_i^{(n-j)}(x)}{dx^{n-j}} = f_i(x), \quad x \in (a_{i-1}, a_i), \quad (i = \overline{1, k}) \tag{2.119}$$

each of which is stated over an individual subinterval. The coefficients $p_{ij}(x)$ of the operators $L_i$ represent continuous functions on $[a_{i-1}, a_i]$, with the leading coefficients $p_{i0}(x)$ being non-zero at any single point on $[a_{i-1}, a_i]$. Each of the right-hand side functions $f_i(x)$ in eqn (2.119) is also continuous on the corresponding subinterval $(a_{i-1}, a_i)$.

The set of equations in (2.119) does not represent a system in the traditional sense. We treat them as a system by imposing the following set of boundary and contact conditions

$$M_q[y_1(a_0), y_1(a_1), y_2(a_1), \ldots, y_k(a_k)] = 0, \quad (q = \overline{1, n \times k}) \tag{2.120}$$

where $M_q$ represent linearly independent forms modeling the boundary and contact conditions imposed at the points $a_i$, $(i = \overline{0, k})$. Two of the relations in eqn (2.120) set up boundary conditions at the end-point $a_0$ and $a_k$, the rest state contact conditions at the internal points $a_i$, $(i = \overline{1, k-1})$.

We assume that the homogeneous boundary-value problem corresponding to that stated by eqns (2.119) and (2.120) has only the trivial solution.

To prepare the extension of the notion of Green's function to the setting in eqns (2.119) and (2.120), we introduce a vector-function $\mathbf{Y}(x)$ whose components $Y_i(x)$ are defined in terms of the functions $y_i(x)$ in the following fashion

$$Y_i(x) = \begin{cases} y_i(x), & \text{for } x \in (a_{i-1}, a_i) \\ 0, & \text{for } x \in (a_0, a_k) \setminus (a_{i-1}, a_i) \end{cases}$$

For the right-hand side functions $f_i(x)$ in eqn (2.119), we also introduce a vector-function $\mathbf{F}(x)$ whose components $F_i(x)$ are defined as

$$F_i(x) = \begin{cases} f_i(x), & \text{for } x \in (a_{i-1}, a_i) \\ 0, & \text{for } x \in (a_0, a_k) \setminus (a_{i-1}, a_i) \end{cases}$$

We are now in a position to formally extend the definition of Green's function so as to make it valid for the multi-point posed boundary-value problems in eqns (2.119) and (2.120).

**Definition:** If for any allowable vector-function $\mathbf{F}(x)$, the vector-function $\mathbf{Y}(x)$ is expressed in the integral form

$$\mathbf{Y}(x) = -\int_{a_0}^{a_k} G(x, s)\mathbf{F}(s)\, ds \tag{2.121}$$

then let the kernel matrix

$$G(x, s) = (g_{ij}(x, s))_{i,j=\overline{1,k}}$$

in eqn (2.121) be referred to as the *matrix of Green's type* for the homogeneous multi-point posed boundary-value problem corresponding to that of eqns (2.119) and (2.120). Note that the first subscript $i$ in the component $g_{ij}(x, s)$ matches the domain of the $x$ variable, $x \in [a_{i-1}, a_i]$, while the second subscript $j$ matches the domain of the $s$ variable, $s \in [a_{j-1}, a_j]$.

For any fixed value of $s$, the components $g_{ij}(x, s)$ hold the following properties:

1. For $i \neq j$ (meaning that the domains of $x$ and $s$ never overlap), the functions $g_{ij}(x, s)$ are continuous along with their derivatives with respect to $x$ of up to the $n$-th order included.
2. For $i = j$ ($x$ and $s$ share the domain), when $x \neq s$, $g_{ii}(x, s)$ are also continuous along with their derivatives with respect to $x$ of up to the $n$-th order included, but as $x = s$, $g_{ii}(x, s)$ are continuous along with their derivatives with respect to $x$ of up to the $(n-2)$-nd order included, whereas their $(n-1)$-st derivatives make a jump of discontinuity, the magnitude of which equals $-p_{i0}^{-1}(s)$.
3. For $x \neq s$, $g_{ij}(x, s)$, as functions of $x$, satisfy the homogeneous equations
$$L_i[g_{ij}(x, s)] = 0, \quad x \in (a_{i-1}, a_i), \quad (i = \overline{1, k})$$
in the domain of $x$.
4. $g_{ij}(x, s)$ satisfy the boundary and contact conditions in (2.120) i.e.:
$$M_q[g_{ij}(a_0, s), g_{ij}(a_1, s), \ldots, g_{ij}(a_k, s)] = 0, \quad (q = \overline{1, n \times k})$$
in which they are involved.

### 2.4.2 Particular examples

In what follows, we will present several particular examples showing how matrices of Green's type can practically be constructed.

**Example 4.1:** Start with the simplest three-point posed boundary-value problem written as

$$\frac{d^2y_1(x)}{dx^2} = f_1(x), \quad x \in (-1, 0) \tag{2.122}$$

$$\frac{d^2y_2(x)}{dx^2} = f_2(x), \quad x \in (0, 1) \tag{2.123}$$

$$y_1(-1) = 0, \quad y_2(1) = 0 \tag{2.124}$$

$$y_1(0) = y_2(0), \quad \frac{dy_1(0)}{dx} = \lambda\frac{dy_2(0)}{dx} \tag{2.125}$$

This problem might, in particular, be interpreted as a model for steady-state heat conduction in a compound bar consisting of two physically homogeneous segments built of different materials. Parameter $\lambda$ represents here the ratio $\lambda_2/\lambda_1$ of the heat conductivities of the materials of which the bar is composed.

The homogeneous problem corresponding to that of eqns (2.122)–(2.125) has only the trivial solution (see Exercise 2.9(a)). Following Lagrange's method of variation of parameters, we represent the general solution of eqn (2.122) as

$$y_1(x) = C_1(x) + xC_2(x) \tag{2.126}$$

This yields the following system of linear algebraic equations

$$\begin{pmatrix} 1 & x \\ 0 & 1 \end{pmatrix} \times \begin{pmatrix} C_1'(x) \\ C_2'(x) \end{pmatrix} = \begin{pmatrix} 0 \\ f_1(x) \end{pmatrix}$$

with a well-posed coefficient matrix. From this, it follows that

$$C_1'(x) = -xf_1(x), \quad C_2'(x) = f_1(x)$$

Expressions for $C_1(x)$ and $C_2(x)$ are obtained by a straightforward integration as

$$C_1(x) = -\int_{-1}^{x} sf_1(s)\,ds + M_1, \quad C_2(x) = \int_{-1}^{x} f_1(s)\,ds + M_2$$

Substituting these expressions of $C_1(x)$ and $C_2(x)$ into eqn (2.126) and combining the integral terms, we obtain

$$y_1(x) = \int_{-1}^{x} (x-s)f_1(s)\,ds + M_1 + M_2x, \quad x \in [-1, 0] \tag{2.127}$$

Analogously for $y_2(x)$, we obtain

$$y_2(x) = \int_{0}^{x} (x-s)f_2(s)\,ds + N_1 + N_2x, \quad x \in [0, 1] \tag{2.128}$$

The boundary and contact conditions in eqns (2.124) and (2.125) applied to $y_1(x)$ and $y_2(x)$ will be used to compute the values of $M_1$, $M_2$, $N_1$, and $N_2$. The boundary conditions in eqn (2.124) yield

$$M_1 - M_2 = 0$$

$$N_1 + N_2 = \int_{0}^{1} (s-1)f_2(s)\,ds$$

while the contact conditions in eqn (2.125) result in

$$M_1 - N_1 = \int_{-1}^{0} sf_1(s)\,ds$$

$$\lambda N_2 - M_2 = \int_{-1}^{0} f_1(s)\,ds$$

These relations form a well-posed system of linear algebraic equations in $M_1$, $M_2$, $N_1$, and $N_2$, whose solution is

$$M_1 = M_2 = \frac{1}{1+\lambda}\int_{-1}^{0}(\lambda s - 1) f_1(s)\, ds + \frac{\lambda}{1+\lambda}\int_{0}^{1}(s-1) f_2(s)\, ds$$

$$N_1 = -\frac{1}{1+\lambda}\int_{-1}^{0}(s+1) f_1(s)\, ds + \frac{\lambda}{1+\lambda}\int_{0}^{1}(s-1) f_2(s)\, ds$$

and

$$N_2 = \frac{1}{1+\lambda}\int_{-1}^{0}(s+1) f_1(s)\, ds + \frac{1}{1+\lambda}\int_{0}^{1}(s-1) f_2(s)\, ds$$

Hence, substituting these values of the coefficients into eqns (2.127) and (2.128), after elementary transformations, one obtains

$$y_1(x) = \int_{-1}^{0}\frac{(x+1)(\lambda s-1)}{1+\lambda} f_1(s)\, ds + \int_{-1}^{x}(x-s) f_1(s)\, ds + \int_{0}^{1}\frac{\lambda(x+1)(s-1)}{1+\lambda} f_2(s)\, ds \tag{2.129}$$

and

$$y_2(x) = \int_{-1}^{0}\frac{(x-1)(s+1)}{1+\lambda} f_1(s)\, ds + \int_{0}^{x}(x-s) f_2(s)\, ds + \int_{0}^{1}\frac{(x+\lambda)(s-1)}{1+\lambda} f_2(s)\, ds \tag{2.130}$$

Combining the first and the second integrals from eqn (2.129), we obtain

$$y_1(x) = -\int_{-1}^{0} g_{11}(x,s) f_1(s)\, ds - \int_{0}^{1} g_{12}(x,s) f_2(s)\, ds \tag{2.131}$$

while combining the second and the third integrals in eqn (2.130), we have

$$y_2(x) = -\int_{-1}^{0} g_{21}(x,s) f_1(s)\, ds - \int_{0}^{1} g_{22}(x,s) f_2(s)\, ds \tag{2.132}$$

where the kernel functions $g_{ij}(x,s)$ are obtained as

$$g_{11}(x,s) = \begin{cases} \beta(x+1)(1-\lambda s), & \text{for } -1 \le x \le s < 0 \\ \beta(s+1)(1-\lambda x), & \text{for } -1 < s \le x \le 0 \end{cases}$$

$$g_{12}(x,s) = \lambda\beta(x+1)(1-s), \quad \text{for } -1 \le x \le 0 < s < 1$$

$$g_{21}(x,s) = \beta(1-x)(s+1), \text{ for } -1 < s < 0 \le x \le 1$$

and

$$g_{22}(x,s) = \begin{cases} \beta(x+\lambda)(1-s), & \text{for } 0 \le x \le s < 1 \\ \beta(s+\lambda)(1-x), & \text{for } 0 < s \le x \le 1 \end{cases}$$

with $\beta = (1+\lambda)^{-1}$.

In accordance with the approach suggested earlier in this section, we introduce the vector $\mathbf{Y}(x)$ whose components are defined as

$$Y_1(x) = \begin{cases} y_1(x), & \text{for } x \in (-1, 0) \\ 0, & \text{for } x \in (0, 1) \end{cases} \qquad Y_2(x) = \begin{cases} 0, & \text{for } x \in (-1, 0) \\ y_2(x), & \text{for } x \in (0, 1) \end{cases}$$

and the vector $\mathbf{F}(x)$ with components defined as

$$F_1(x) = \begin{cases} f_1(x), & \text{for } x \in (-1, 0) \\ 0, & \text{for } x \in (0, 1) \end{cases} \qquad F_2(x) = \begin{cases} 0, & \text{for } x \in (-1, 0) \\ f_2(x), & \text{for } x \in (0, 1) \end{cases}$$

In terms of the vectors $\mathbf{Y}(x)$ and $\mathbf{F}(x)$, the integrals from eqns (2.131) and (2.132) can be rewritten as a single integral

$$\mathbf{Y}(x) = -\int_{-1}^{1} G(x, s)\mathbf{F}(s)\, ds$$

Thus, from the definition introduced in this section, it follows that the expressions of $g_{ij}(x, s)$ just derived can be referred to as the entries of the matrix of Green's type $G(x, s)$ for the three-point posed homogeneous problem associated with that of eqns (2.122)–(2.125). ▲

**Example 4.2:** Consider a system of Cauchy–Euler equations

$$\frac{d}{dx}\left(x\frac{dy_1(x)}{dx}\right) - \frac{1}{x}y_1(x) = f_1(x), \quad x \in (0, a) \tag{2.133}$$

$$\frac{d}{dx}\left(x\frac{dy_2(x)}{dx}\right) - \frac{1}{x}y_2(x) = f_2(x), \quad x \in (a, \infty) \tag{2.134}$$

with the boundary and contact conditions imposed as

$$|y_1(0)| < \infty, \quad |y_2(\infty)| < \infty \tag{2.135}$$

$$y_1(a) = y_2(a), \quad \frac{dy_1(a)}{dx} = \lambda\frac{dy_2(a)}{dx} \tag{2.136}$$

Note that two remarkable scientists, after whom the governing equations in this example were named (the French mathematician A. Cauchy (1789–1857) and the Swiss mathematician and physicist L. Euler (1707–1783)), are well-known for their priceless contribution to many branches of contemporary mathematics and engineering science.

This formulation is presented to show how Lagrange's method works for boundary-value problems with singular points for governing equations and stated over unbounded domains.

Exercise 2.9(b) shows that the homogeneous problem associated with that in eqns (2.133)–(2.136) has only the trivial solution, justifying, consequently, the existence and uniqueness of its matrix of Green's type.

Clearly, the functions $Y_1(x) \equiv x$ and $Y_2(x) \equiv x^{-1}$ constitute a fundamental set of solutions for the homogeneous equation corresponding to that in (2.133). Therefore, tracing out our procedure, we express the general solution of eqn (2.133) in the form

$$y_1(x) = C_1(x)x + C_2(x)x^{-1} \tag{2.137}$$

This yields the following well-posed system of linear algebraic equations

$$\begin{pmatrix} x & x^{-1} \\ 1 & -x^{-2} \end{pmatrix} \times \begin{pmatrix} C_1'(x) \\ C_2'(x) \end{pmatrix} = \begin{pmatrix} 0 \\ f_1(x)/x \end{pmatrix}$$

in $C_1'(x)$ and $C_2'(x)$, whose solution is

$$C_1'(x) = \frac{f_1(x)}{2x}, \quad C_2'(x) = -\frac{x\, f_1(x)}{2}$$

Integrating these relations and substituting the values of $C_1(x)$ and $C_2(x)$ into eqn (2.137) provides

$$y_1(x) = \int_0^x \frac{x^2 - s^2}{2xs} f_1(s)\, ds + D_{11}x + D_{12}x^{-1} \tag{2.138}$$

Analogously, for $y_2(x)$ one obtains

$$y_2(x) = \int_a^x \frac{x^2 - s^2}{2xs} f_2(s)\, ds + D_{21}x + D_{22}x^{-1} \tag{2.139}$$

Notice that the lower limits of the above two integrals are different representing the left-end points of the intervals $(0, a)$ and $(a, \infty)$, respectively.

The constants of integration in eqns (2.138) and (2.139) are to be obtained by the boundary and contact conditions from eqns (2.135) and (2.136). The first condition in eqn (2.135) requires $D_{12} = 0$, since $x^{-1}$ is unbounded as $x$ approaches zero. To satisfy the second condition in eqn (2.135), we regroup the terms in eqn (2.139) in the following manner

$$y_2(x) = \left( \int_a^x \frac{1}{2s} f_2(s)\, ds + D_{21} \right) x + \left( - \int_a^x \frac{s}{2} f_2(s)\, ds + D_{22} \right) x^{-1} \tag{2.140}$$

By observation, it can easily be seen that, as $x$ goes to infinity, the coefficient of $x$ in eqn (2.140) ought to equal zero, resulting in

$$D_{21} = - \int_a^\infty \frac{1}{2s} f_2(s)\, ds$$

Recalling the values of $D_{12}$ and $D_{21}$ just found, we write the first condition in eqn (2.136) in the form

$$D_{11}a - D_{22}a^{-1} = - \int_0^a \frac{a^2 - s^2}{2as} f_1(s)\, ds - \int_a^\infty \frac{a}{2s} f_2(s)\, ds \tag{2.141}$$

To properly treat the second condition in eqn (2.136), we first differentiate $y_1(x)$ and $y_2(x)$, providing

$$y_1'(x) = \int_0^x \frac{x^2+s^2}{2sx^2} f_1(s)\, ds + D_{11}$$

and

$$y_2'(x) = \int_a^x \frac{x^2+s^2}{2sx^2} f_2(s)\, ds + D_{21} - D_{22}x^{-2}$$

Hence, the second condition in eqn (2.136) yields

$$D_{11} + \lambda D_{22}a^{-2} = -\int_0^a \frac{a^2+s^2}{2sa^2} f_1(s)\, ds - \int_a^\infty \frac{\lambda}{2s} f_2(s)\, ds \tag{2.142}$$

Equations (2.141) and (2.142) form a well-posed system of linear algebraic equations in $D_{11}$ and $D_{22}$, whose solution is

$$D_{11} = -\int_a^\infty \frac{\lambda}{(1+\lambda)s} f_2(s)\, ds - \int_0^a \frac{(a^2+s^2)+\lambda(a^2-s^2)}{2(1+\lambda)a^2 s} f_1(s)\, ds$$

and

$$D_{22} = -\int_0^a \frac{(a^2+s^2)-\lambda(a^2-s^2)}{4\lambda s} f_1(s)\, ds$$

Substituting the values of $D_{ij}$, $(i, j = 1, 2)$ just computed in eqns (2.138) and (2.139), we obtain the solution of the problem posed by eqns (2.133)–(2.136) as

$$\begin{aligned} y_1(x) = &-\int_0^a \frac{x[(a^2+s^2)+\lambda(a^2-s^2)]}{2(1+\lambda)a^2 s} f_1(s)\, ds \\ &+ \int_0^x \frac{x^2-s^2}{2xs} f_1(s)\, ds - \int_a^\infty \frac{\lambda x}{(1+\lambda)s} f_2(s)\, ds \end{aligned}$$

and

$$\begin{aligned} y_2(x) = &-\int_0^a \frac{(a^2+s^2)-\lambda(a^2-s^2)}{4\lambda xs} f_1(s)\, ds \\ &+ \int_a^x \frac{x^2-s^2}{2xs} f_2(s)\, ds - \int_a^\infty \frac{x}{2s} f_2(s)\, ds \end{aligned}$$

From these integral representations for $y_1(x)$ and $y_2(x)$, in accordance with the definition given at the beginning of this section, the entries $g_{ij}(x, s)$ of the matrix of Green's type to the homogeneous boundary-value problem associated with that occurring in eqns (2.133)–(2.136) are finally found to be as follows:

$$g_{11}(x, s) = \begin{cases} x[(a^2+s^2)+\lambda(a^2-s^2)][2(1+\lambda)a^2 s]^{-1}, & \text{for } 0 \le x \le s < a \\ s[(a^2+x^2)+\lambda(a^2-x^2)][2(1+\lambda)a^2 x]^{-1}, & \text{for } 0 < s \le x \le a \end{cases}$$

$$g_{12}(x, s) = \lambda x[(1+\lambda)s]^{-1}, \quad \text{for } 0 \le x \le a < s < \infty$$

$$g_{21}(x,s) = [(a^2+s^2) - \lambda(a^2-s^2)](4\lambda xs)^{-1}, \quad \text{for } 0 < s < a \leq x < \infty$$

$$g_{22}(x,s) = \begin{cases} x(2s)^{-1}, & \text{for } a \leq x \leq s < \infty \\ s(2x)^{-1}, & \text{for } a < s \leq x < \infty \end{cases}$$

Before we start with the next example, it is worth noting that in the formulations that have been discussed so far in this section, we considered multi-point posed boundary-value problems where domains of independent variables consist of a series of segments. The four-point posed problem to be considered in Example 4.3 that follows, is different. Three segments are joined in an assembly by allowing their left-end points to contact in a way shown in Figure 2.1. ▲

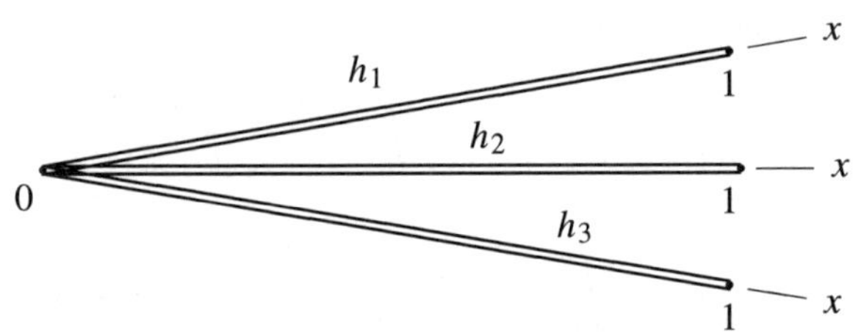


Figure 2.1: Heat conduction in an assembly of rods

**Example 4.3:** Let us consider the following problem

$$\frac{d^2 y_i(x)}{dx^2} = -f_i(x), \quad x \in (0,1), \quad i = 1,2,3 \tag{2.143}$$

$$y_1(0) = y_2(0) = y_3(0), \quad h_1 y_1'(0) + h_2 y_2'(0) + h_3 y_3'(0) = 0 \tag{2.144}$$

$$y_1(1) = 0, \quad y_2(1) = 0, \quad y_3(1) = 0 \tag{2.145}$$

This formulation can, for example, be associated with a steady-state heat conduction in an assembly of three rods each of unit length as shown in Figure 2.1. The rods are assumed to be made of conductive materials whose heat conductivities are $h_1$, $h_2$, and $h_3$. The relations in eqn (2.144) in a case of such interpretation can be referred to as conditions of the ideal thermal contact.

In what follows, we will show how the technique described earlier in this section can be applied to the construction of matrices of Green's type for problems of the kind in eqns (2.143)–(2.145).

Exercise 2.9(c) shows that the homogeneous problem associated with that in eqns (2.143)–(2.145) has only the trivial solution, justifying, consequently, the existence and uniqueness of its matrix of Green's type.

Clearly, the general solution of eqn (2.143) is

$$y_i(x) = C_i(x) + D_i(x)x, \quad i = 1, 2, 3$$

By the method of variation of parameters, this reduces to the following integral form

$$y_i(x) = \int_0^x (s - x) f_i(s)\, ds + M_i + N_i x, \quad i = 1, 2, 3 \qquad (2.146)$$

Satisfying the first group $y_1(0) = y_2(0) = y_3(0)$ of conditions in eqn (2.144), we derive the following two equations in $M_1$, $M_2$ and $M_3$

$$M_1 = M_2 = M_3 \qquad (2.147)$$

while the second condition from eqn (2.144) yields

$$h_1 N_1 + h_2 N_2 + h_3 N_3 = 0 \qquad (2.148)$$

The boundary conditions in eqn (2.145) finally provide three additional relations for $M_i$ and $N_i$

$$M_i + N_i = \int_0^1 (1 - s) f_i(s)\, ds, \quad i = 1, 2, 3 \qquad (2.149)$$

Relations (2.147)–(2.149) form a well-posed system of six linear algebraic equations in six unknowns, which are found as

$$M_1 = M_2 = M_3 = H \int_0^1 (1 - s)[h_1 f_1(s) + h_2 f_2(s) + h_3 f_3(s)]\, ds$$

$$N_1 = H \int_0^1 (1 - s)[(h_2 + h_3) f_1(s) - h_2 f_2(s) - h_3 f_3(s)]\, ds$$

$$N_2 = H \int_0^1 (1 - s)[(h_1 + h_3) f_2(s) - h_1 f_1(s) - h_3 f_3(s)]\, ds$$

and

$$N_3 = H \int_0^1 (1 - s)[(h_1 + h_2) f_3(s) - h_1 f_1(s) - h_2 f_2(s)]\, ds$$

where $H = (h_1 + h_2 + h_3)^{-1}$.

Substituting these in eqn (2.146), one obtains the solution of the problem formulated by eqns (2.143)–(2.145) in the matrix form

$$\begin{pmatrix} y_1(x) \\ y_2(x) \\ y_3(x) \end{pmatrix} = \int_0^1 \begin{pmatrix} g_{11}(x,s) & g_{12}(x,s) & g_{13}(x,s) \\ g_{21}(x,s) & g_{22}(x,s) & g_{23}(x,s) \\ g_{31}(x,s) & g_{32}(x,s) & g_{33}(x,s) \end{pmatrix} \begin{pmatrix} f_1(s) \\ f_2(s) \\ f_3(s) \end{pmatrix} ds \tag{2.150}$$

The entries $g_{ij}(x,s)$ of the kernel-matrix in the above integral representation are expressed as follows

$$g_{11}(x,s) = \begin{cases} H(1-s)[h_1 + x(h_2+h_3)], & \text{for } x \le s \\ H(1-x)[h_1 + s(h_2+h_3)], & \text{for } s \le x \end{cases}$$

$$g_{12}(x,s) = Hh_2(1-s)(1-x), \quad g_{13}(x,s) = Hh_3(1-s)(1-x)$$

$$g_{22}(x,s) = \begin{cases} H(1-s)[h_2 + x(h_1+h_3)], & \text{for } x \le s \\ H(1-x)[h_2 + s(h_1+h_3)], & \text{for } s \le x \end{cases}$$

$$g_{21}(x,s) = Hh_1(1-s)(1-x), \quad g_{23}(x,s) = Hh_3(1-s)(1-x)$$

$$g_{31}(x,s) = Hh_1(1-s)(1-x), \quad g_{32}(x,s) = Hh_2(1-s)(1-x)$$

and

$$g_{33}(x,s) = \begin{cases} H(1-s)[h_3 + x(h_1+h_2)], & \text{for } x \le s \\ H(1-x)[h_3 + s(h_1+h_2)], & \text{for } s \le x \end{cases}$$

From the definition that has been introduced in the opening part of this section, it follows that the kernel-matrix of the integral in eqn (2.150) represents the matrix of Green's type to the homogeneous boundary-contact value problem corresponding to that in eqns (2.143)–(2.145). This implies that this matrix can be interpreted as the influence function of a point source for the entire assembly of rods shown in Figure 2.1. Hence, the entry $g_{ij}(x,s)$ of this matrix simulates the response of the $i$-th rod in the assembly to a point source acting at an arbitrary point $s$ of the $j$-th rod. ▲

It is evident that matrices of Green's type, similarly to Green's functions in the case of a single equation, can naturally be utilized for the solution of boundary-contact value problems for nonhomogeneous systems of differential equations subject to homogeneous boundary and contact conditions. Exercise 2.11 is offered in the End Chapter Exercises section to give the reader a possibility to go through such a solution in detail. Based on that, the reader can later develop a computer-algebra routine that could be used for a fast obtaining of matrices of Green's type for a given assembly configuration and the set of boundary and contact conditions imposed.

## 2.5 Matrix of Green's Type Formalism Extended

From what we have learned in the previous section, the applicability of the matrix of Green's type formalism developed is limited to a *sandwich* type of a piecewise homogeneity of a material of which an assembly is composed. To make the range of possible applications of this formalism broader, we extend it to multi-point posed boundary-contact value problems of a more complex type. The framework of graph theory is used for such an extension.

Sets of linear ordinary differential equations are considered, the ones formulated on finite weighted graphs in such a special way that every equation governs a single unknown function and is stated on a single edge of the graph. The individual equations of the set are put into a system format by subjecting contact and boundary conditions at the vertices and end-points of the graph. Based on such a statement, a novel definition of the matrix of Green's type is introduced. Existence and uniqueness of such matrices are discussed and two methods for their practical construction are proposed. Several particular examples from mechanics are considered.

Computational utilization of the Green's function approach to problems of applied mathematical physics has been recommended and used by many authors (see, for example, [9, 10, 14, 24, 29, 31. 34, 35, 41, 43, 45, 47-49, 51, 52]). However, as we have already mentioned in this text, the practical use of these functions for actual computations in engineering and science is substantially limited because of a lack of their appropriate representations available in the literature.

Use of the Green's function formalism is justified for settings where governing differential equations have continuous coefficients. However, in [44] attempts were undertaken to extend this formalism to boundary-value problems of continuum mechanics, formulated throughout piecewise homogeneous regions, yielding discontinuity of the coefficients in governing differential equations. An effort has been put forth to implement this formalism for treating the so-called multi-point posed boundary-contact value problems that model various settings in continuum mechanics for piecewise homogeneous media.

The first attempts have been undertaken in [43] to introduce the notion of a matrix of Green's type. In Section 2.4, we followed the concept which was proposed in [43]. It is worth noting, however, that the implementation range of that notion is limited to the sandwich type of the material inhomogeneity. Our intention in the present section is to introduce the notion of a matrix of Green's type in a different way. The objective is to provide the extension of the Green's function formalism to multi-point posed boundary-contact value problems occurring in complex assemblies consisting of different homogeneous elements.

For the notational convenience in what follows, boundary-contact value problems for governing systems of differential equations are set up on finite weighted graphs. This allows a considerably systematic analysis of a variety of problems stated in assemblies of one-dimensional elements.

A finite weighted graph $R$ is considered. For terminological convenience, vertices of degree one will be referred to as the *end-points*. Let the graph contain $m$

end-points, $E_h$, $(h = \overline{1, m})$ and $r$ vertices, $V_k$, $(k = \overline{1, r})$ of degree $d_k$ joined by $n$ edges, $e_i$, $(i = \overline{1, n})$ (see Figure 2.2). Let also positive real numbers $l_i$, $(i = \overline{1, n})$, each representing the length of edge $e_i$, be regarded as its weight.

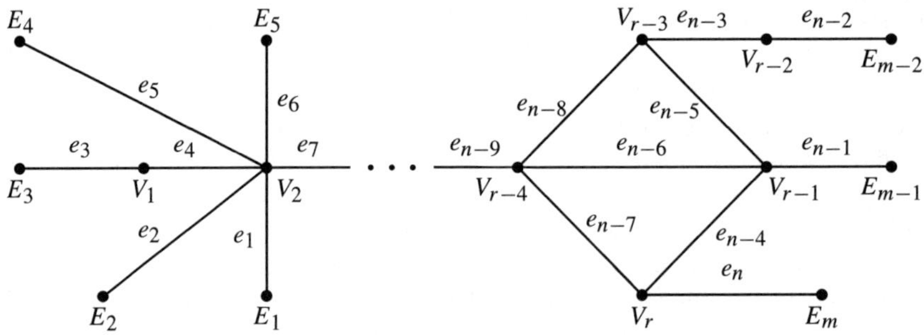


Figure 2.2: Graph $R$ hosting a system of equations

Suppose that every edge $e_i$ of $R$ (every element of the assembly) is occupied with a conductive material (of either thermal or electrical or any other relevant nature) whose conductivity $p_i(x)$ is a continuously differentiable function of the longitudinal coordinate $x$.

Let $u_i(x)$ represent the unknown function (temperature, electric potential, etc.) to be determined throughout the edge $e_i$ of $R$. We will determine the set of these functions by the following set of differential equations

$$\frac{d}{dx}\left(p_i(x)\frac{du_i(x)}{dx}\right) + q_i(x)u_i(x) = -f_i(x), \quad x \in (0, l_i), \quad (i = \overline{1, n}) \tag{2.151}$$

These individual equations are put into a system format by assigning the set of contact conditions

$$u_1(V_k) = \cdots = u_{d_k}(V_k), \quad \sum_{h=1}^{d_k} p_h(V_k)\frac{du_h(V_k)}{dx} = 0, \quad (k = \overline{1, r}) \tag{2.152}$$

at each of the vertices $V_k$, with $d_k$ being their degrees. Notice that for the notational convenience, in formulating these conditions, we use a 'local' numbering of the edges incident to the vertex $V_k$. It can easily be seen that the number of the contact conditions assigned at each of the vertices equals the degree of the vertex. Clearly, the contact conditions in eqn (2.152) model conservation of energy at every vertex $V_k$ of $R$. In addition, the boundary conditions

$$\alpha_h\frac{du_i(E_h)}{dx} + \beta_h u_i(E_h) = 0, \quad (h = \overline{1, m}) \tag{2.153}$$

are subjected at each of the end-points $E_h$ of $R$. This implies that the functions $u_i(x)$ in the above equation are defined on the end edges $e_i$ incident to $E_h$.

Clearly, the number of contact conditions implied at a vertex equals the degree of the vertex, while a single boundary condition is implied at each end-point. This makes the total number of conditions of uniqueness imposed by eqns (2.152) and (2.153)

$$\sum_{k=1}^{r} d_k + m = 2n$$

meaning that the setting in eqns (2.151)–(2.153) is well-posed.

In this section, we will be focusing on the influence function (matrix) which represents the response of the entire assembly to a unit energy source acting at an arbitrary point $s$ within an arbitrary edge of $R$. Notice that the emphasis will be on boundary-value problems of the type in eqns (2.151)–(2.153). However, the results of this section can readily be extended to problems formulated for differential equations of higher order. This point will be addressed later in this text while problems for multi-span beams are considered.

We are now in a position to extend the conventional definition of the Green's function so as to make it valid for the multi-point posed boundary-value problems of the type in eqns (2.151)–(2.153).

**Definition:** An $n \times n$ matrix $G(x, s)$, whose entries $g_{ij}(x, s)$ are defined for $x \in e_i$ and $s \in e_j$ on $R$, is referred to as the *matrix of Green's type* of the homogeneous boundary-value problem corresponding to that posed by eqns (2.151)–(2.153), if for any fixed value of $s$, the entries $g_{ij}(x, s)$ hold the following properties:

1. As $x \neq s$, the entries $g_{ii}(x, s)$ of the principal diagonal $(i = j)$ represent continuous functions of $x$ on $e_i$, they have continuous partial derivatives with respect to $x$ of up to the second order included, and satisfy the homogeneous equations corresponding to those in (2.151);
2. As $x = s$, the entries $g_{ii}(x, s)$ of the principal diagonal are continuous functions of $x$, whereas their partial derivatives of the first order with respect to $x$ are discontinuous functions, providing

$$\lim_{x \to s^+} \frac{\partial g_{ii}(x, s)}{\partial x} - \lim_{x \to s^-} \frac{\partial g_{ii}(x, s)}{\partial x} = -\frac{1}{p_i(s)}$$

   and

$$\lim_{s \to x^+} \frac{\partial g_{ii}(x, s)}{\partial x} - \lim_{s \to x^-} \frac{\partial g_{ii}(x, s)}{\partial x} = \frac{1}{p_i(s)}$$

3. The peripheral $(i \neq j)$ entries $g_{ij}(x, s)$ of $G(x, s)$ are continuous functions of $x$ for any value of $s \in e_j$, they have continuous partial derivatives with respect to $x$ of up to the second order included, and satisfy the homogeneous equations corresponding to those in (2.151);
4. All entries $g_{ij}(x, s)$ of $G(x, s)$ satisfy the contact and the end conditions (in which they are involved) in eqns (2.152) and (2.153), in the sense that each of these conditions is satisfied for $s$ belonging to any of the edges $e_j$, $(j = \overline{1, n})$.

In the discussion that follows, the arguments $x$ and $s$ in the matrix of Green's type are referred to (analogously to those in the Green's function) as the *observation (field) point* and the *source point*, respectively.

The following theorem can be formulated to stipulate the existence and uniqueness of the matrix of Green's type for the homogeneous boundary-value problem corresponding to that posed by eqns (2.151)–(2.153).

**Theorem 2.5:** If the multi-point posed boundary-value problem stated by eqns (2.151)–(2.153) has a unique solution (that is, the corresponding homogeneous problem has only the trivial solution), then there exists a unique matrix of Green's type $G(x, s)$ of the corresponding homogeneous problem.

**Proof:** Let $u_{i1}(x)$ and $u_{i2}(x)$, $(i = \overline{1, n})$ represent pairs of linearly independent on $e_i$ particular solutions (fundamental sets of solutions) of the homogeneous equations corresponding to those in (2.151). If so, then, by virtue of the defining property 1, the diagonal entries $g_{ii}(x, s)$ of $G(x, s)$ can be sought in the form

$$g_{ii}(x, s) = \begin{cases} a_{i1}(s)u_{i1}(x) + a_{i2}(s)u_{i2}(x), & \text{for } x \le s \\ b_{i1}(s)u_{i1}(x) + b_{i2}(s)u_{i2}(x), & \text{for } x \ge s \end{cases} \tag{2.154}$$

whereas, in compliance with the defining property 3, the peripheral $(i \neq j)$ entries $g_{ij}(x, s)$ of $G(x, s)$ can be written as

$$g_{ij}(x, s) = c_{ij}(s)u_{i1}(x) + d_{ij}(s)u_{i2}(x) \tag{2.155}$$

The coefficients $a_{i1}(s)$, $a_{i2}(s)$, $b_{i1}(s)$, $b_{i2}(s)$, $c_{ij}(s)$, and $d_{ij}(s)$ in the above representations are to be determined upon applying the remaining defining properties of the matrix of Green's type. Notice that the total number of these coefficients equals $2n(n + 1)$ while the total number of the relations provided by properties 2 and 4 equals also $2n(n + 1)$. This is a "good news" as to the well-posedness of the problem.

By virtue of property 2, one obtains $n$ well-posed systems of linear algebraic equations

$$\begin{pmatrix} u_{i1}(s) & u_{i2}(s) \\ u'_{i1}(s) & u'_{i2}(s) \end{pmatrix} \times \begin{pmatrix} C_{i1}(s) \\ C_{i2}(s) \end{pmatrix} = \begin{pmatrix} 0 \\ p_i^{-1}(s) \end{pmatrix}, \quad (i = \overline{1, n}) \tag{2.156}$$

in two unknowns each, of the total amount of $2n$ equations in $2n$ unknowns $C_{i1}(s)$ and $C_{i2}(s)$, $(i = \overline{1, n})$. These unknowns are expressed in terms of the coefficients of $g_{ii}(x, s)$ in eqn (2.154) as

$$C_{ik}(s) = b_{ik}(s) - a_{ik}(s), \quad k = 1, 2 \tag{2.157}$$

The well-posedness of the systems in eqn (2.156) follows from the fact that the determinants of their coefficient matrices represent Wronskian of the linearly independent functions $u_{i1}(x)$ and $u_{i2}(x)$. Hence, the unique values of $C_{i1}(s)$ and $C_{i2}(s)$ can readily be obtained. Subsequently, in compliance with eqn (2.157), the coefficients $a_{i1}(s)$ and $a_{i2}(s)$ can uniquely be expressed in terms of $b_{i1}(s)$ and $b_{i2}(s)$ and vice versa.

Thus, the number of undetermined coefficients in eqns (2.154) and (2.155) reduces to $2n^2$. And they can ultimately be found by applying the defining property 4. Indeed, by satisfying the entire set of boundary and contact conditions posed by eqns (2.152) and (2.153) $n$ times (once for each location of the source point $s \in e_j$, $j = \overline{1, n}$), we finally obtain a nonhomogeneous system of $2n^2$ linear algebraic equations in $2n^2$ unknowns. The coefficient matrix of this system reduces to the following partitioned diagonal form

$$M = \begin{pmatrix} A_{11} & 0 & . & . & . & 0 \\ 0 & A_{22} & . & . & . & 0 \\ . & . & . & . & . & . \\ 0 & 0 & . & . & . & A_{nn} \end{pmatrix}$$

in which $A_{ii}$ $(i = \overline{1, n})$ represent $2n \times 2n$ sub-matrices whose regularity follows from the well-posedness of the original boundary-value problem in eqns (2.151)–(2.153). The peripheral sub-matrices of $M$ represent the null $2n \times 2n$ matrices. Thus, $M$ represents a non-singular matrix providing all of the coefficients of the representations in eqns (2.154) and (2.155) and can be uniquely found.

This completes the proof of Theorem 2.5 because, once the values of the coefficients $a_{i1}(s)$, $a_{i2}(s)$, $b_{i1}(s)$, $b_{i2}(s)$, $c_{ij}(s)$, and $d_{ij}(s)$ are found, one can immediately obtain explicit representations of the entries of $G(x, s)$ by substituting those values into eqns (2.154) and (2.155). Notice that this proof suggests the procedure for the actual construction of matrices of Green's type for boundary-value problems posed on graphs. ■

Another effective procedure of obtaining matrices of Green's type for homogeneous boundary-value problems of the type posed on graphs by eqns (2.151)–(2.153) is also brought forward in this section. To describe it let us introduce a vector-function $\mathbf{U}(x)$ whose components $U_i(x)$, $(i = \overline{1, n})$ are defined in terms of the solutions $u_i(x)$ of eqn (2.151) as

$$U_i(x) = \begin{cases} u_i(x), & \text{for } x \in e_i \\ 0, & \text{for } x \in R \setminus e_i \end{cases} \tag{2.158}$$

We also introduce a vector-function $\mathbf{F}(x)$ whose components $F_i(x)$ are defined in terms of the right-hand side functions $f_i(x)$ of eqn (2.151) in the form

$$F_i(x) = \begin{cases} f_i(x), & \text{for } x \in e_i \\ 0, & \text{for } x \in R \setminus e_i \end{cases} \tag{2.159}$$

The following theorem is formulated and proved to determine the solution of the boundary-value problem posed by eqns (2.151)–(2.153) in terms of the matrix of Green's type of the corresponding homogeneous problem.

**Theorem 2.6:** If $G(x, s)$ represents the matrix of Green's type of the homogeneous boundary-value problem corresponding to that in eqns (2.151)–(2.153), then the solution of the problem posed by eqns (2.151)–(2.153) on $R$ can be written as

$$\mathbf{U}(x) = \int_R G(x, s)\mathbf{F}(s)\, dR(s), \qquad x \in R \tag{2.160}$$

where the integration is carried out over the entire graph $R$. The converse is also true. That is, if the solution of the problem posed by eqns (2.151)–(2.153) on $R$ is obtained in the form of the integral in eqn (2.160), then the kernel $G(x, s)$ of that integral represents the matrix of Green's type for the homogeneous boundary-value problem corresponding to that in eqns (2.151)–(2.153).

**Proof:** By virtue of the relations in eqns (2.158) and (2.159), the integral in eqn (2.160) can be read off in the scalar form as

$$u_i(x) = \sum_{j=1}^{n} \int_{e_j} g_{ij}(x, s) f_j(s)\, de_j(s), \quad i = \overline{1, n}$$

which can be rewritten in terms of the local coordinates as

$$u_i(x) = \sum_{j=1}^{n} \int_0^{l_j} g_{ij}(x, s) f_j(s)\, ds, \quad x \in [0, l_i], \quad i = \overline{1, n} \tag{2.161}$$

Since the diagonal $g_{ii}(x, s)$ and the peripheral $g_{ij}(x, s)$ entries of the matrix of Green's type are defined in different manner (see eqns (2.154) and (2.155)), we isolate the $i$-th term of the finite sum in eqn (2.161)

$$u_i(x) = \sum_{j=1}^{i-1} \int_0^{l_j} g_{ij}(x, s) f_j(s)\, ds + \int_0^{l_i} g_{ii}(x, s) f_i(s)\, ds$$
$$+ \sum_{j=i+1}^{n} \int_0^{l_j} g_{ij}(x, s) f_j(s)\, ds, \quad x \in [0, l_i], \quad i = \overline{1, n}$$

Since the diagonal entries of $G(x, s)$ are defined in pieces, we break down the integral containing $g_{ii}(x, s)$ into two integrals as shown

$$u_i(x) = \sum_{j=1}^{i-1} \int_0^{l_j} g_{ij}(x, s) f_j(s)\, ds + \int_0^{x} g_{ii}^{-}(x, s) f_i(s)\, ds$$
$$+ \int_x^{l_i} g_{ii}^{+}(x, s) f_i(s)\, ds$$
$$+ \sum_{j=i+1}^{n} \int_0^{l_j} g_{ij}(x, s) f_j(s)\, ds, \quad x \in [0, l_i], \quad i = \overline{1, n}$$

where $g_{ii}^{-}(x, s)$ and $g_{ii}^{+}(x, s)$ represent the lower and the upper branches of the diagonal entries of $G(x, s)$, which are valid for $x \geq s$ and $x \leq s$, respectively (see eqn (2.154)).

To properly differentiate the functions $u_i(x)$ in the above equation, we recall the defining properties of the entries of $G(x, s)$ and notice also that the above expression contains integrals involving parameter and having variable limits. With this in mind, one obtains

$$\frac{du_i(x)}{dx} = \sum_{j=1}^{i-1} \int_0^{l_j} \frac{\partial g_{ij}(x,s)}{\partial x} f_j(s)\,ds + \int_0^x \frac{\partial g_{ii}^-(x,s)}{\partial x} f_i(s)\,ds$$
$$+ g_{ii}(x,x^-) f_i(x) + \int_x^{l_i} \frac{\partial g_{ii}^+(x,s)}{\partial x} f_i(s)\,ds - g_{ii}(x,x^+) f_i(x)$$
$$+ \sum_{j=i+1}^{n} \int_0^{l_j} \frac{\partial g_{ij}(x,s)}{\partial x} f_j(s)\,ds, \quad x \in [0, l_i], \quad i = \overline{1,n}$$

The sum of the two non-integral terms

$$g_{ii}(x,x^-) f_i(x) - g_{ii}(x,x^+) f_i(x)$$

equals zero because, according to the definition of the matrix of Green's type, its diagonal entries are continuous as $x = s$. This yields

$$\frac{du_i(x)}{dx} = \sum_{j=1}^{i-1} \int_0^{l_j} \frac{\partial g_{ij}(x,s)}{\partial x} f_j(s)\,ds + \int_0^x \frac{\partial g_{ii}^-(x,s)}{\partial x} f_i(s)\,ds$$
$$+ \int_x^{l_i} \frac{\partial g_{ii}^+(x,s)}{\partial x} f_i(s)\,ds$$
$$+ \sum_{j=i+1}^{n} \int_0^{l_j} \frac{\partial g_{ij}(x,s)}{\partial x} f_j(s)\,ds, \quad x \in [0, l_i], \quad i = \overline{1,n}$$

or in a compact form

$$\frac{du_i(x)}{dx} = \sum_{j=1}^{n} \int_0^{l_j} \frac{\partial g_{ij}(x,s)}{\partial x} f_j(s)\,ds, \quad x \in [0, l_i], \quad i = \overline{1,n} \tag{2.162}$$

Hence, the first order derivatives of the integral representations of $u_i(x)$ can be obtained by a straightforward differentiation of their integrands. Consequently, these representations of $u_i(x)$ satisfy the boundary conditions in eqns (2.152) and (2.153) because so do the entries of $G(x, s)$.

To find out whether $u_i(x)$, as shown in eqn (2.161), satisfy the governing differential equations, we obtain the second derivatives of $u_i(x)$

$$\frac{d^2u_i(x)}{dx^2} = \sum_{j=1}^{i-1} \int_0^{l_j} \frac{\partial^2 g_{ij}(x, s)}{\partial x^2} f_j(s)\, ds + \int_0^x \frac{\partial^2 g_{ii}^-(x, s)}{\partial x^2} f_i(s)\, ds$$
$$+ \frac{\partial g_{ii}(x, x^-)}{\partial x} f_i(x) + \int_x^{l_i} \frac{\partial^2 g_{ii}^+(x, s)}{\partial x^2} f_i(s)\, ds - \frac{\partial g_{ii}(x, x^+)}{\partial x} f_i(x)$$
$$+ \sum_{j=i+1}^{n} \int_0^{l_j} \frac{\partial^2 g_{ij}(x, s)}{\partial x^2} f_j(s)\, ds, \quad x \in [0, l_i], \quad i = \overline{1, n}$$

In compliance with property 2 of the definition of $G(x, s)$, we have

$$\frac{\partial g_{jj}(x, x^-)}{\partial x} f_i(x) - \frac{\partial g_{jj}(x, x^+)}{\partial x} f_i(x) = -\frac{f_i(x)}{p_i(s)}$$

And for the second derivative of $u_i(x)$, we finally obtain its compact representation written as

$$\frac{d^2u_i(x)}{dx^2} = \sum_{j=1}^{n} \int_0^{l_j} \frac{\partial^2 g_{ij}(x, s)}{\partial x^2} f_j(s)\, ds - \frac{f_i(x)}{p_i(x)}, \quad x \in [0, l_i], \quad i = \overline{1, n} \tag{2.163}$$

Upon substituting the values of $u_i(x)$ and their derivatives from eqns (2.161)–(2.163) into eqn (2.151), we finally obtain

$$\sum_{j=1}^{n} \int_0^{l_j} L[g_{ij}(x, s)] f_j(s)\, ds - f_i(x) = -f_i(x), \quad x \in (0, l_i)$$

where $L$ represents the differential operator of eqn (2.151).

Thus, the integral representations in eqn (2.161) satisfy the governing differential equations because the entries of the matrix of Green's type satisfy the homogeneous equations corresponding to those in eqn (2.151). That is, $L[g_{ij}(x, s)] = 0$, vanishing the integral terms in the above equation. Hence, the theorem has been proven. ■

Theorem 2.6 clearly suggests that, once the solution to the original problem posed by eqns (2.151)–(2.153) is expressed in terms of the integral in eqn (2.160), the kernel $G(x, s)$ of that integral represents the matrix of Green's type of the corresponding homogeneous problem.

A version of the method of variation of parameters is proposed below to obtain an integral representation of the form in eqn (2.160) for the solution of the nonhomogeneous boundary-value problem posed by eqn (2.151)–(2.153).

In doing so, we again recall the fundamental sets of solutions $u_{i1}(x)$ and $u_{i2}(x)$ of the homogeneous equations corresponding to those in (2.151). The general solution $u_i(x)$ of eqn (2.151) is sought as follows

$$u_i(x) = D_{i1}(x)u_{i1}(x) + D_{i2}(x)u_{i2}(x), \quad i = \overline{1, n} \tag{2.164}$$

Based on this and following the standard procedure of the method of variation of parameters, one obtains the well-posed systems of linear algebraic equations

$$\begin{pmatrix} u_{i1}(x) & u_{i2}(x) \\ u'_{i1}(x) & u'_{i2}(x) \end{pmatrix} \times \begin{pmatrix} D'_{i1}(x) \\ D'_{i2}(x) \end{pmatrix} = \begin{pmatrix} 0 \\ -f_i(x)/p_i(x) \end{pmatrix}, \quad i = \overline{1, n}$$

in the derivatives of the coefficients $D_{i1}(x)$ and $D_{i2}(x)$ of $u_i(x)$. From this system it follows that

$$D'_{i1}(x) = \frac{u_{i2}(x)f_i(x)}{p_i(x)W_i(x)}, \quad D'_{i2}(x) = -\frac{u_{i1}(x)f_i(x)}{p_i(x)W_i(x)}, \quad i = \overline{1, n}$$

where $W_i(x) = u_{i1}(x)u'_{i2}(x) - u_{i2}(x)u'_{i1}(x)$ represent the Wronskian of the fundamental set of solutions $u_{i1}(x)$ and $u_{i2}(x)$.

Integration of the derivatives $D'_{i1}(x)$ and $D'_{i2}(x)$ yields

$$D_{i1}(x) = \int_0^x \frac{u_{i2}(s)f_i(s)}{p_i(s)W_i(s)}\, ds + E_{i1}, \quad i = \overline{1, n}$$

and

$$D_{i2}(x) = -\int_0^x \frac{u_{i1}(s)f_i(s)}{p_i(s)W_i(s)}\, ds + E_{i2}, \quad i = \overline{1, n}$$

where $E_{i1}$ and $E_{i2}$ represent undetermined coefficients. Upon substituting $D_{i1}(x)$ and $D_{i2}(x)$ just found into eqn (2.164), the latter can be rewritten as

$$u_i(x) = u_{i1}(x)\int_0^x \frac{u_{i2}(s)f_i(s)}{p_i(s)W_i(s)}\, ds - u_{i2}(x)\int_0^x \frac{u_{i1}(s)f_i(s)}{p_i(s)W_i(s)}\, ds$$
$$+ E_{i1}u_{i1}(x) + E_{i2}u_{i2}(x), \quad i = \overline{1, n}$$

By combining the integral terms in the above equation, the general solution of eqn (2.151) is finally obtained in the form

$$u_i(x) = \int_0^x \frac{u_{i1}(x)u_{i2}(s) - u_{i2}(x)u_{i1}(s)}{p_i(s)W_i(s)} f_i(s)\, ds$$
$$+ E_{i1}u_{i1}(x) + E_{i2}u_{i2}(x), \quad x \in (0, l_i), \quad i = \overline{1, n} \tag{2.165}$$

The undetermined coefficients $E_{i1}$ and $E_{i2}$, of a total number of $2n$, can be obtained upon satisfying the contact and boundary conditions in eqns (2.152) and (2.153), whose total number equals also $2n$. This yields a well-posed system of linear algebraic equations which leads finally to the integral representation of the form in eqn (2.160) for the solution of the problem under consideration. The kernel of that integral represents the matrix of Green's type of the problem.

Our approach enables us to naturally apply the matrix of Green's type formalism to a problem formulated for the medium whose property is a discontinuous function of the spatial variable.

**Example 5.1:** Construct the matrix of Green's type for the steady-state heat conduction in an assembly of rods (see Figure 2.3), each of which is composed of a homogeneous material whose heat conductivity is $p_i$.

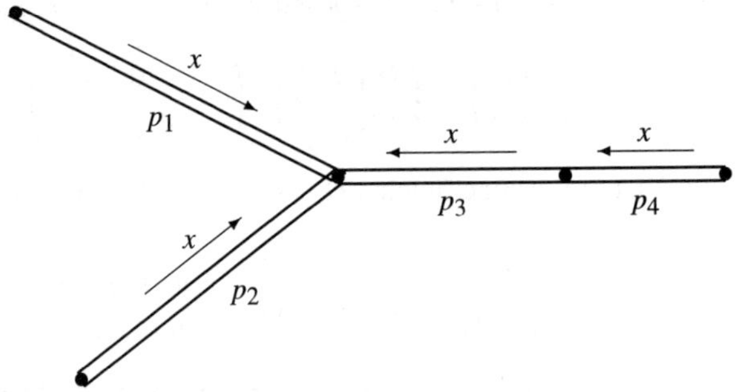


Figure 2.3: An assembly of heat conductive rods

On the weighted graph associated with the above assembly, we formulate the following multi-point posed boundary-value problem

$$p_i \frac{d^2 u_i(x)}{dx^2} = -f_i(x), \quad x \in (0, l_i), \quad i = \overline{1,4} \tag{2.166}$$

$$u_1(l_1) = u_2(l_2) = u_3(l_3) \tag{2.167}$$

$$p_1 \frac{du_1(l_1)}{dx} + p_2 \frac{du_2(l_2)}{dx} + p_3 \frac{du_3(l_3)}{dx} = 0 \tag{2.168}$$

$$u_3(0) = u_4(l_4) \tag{2.169}$$

$$p_3 \frac{du_3(0)}{dx} - p_4 \frac{du_4(l_4)}{dx} = 0 \tag{2.170}$$

$$u_1(0) = u_2(0) = u_4(0) = 0 \tag{2.171}$$

that describes the steady-state heat conduction phenomenon in the assembly. Here $l_i$, $(i = \overline{1,4})$ represent the lengths of the rods.

In compliance with the procedure of the method of variation of parameters, we seek the general solution of eqn (2.166) in the form

$$u_{i,g}(x) = D_{i1}(x) + D_{i2}(x)x, \quad i = \overline{1,4}$$

that ultimately reduces in this case (see eqn (2.165)) to

$$u_i(x) = \int_0^x \frac{s-x}{p_i} f_i(s)\, ds + E_{i1} + E_{i2}x, \quad x \in (0, l_i), \quad i = \overline{1,4} \tag{2.172}$$

The undetermined coefficients $E_{i1}$ and $E_{i2}$, $(i = \overline{1,4})$ in eqn (2.172) are to be determined upon satisfying the contact and boundary conditions posed by eqns (2.167)–(2.171). The conditions in eqn (2.171) yield, in particular, $E_{11} = E_{21} = E_{41} = 0$. For the rest of the coefficients, one obtains a well-posed system of linear algebraic equations written as

$$\begin{pmatrix} l_1 & -l_2 & 0 & 0 & 0 \\ l_1 & 0 & -1 & -l_3 & 0 \\ p_1 & p_2 & 0 & p_3 & 0 \\ 0 & 0 & 1 & 0 & -l_4 \\ 0 & 0 & 0 & p_3 & -p_4 \end{pmatrix} \times \begin{pmatrix} E_{12} \\ E_{22} \\ E_{31} \\ E_{32} \\ E_{42} \end{pmatrix} = \begin{pmatrix} A_2 - A_1 \\ A_3 - A_1 \\ B_1 + B_2 + B_3 \\ A_4 \\ -B_4 \end{pmatrix} \tag{2.173}$$

where

$$A_i = \int_0^{l_i} \frac{s - l_i}{p_i} f_i(s)\, ds, \quad B_i = \int_0^{l_i} f_i(s)\, ds, \quad i = \overline{1, 4}$$

For the sake of simplicity, we assume in what follows that the edges of the graph have equal lengths, that is $l_1 = l_2 = l_3 = l_4 = l$. The determinant $\Delta$ of the coefficient matrix of the system in eqn (2.173) is found, in this case, in the form

$$\Delta = l^2[(p_1 + p_2)(p_3 + p_4) + p_3 p_4]$$

When solving the system in eqn (2.173) and substituting thereupon values of the coefficients $E_{i1}$ and $E_{i2}$ found into eqn (2.172), we finally obtain

$$\begin{aligned} u_1(x) = & \int_0^l \frac{x}{\Delta^* p_1}\{\Delta^* - s[p_2(p_3 + p_4) + p_3 p_4]\} f_1(s)\, ds \\ & + \int_0^x \frac{s - x}{p_1} f_1(s)\, ds + \int_0^l \frac{xs}{\Delta^*}(p_3 + p_4) f_2(s)\, ds \\ & + \int_0^l \frac{x}{\Delta^*}(lp_3 + sp_4) f_3(s)\, ds + \int_0^l \frac{xs}{\Delta^*} p_3 f_4(s)\, ds \end{aligned} \tag{2.174}$$

$$\begin{aligned} u_2(x) = & \int_0^l \frac{xs}{\Delta^*}(p_3 + p_4) f_1(s)\, ds \\ & + \int_0^x \frac{s - x}{p_2} f_2(s)\, ds \\ & + \int_0^l \frac{x}{\Delta^* p_2}\{\Delta^* - s[p_1(p_3 + p_4) + p_3 p_4]\} f_2(s)\, ds \\ & + \int_0^l \frac{x}{\Delta^*}(lp_3 + sp_4) f_3(s)\, ds + \int_0^l \frac{xs}{\Delta^*} p_3 f_4(s)\, ds \end{aligned} \tag{2.175}$$

$$u_3(x) = \int_0^l \frac{s}{\Delta^*}(lp_3 + xp_4) f_1(s)\, ds + \int_0^l \frac{s}{\Delta^*}(lp_3 + xp_4) f_2(s)\, ds$$
$$+ \int_0^l \frac{1}{\Delta^* p_3}[l(p_1 + p_2 + p_3) - s(p_1 + p_2)](lp_3 + xp_4) f_3(s)\, ds$$
$$+ \int_0^x \frac{s - x}{p_3} f_3(s)\, ds + \int_0^l \frac{s}{\Delta^*}[l(p_1 + p_2 + p_3) - x(p_1 + p_2)] f_4(s)\, ds \tag{2.176}$$

and

$$u_4(x) = \int_0^l \frac{xs}{\Delta^*} p_3 f_1(s)\, ds + \int_0^l \frac{xs}{\Delta^*} p_3 f_2(s)\, ds$$
$$+ \int_0^l \frac{x}{\Delta^*}[l(p_1 + p_2 + p_3) - s(p_1 + p_2)] f_3(s)\, ds$$
$$+ \int_0^x \frac{s - x}{p_4} f_4(s)\, ds + \int_0^l \frac{x}{\Delta^* p_4}[\Delta^* - sp_3(p_1 + p_2)] f_4(s)\, ds \tag{2.177}$$

where $\Delta^* = \Delta/l$.

Since the solution to the boundary-contact value problem posed by eqns (2.167)–(2.171) is expressed in the form of the integral in eqn (2.160), the entries $g_{ij}(x, s)$ of the matrix of Green's type $G(x, s)$ of the corresponding homogeneous problem can be read off from the integral representations in eqns (2.174)–(2.177). The entries $g_{i1}(x, s)$ of the first column, for example, of $G(x, s)$ are exhibited as

$$g_{11}(x, s) = \frac{1}{\Delta^* p_1} \begin{cases} x\{\Delta^* - s[p_2(p_3 + p_4) + p_3 p_4]\}, & \text{for } x \le s \\ s\{\Delta^* - x[p_2(p_3 + p_4) + p_3 p_4]\}, & \text{for } x \ge s \end{cases}$$

$$g_{21}(x, s) = \frac{xs}{\Delta^*}(p_3 + p_4), \quad g_{31}(x, s) = \frac{s}{\Delta^*}(lp_3 + xp_4)$$

$$g_{41}(x, s) = \frac{xs}{\Delta^*} p_3$$

These specify the response of the assembly of rods to a unit point source acting at a source point $s$ arbitrarily located in the rod number one. The rest of the entries of the matrix of Green's type $G(x, s)$ specifying the response to a unit source acting at other rods, could, if required, also be directly obtained from the above integral representations. ▲

## 2.6 End Chapter Exercises

**2.1** Construct Green's functions for the following boundary-value problems on the indicated interval:

**a)** $y''(x) = 0$, with $y(0) = 0$ and $y'(a) = 0$;

**b)** $y''(x) = 0$, with $y(0) = 0$ and $y'(a) + hy(a) = 0$, $(h \geq 0)$. Show that if $h = 0$, then the Green's function for this problem reduces to that in Exercise 2.1, part **a)**;

**c)** $y''(x) = 0$, with $y'(0) - h_1 y(0) = 0$ and $y'(a) + h_2 y(a) = 0$, when $h_1$ and $h_2$ both are not zero. Show that if $h_1 = 0$, then the Green's function for this problem reduces to that of Example 1.3 in Section 2.1;

**d)** $((mx + p)y'(x))' = 0$, with $y(0) = 0$ and $y(a) = 0$, when $m > 0$ and $p > 0$;

**e)** $(\exp(\beta x)y'(x))' = 0$, with $y(0) = 0$ and $y(a) = 0$;

**f)** $(\exp(\beta x)y'(x))' = 0$, with $y(0) = 0$ and $y'(a) = 0$;

**g)** $y''(x) + k^2 y(x) = 0$, with $y(0) = 0$ and $y(a) = 0$;

**h)** $y^{IV}(x) = 0$ for $x \in (0, 1)$, with $y(0) = y'(0) = 0$ and $y''(1) = y'''(1) = 0$.

**2.2** Determine whether the following equations are in a self-adjoint form:

**a)** $y''(x) + k^2 y(x) = 0$;

**b)** $x^2 y''(x) + 2xy'(x) - (x^2 - 1)y(x) = 0$;

**c)** $x^2 y''(x) - 2xy'(x) + y(x) = 0$;

**d)** $y''(x) + 3y'(x) + 9y(x) = 0$;

**e)** $\sin^2(x)y''(x) + \sin(2x)y'(x) - y(x) = 0$.

**2.3** By introducing integrating factors, reduce the following differential equations to a self-adjoint form:

**a)** $y''(x) - 2y'(x) + 4y(x) = 0$;

**b)** $y''(x) + xy'(x) - x^2 y(x) = 0$;

**c)** $x^2 y''(x) - xy'(x) + y(x) = 0$;

**d)** $x^2 y''(x) + xy'(x) - y(x) = 0$.

**2.4** Determine whether the following boundary-value problems are self-adjoint:

**a)** $y''(x) + y(x) = 0$, with $y(a) = 0$ and $y'(b) + hy(b) = 0$;

**b)** $y''(x) - y(x) = 0$, with $y'(a) = 0$ and $y'(b) + hy(b) = 0$;

**c)** $x\, y''(x) + y'(x) - y(x) = 0$, $y'(a) + h_1 y(a) = 0$, $y'(b) + h_2 y(b) = 0$;

**d)** $xy''(x) + y'(x) = 0$, with $y(a) = y(b)$ and $ay'(a) = by'(b)$;

**e)** $(x - a)y''(x) + y'(x) - y(x) = 0$, with $|y(a)| < \infty$ and $y'(b) + hy(b) = 0$;

**f)** $y''(x) + y(x) = 0$, with $y'(0) + y(0) + y(a) = 0$ and $y'(0) - y(0) + y'(a) = 0$.

**2.5** Construct the Green's function for the following problem

$$y''(x) + 3y'(x) - 10y(x) = 0, \quad y(0) = 0, \quad |y(\infty)| < \infty$$

Reduce then this problem to a self-adjoint form and construct the Green's function again. Observe how this affects the symmetry of the Green's function.

**2.6** Construct the Green's function for the problem

$$y''(x) + k^2 y(x) = 0, \quad y(0) = 0, \quad y'(1) = 0$$

by the approach discussed in the proofs of Theorem 2.2. Compare this to the method used in Theorem 2.4.

**2.7** Use Lagrange's method to construct Green's functions for the following boundary-value problems:

**a)** $(xy'(x))' = 0$, with $|y(0)| < \infty$, $y(a) = 0$;
**b)** $y''(x) - k^2 y(x) = 0$, with $y'(0) - hy(0) = 0$, $y(a) = 0$;
**c)** $y^{IV}(x) = 0$, with $y(0) = y'(0) = 0$, $y(1) = y'(1) = 0$;
**d)** $y^{IV}(x) = 0$, with $y(0) = y''(0) = 0$, $y(1) = y''(1) = 0$.

**2.8** Based on Theorem 2.4, compute solutions for the following boundary-value problems by utilizing corresponding Green's functions:

**a)** $y''(x) + y(x) = \exp(2x)$, with $y(a) = 0$, $y'(b) + hy(b) = 0$;
**b)** $y''(x) - y(x) = x$, with $y'(a) = 0$, $y'(b) + hy(b) = 0$;
**c)** $y''(x) + 2y'(x) + y(x) = \sin x$, $y'(a) = 0$, $y'(b) = 0$;
**d)** $y^{IV}(x) = x^2$, with $y(0) = y'(0) = 0$, $y(1) = y'(1) = 0$;
**e)** $y^{IV}(x) - 2y''(x) + y(x) = 1$, $y(0) = y'(0) = 0$, $|y(\infty)| < \infty$, $|y'(\infty)| < \infty$.

**2.9** Determine whether the following multi-point posed boundary-value problems have only the trivial solution:

**a)** Homogeneous problem corresponding to that in eqns (2.122)–(2.125);
**b)** Homogeneous problem corresponding to that in eqns (2.133)–(2.136);
**c)** Homogeneous problem corresponding to that in eqns (2.143)–(2.145).

**2.10** Construct matrix of Green's type for the following multi-point posed boundary-value problem:

**a)** $y_1''(x) = 0$ for $x \in (-a, 0)$ and $y_2''(x) - k^2 y_2(x) = 0$ for $x \in (0, \infty)$ with $y_1(-a) = 0$, $|y_2(\infty)| < \infty$, $y_1(0) = y_2(0)$, $y_1'(0) = \lambda y_2'(0)$;

**b)** $y_1''(x) - k^2 y_1(x) = 0, x \in (-a, 0)$ and $y_2''(x) - k^2 y_2(x) = 0, x \in (0, \infty)$ with $y_1(-a) = 0$, $|y_2(\infty)| < \infty$, $y_1(0) = y_2(0)$, $y_1'(0) = \lambda y_2'(0)$;

**c)** $y_1''(x) + k^2 y_1(x) = 0$ for $x \in (-a, 0)$ and $y_2''(x) + k^2 y_2(x) = 0$ for $x \in (0, a)$ with $y_1(-a) = 0$, $y_2(a) = 0$, $y_1(0) = y_2(0)$, $y_1'(0) = \lambda y_2'(0)$;

**d)** $y_i''(x) - k^2 y_i(x) = 0, \quad (i = 1, 2, 3)$ for $x \in (0, \infty)$ with $y_1(0) = y_2(0) = y_3(0), \quad h_1 y_1'(0) + h_2 y_2'(0) + h_3 y_3'(0) = 0, \quad |y_i(\infty)| < \infty$, $(i = 1, 2, 3)$.

**2.11** Solve the four-point posed boundary-value problem posed by eqns (2.143)–(2.145) for $f_1(x) \equiv \sin(\pi x)$, $f_2(x) = f_3(x) \equiv 0$.

# Chapter 3

# Kirchhoff Beam Problems

Among various topics of applied mechanics that could be successfully approached with the influence (Green's) function method, the bending of elastic beams represents one of the most beneficial areas. In this chapter, the reader will find a variety of applications of the method to some basic classical beam problems that are traditionally discussed in *mechanics of materials, strength of materials, structural analysis*, and relevant courses. Some nontraditional implementations of this method in elastic beam problems are also discussed.

Application of influence functions of a point force to the development of solution algorithms for some classes of beam problems within the scope of Kirchhoff theory could be linked to the so-called *singularity method* and to the concept of *influence lines* (see, for example, [12, 21, 27, 55]). The methodology of the singularity method was developed many years ago. However, it has never been systematically used in the beam theory and has not been suggested as a universal approach to all of the problem classes that are discussed in this book.

In this chapter we continue a broad discussion on possible ways of the utilization of the influence function method in computational mechanics. Application of this method to one of the most traditional areas in the field will be brought to the reader's attention first. Namely, the geometrical and physically linear bending of elastic Kirchhoff beams will be considered as they undergo transverse concentrated and distributed forces. Explicit formulae are derived for computing all components of stress-strain state in terms of the influence function.

The discussion in this chapter opens with the establishment of the relation between the notion of influence function of a transverse point force for a beam and the concept of Green's function for the corresponding boundary-value problem that models the bending phenomenon for the beam.

The construction procedures can be found in Section 3.1 for influence functions for a beam of uniform flexural rigidity, with various types of edge conditions imposed. We then show in Section 3.2 various ways of utilizing influence functions for computation of the most important components of the stress-strain state of a beam subject to various types of load. And finally, the influence function formalism is further extended in Section 3.3 to beams resting on a simple (single parameter) elastic foundation.

A specificity of Section 3.5 is in its referential nature. The section assembles together an extensive catalogue of influence functions of a transverse point force for single-span Kirchhoff beams of uniform flexural rigidity. More than two dozens of different combinations of edge conditions are covered. Many of these influence functions are not available in the existing literature on elastic beams and published for the first time in this text.

## 3.1 Single-span Beams

The classical Kirchhoff theory [12, 21] assumes that bending of a beam is linear in both physical and geometrical sense. Physical linearity means that the material, of which the beam is made, follows the Hooke's law (the linear portion of the stress-strain curve) [21, 27, 55], while geometric linearity implies that the maximum of the deflection is *small* compared to the beam's length. Mathematical sense of the geometric linearity is that the derivative of the deflection function is supposed to be negligibly small compared to the unity.

### 3.1.1 Statement of basic problems

For a beam of length $a$, Kirchhoff theory yields the displacement formulation which is a boundary-value problem for the so-called Euler–Bernoulli equation

$$\frac{d^2}{dx^2}\left(EI(x)\frac{d^2w(x)}{dx^2}\right) = -q(x), \quad x \in (0, a) \tag{3.1}$$

where $w(x)$ represents the beam's deflection, $E(x)$ is the modulus of elasticity of the material of which the beam is made, $I(x)$ is the moment of inertia of the cross-section $x$. The product $EI(x)$ is usually referred to as the *flexural rigidity* of the beam, and $q(x)$ is a transverse load applied to the beam.

As follows from the general theory of ODE, the total number of four boundary conditions ought to be imposed at the end-points of the interval $[0, a]$ to ensure a unique solution to eqn (3.1). This follows from the fact that the general solution of the above equation and of the corresponding homogeneous equation

$$\frac{d^2}{dx^2}\left(EI(x)\frac{d^2w(x)}{dx^2}\right) = 0 \tag{3.2}$$

contains four arbitrary constants which are supposed to be defined from boundary conditions.

In standard beam problem settings, we usually impose two conditions at each end-point of a beam. Boundary conditions in such settings naturally follow from the manner the beam's edges are fixed. At this point in our presentation, we are going to introduce a variety of different types of boundary conditions that are physically feasible for a beam and which will be dealt with in this text.

But before going any further, we recall expressions for the force components occurring in the cross-section $x$ as they are written in terms of the deflection function $w(x)$, because those components along with the deflection function itself are required to formulate various types of boundary conditions. The bending moment, for example, is defined in terms of the deflection function as

$$M(x) = EI(x)\frac{d^2w(x)}{dx^2} \tag{3.3}$$

while the shear force is expressed as

$$Q(x) = \frac{d}{dx}\left(EI(x)\frac{d^2w(x)}{dx^2}\right) \tag{3.4}$$

Consider first the so-called cantilever beam whose edge $x = 0$ is built into a wall while the edge $x = a$ is free of tension as shown in Figure 3.1

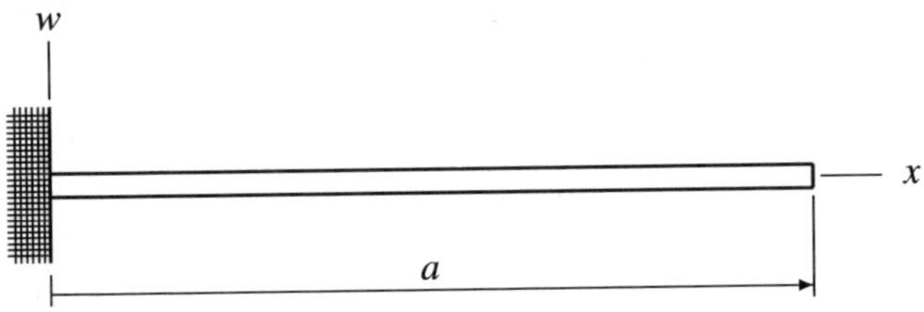


Figure 3.1: Cantilever beam

From this setting, it follows that the deflection function and also its first order derivative ought to be zero as $x = 0$. That is

$$w(0) = 0 \quad \text{and} \quad \frac{dw(0)}{dx} = 0 \tag{3.5}$$

The second condition in eqn (3.5) is evident, because if it does not hold then the beam actually breaks at the place where it is built into the wall. Boundary conditions imposed at the right-hand edge imply that the bending moment $M(x)$ as well as the shear force $Q(x)$ are zero as $x = a$. That is

$$\frac{d^2w(a)}{dx^2} = 0 \quad \text{and} \quad \frac{d^3w(a)}{dx^3} = 0 \tag{3.6}$$

The first relation in eqn (3.6) is evident since it directly follows from the relation $M(a) = 0$ (see eqn (3.3)), while the second relation can be justified by applying the product rule of differentiation to the expression for $Q(x)$ in eqn (3.4)

$$Q(x) = \frac{d(EI(x))}{dx}\frac{d^2w(x)}{dx^2} + EI(x)\frac{d^3w(x)}{dx^3}$$

and realizing that the term containing $d^2w/dx^2$ vanishes in light of the first relation in (3.6).

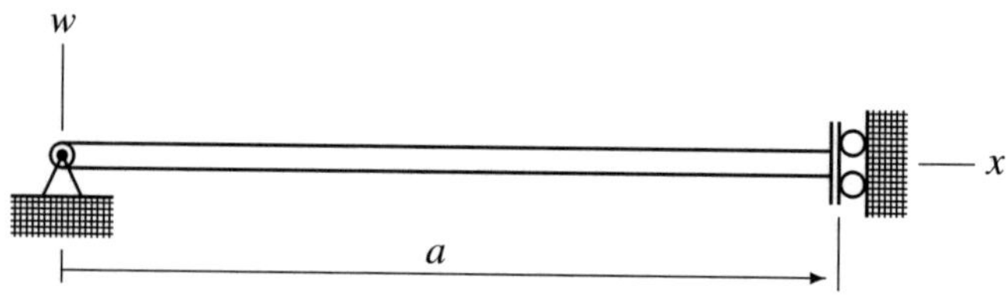


Figure 3.2: Simply supported and sliding beam

Another two standard boundary conditions are introduced by Figure 3.2 that depicts a beam whose left-hand edge is hinged (we say *simply supported*) while the right-hand edge is freely sliding along a wall without detachment.

This setting implies that the deflection and the bending moment ought to equal zero as $x = 0$. That is

$$w(0) = 0 \quad \text{and} \quad \frac{d^2w(0)}{dx^2} = 0 \tag{3.7}$$

while the way by which the right-hand edge is fixed suggests that the slope of the deflection function along with the shear force equal zero as $x = a$. That is

$$\frac{dw(a)}{dx} = 0 \quad \text{and} \quad \frac{d}{dx}\left(EI(a)\frac{d^2w(a)}{dx^2}\right) = 0 \tag{3.8}$$

Note that the boundary conditions in eqn (3.8) also occur if the stress-strain state of a beam of length $2a$ is symmetric about the line $x = a$. In such a case we can consider only half-interval $[0, a]$ and impose the conditions in eqn (3.8) to the right-hand end-point $x = a$.

The boundary conditions in eqns (3.5)–(3.8) are most widely spread in applications, but are not the only conditions that may occur in beam problems. Also of a great practical importance in engineering science are, for example, beams whose edges are hinged and supported by elastic springs (we say *elastically supported*) in a way shown in Figure 3.3.

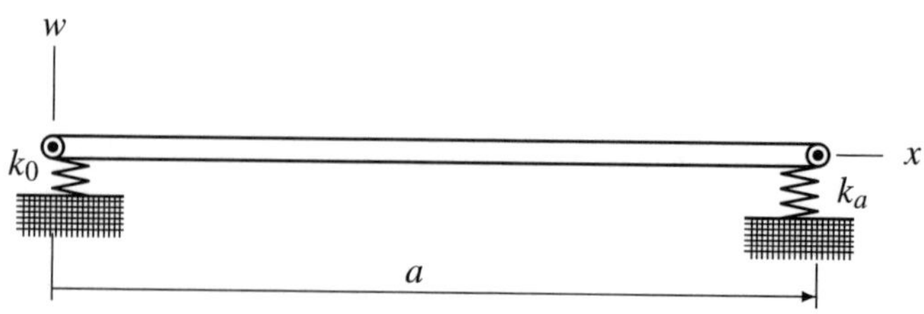


Figure 3.3: Hinged and elastically supported beam

Mechanical interpretation of the boundary conditions shown in Figure 3.3 is that the bending moment is equal to zero, while the shear force is directly proportional to the deflection (the Hooke's law) at both edges. This reads mathematically as

$$\frac{d^2w(0)}{dx^2} = 0 \quad \text{and} \quad EI(0)\frac{d^3w(0)}{dx^3} + k_0 w(0) = 0, \quad k_0 > 0 \tag{3.9}$$

while as $x = a$, we have

$$\frac{d^2w(a)}{dx^2} = 0 \quad \text{and} \quad EI(a)\frac{d^3w(a)}{dx^3} - k_a w(a) = 0, \quad k_a > 0 \tag{3.10}$$

The appearance of the shear force-related terms in the second relations of eqns (3.9) and (3.10) seemingly contradicts the expression for the shear force from eqn (3.4). However, this is not a contradiction. Indeed, similarly to the transformation in eqn (3.6), one arrives at the actual form of the second relations in eqns (3.9) and (3.10) after the product rule of differentiation is applied to eqn (3.4) and the first relation $d^2w/dx^2 = 0$ is taken into account.

Note that if both the elastic constants $k_0$ and $k_a$ approach infinity, then the relations in eqns (3.9) and (3.10) transform into conditions of simple support. The hypothetical case when both $k_0$ and $k_a$ approach zero is clearly meaningless from the practice stand point. It is interesting to note a differential that appears in the signs in the second conditions of eqns (3.9) and(3.10). The point is that the derivatives in eqn (3.9) are taken with the $x$ variable *increasing*, whereas when we differentiate in eqn (3.10), the $x$ variable is *decreasing*.

The beam shown in Figure 3.4 brings another two types of boundary conditions to a beam problem that might occur in engineering practice

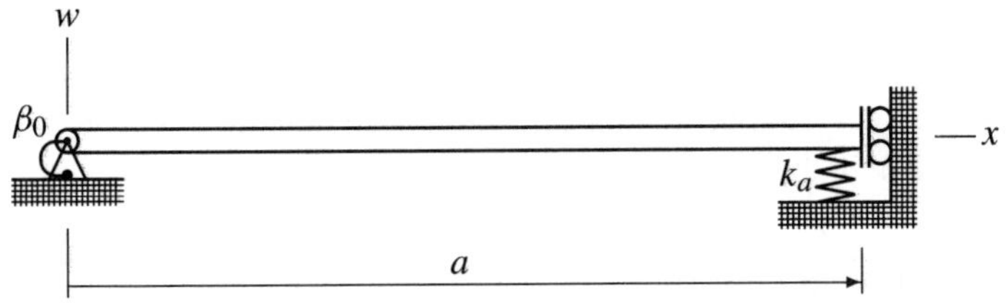


Figure 3.4: Elastically hinged and sliding beam

The left-hand edge of this beam is fixed so that the deflection is zero and the bending moment is directly proportional to the rotation angle (the derivative of the deflection function) of the edge cross-section. This yields

$$w(0) = 0 \quad \text{and} \quad EI(0)\frac{d^2w(0)}{dx^2} - \beta_0 \frac{dw(0)}{dx} = 0, \quad \beta_0 \geq 0 \tag{3.11}$$

As to the right-hand edge, it is allowed to slide along a vertical wall with no detachment, but the sliding is restricted by a spring whose elastic constant is $k_a$.

This formalizes into the following conditions

$$\frac{dw(a)}{dx} = 0 \quad \text{and} \quad \frac{d}{dx}\left(EI(a)\frac{d^2w(a)}{dx^2}\right) - k_a w(a) = 0, \quad k_a \geq 0 \tag{3.12}$$

Note that if the elastic constant $\beta_0$ equals zero, then the relations in eqn (3.11) convert into conditions of simple support, whereas if $\beta_0$ is taken to infinity, then they transform into conditions of clamp. If the elastic constant $k_a$ goes to infinity, then relations in eqn (3.12) convert to conditions of symmetry (see eqn (3.8)), whereas as $k_a = 0$, we have conditions of clamp.

In addition to the types of boundary conditions listed in eqns (3.5)–(3.12), a beam's edge could be fixed in such a way that the bending moment is directly proportional to the rotation angle (the derivative of the deflection function), while the shear force is either zero (the deflection is not limited) or directly proportional to the deflection at the edge cross-section.

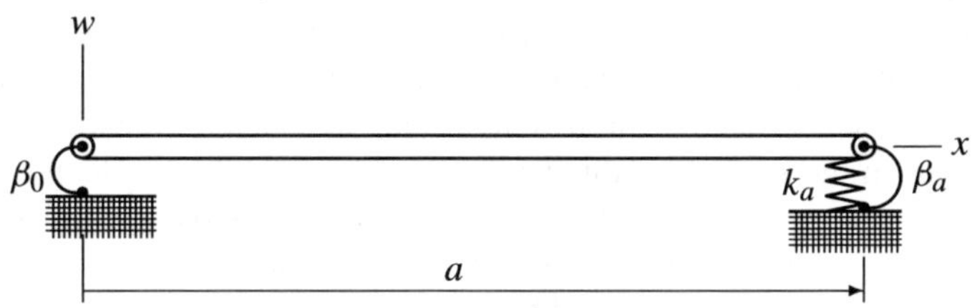


Figure 3.5: Elastically hinged and supported beam

The end conditions of these types take place at the edges of the beam depicted in Figure 3.5. And the boundary condition simulating this type of end conditions are written in the form

$$EI(0)\frac{d^2w(0)}{dx^2} - \beta_0\frac{dw(0)}{dx} = 0 \quad \text{and} \quad \frac{d}{dx}\left(EI(0)\frac{d^2w(0)}{dx^2}\right) = 0 \tag{3.13}$$

and

$$EI(a)\frac{d^2w(a)}{dx^2} + \beta_a\frac{dw(a)}{dx} = 0 \quad \text{and} \quad \frac{d}{dx}\left(EI(a)\frac{d^2w(a)}{dx^2}\right) - k_a w(a) = 0 \tag{3.14}$$

where the parameters $\beta_0$, $\beta_a$, $k_0$ and $k_a$ represent non-negative constants.

Note that not every combination of the conditions presented by eqns (3.5) through (3.14) is physically feasible for a single beam, because a certain combination of conditions is feasible if it provides only the trivial solution to equation (3.2). We will focus on this point in detail and illustrate it with several examples. For simplicity in the examples that follow, we assume that the flexural rigidity $EI$

of a beam is constant reducing equation (3.2) to the form

$$\frac{d^4w(x)}{dx^4}=0 \tag{3.15}$$

the general solution of which

$$w(x)=C_1+C_2x+C_3x^2+C_4x^3 \tag{3.16}$$

as we have shown in Chapter 2, can be obtained by four successive integrations.

**Example 1.1:** It is evident, for instance, that a long elastic bar whose both edges are free of tension does not represent a beam but is rather a freely moving object. Hence, conditions of eqn (3.6) cannot be imposed at both edges of a beam. Indeed, if they are applied to the general solution of eqn (3.16), then we come up with the following homogeneous system of linear algebraic equations

$$\begin{pmatrix} 0 & 0 & 2 & 0 \\ 0 & 0 & 0 & 6 \\ 0 & 0 & 2 & 6a \\ 0 & 0 & 0 & 6 \end{pmatrix} \times \begin{pmatrix} C_1 \\ C_2 \\ C_3 \\ C_4 \end{pmatrix} = \begin{pmatrix} 0 \\ 0 \\ 0 \\ 0 \end{pmatrix}$$

in $C_j$, $(j=\overline{1,4})$. Since the first column of the above coefficient matrix contains only zero entries, its determinant is zero. This implies that the system has infinitely many solutions making unfeasible the case with conditions of the type in eqn (3.6) imposed at both edges. ▲

**Example 1.2:** The set of boundary conditions in eqns (3.6) and (3.7) is not feasible either for a single beam, because an elastic bar, one edge of which is hinged while another is free of tension does not represent a beam. Indeed, it can freely rotate around the hinged end-point. Mathematics supports this conclusion, since applying conditions (3.6) and (3.7) to (3.16), we obtain the homogeneous system

$$\begin{pmatrix} 1 & 0 & 0 & 0 \\ 0 & 0 & 2 & 0 \\ 0 & 0 & 2 & 6a \\ 0 & 0 & 0 & 6 \end{pmatrix} \times \begin{pmatrix} C_1 \\ C_2 \\ C_3 \\ C_4 \end{pmatrix} = \begin{pmatrix} 0 \\ 0 \\ 0 \\ 0 \end{pmatrix}$$

whose coefficient matrix is singular (its determinant is zero). ▲

**Example 1.3:** The boundary-value problem posed by eqns (3.11) and (3.12) is feasible for a beam. Indeed, if the conditions from these equations are applied to the general solution in eqn (3.16), then we have

$$\begin{pmatrix} 1 & 0 & 0 & 0 \\ 0 & -\beta_0 & 2EI & 0 \\ 0 & 1 & 2a & 3a^2 \\ -k_a & -ak_a & -a^2k_a & 6EI - a^3k_a \end{pmatrix} \times \begin{pmatrix} C_1 \\ C_2 \\ C_3 \\ C_4 \end{pmatrix} = \begin{pmatrix} 0 \\ 0 \\ 0 \\ 0 \end{pmatrix}$$

And the determinant of the coefficient matrix equals in this case

$$-a^3k_a(a\beta_0 + 4EI) - 12EI(a\beta_0 + EI)$$

which is never zero. Indeed, since the parameters $\beta_0$ and $k_a$ are greater than or equal to zero, while $a$ and $EI$ are positive, the above quantity is always negative. So, the boundary-value problem simulating equilibrium of the beam shown in Figure 3.4 is well-posed. ▲

Although in the overwhelming majority of practical cases two boundary conditions are imposed at each end-point of a beam, settings are also possible with three conditions imposed at one end-point and one at the other, and even with all four conditions imposed at one end-point. The latter case represents, in fact, an initial-value problem. Such settings are mathematically feasible although they are not of practical importance.

### 3.1.2 Influence function–Green's function relation

To clear up the physical sense of the concept of Green's function for a boundary-value problem that simulates the elastic equilibrium of a Kirchhoff beam, we introduce the *Dirac delta function* $\delta(x)$ named after a prominent British physicist P.A.M. Dirac (1902–1984) who was the 1933 Nobel Prize winner. Dirac function represents one of the so-called *generalized functions* [4, 32] that play an important role in applied mathematics and engineering sciences.

Dirac delta function is not actually a function in a common sense and its rigorous definition is usually presented in special graduate courses in mathematics and cannot be given within the scope of our text. Instead, we will limit ourselves to a heuristic definition [32] that is formally accepted and successfully used in courses of the engineering curricula.

**Definition:** Dirac delta function $\delta(x)$ of a real variable $x$ can be defined by the following condition

$$\int_a^b f(x)\delta(x-s)\,dx = \begin{cases} 0, & \text{if } s \notin (a,b) \\ \frac{1}{2}f(s), & \text{if } s = a \text{ or } b \\ f(s), & \text{if } s \in (a,b) \end{cases} \tag{3.17}$$

where $f(x)$ represents a function continuous on $[a, b]$.

This definition is not flawless mathematically. Indeed, on one hand, from eqn (3.17), it follows that $\delta(x) = 0$ for any value of $x$ except for $x = 0$. On the other hand, if $f(x) \equiv 1$, then eqn (3.17) yields

$$\int_a^b \delta(x)\,dx = 1$$

But these two conditions are contradictory from the calculus viewpoint, because one of the properties of a definite integral states that the integral of a function, which is zero *almost everywhere* on the segment of integration, must be zero. However, a formal application of the definition in eqn (3.17) allows many convenient generalizations in applied mathematics and is productively used in engineering sciences, where we often deal with discontinuous functions.

Note that in some sources [32] the definition in eqn (3.17) is referred to as the *sifting property* of the Dirac delta function.

Let $g(x, s)$ represent the Green's function for the homogeneous Euler–Bernoulli equation

$$\frac{d^2}{dx^2}\left(EI(x)\frac{d^2w(x)}{dx^2}\right) = 0, \quad x \in (0, a) \tag{3.18}$$

subject to a set of boundary conditions

$$B_{0,i}[w(0)] = 0, \quad B_{a,i}[w(a)] = 0, \quad i = 1, 2 \tag{3.19}$$

which is supposed to be feasible mathematically. This implies that the boundary-value problem in eqns (3.18) and (3.19) has only the trivial solution.

In compliance with Theorem 2.4 from Chapter 2, the solution $w(x)$ of the problem stated by eqns (3.1) and (3.19) can be expressed in terms of $g(x, s)$ as

$$w(x) = \int_0^a g(x, s)q(s)\,ds, \quad x \in [0, a] \tag{3.20}$$

To come up with the physical interpretation of the Green's function $g(x, s)$, let the beam be loaded with a single transverse force of magnitude $P_0$ concentrated at a fixed but arbitrary point $x = s_0$. The right-hand term in eqn (3.1) can be written in this case as a scalar multiple of the Dirac delta function

$$q(x) = P_0\delta(x - s_0)$$

Substituting this in eqn (3.20) and using then the sifting property of the Dirac delta function yields

$$w(x) = \int_0^a g(x, s)P_0\delta(s - s_0)\,ds = P_0 g(x, s_0), \quad x \in [0, a]$$

or

$$g(x, s_0) = \frac{w(x)}{P_0}, \quad x \in [0, a]$$

Thus, the Green's function $g(x, s)$ of the boundary-value problem posed by eqns (3.18) and (3.19), being considered as a function of the observation point $x$, represents the deflection of the beam that is caused by a unit concentrated force applied to an arbitrary point $s_0$. In view of this interpretation, $g(x, s)$ is usually referred to, in mechanics, as the *influence function of a point force* for the beam simulated by eqns (3.18) and (3.19).

This suggests a straight way of obtaining influence functions of a point force for Kirchhoff beams. A boundary-value problem has to be formulated that simulates elastic equilibrium of the beam and then we construct the Green's function for that problem by using the techniques presented in Chapter 2.

### 3.1.3 Influence functions for beams of uniform rigidity

In this section the focus will be on the construction of influence functions of a point force for Kirchhoff beams of a uniform flexural rigidity. As we mentioned earlier, if $EI(x)$ in eqn (3.1) does not vary with $x$, then it can be taken out of the differentiation sign. This substantially simplifies eqn (3.1) reducing it to

$$\frac{d^4w(x)}{dx^4} = -\frac{q(x)}{EI}, \quad x \in (0, a) \tag{3.21}$$

From Chapter 2, we recall the modification of the classical method for the construction of Green's functions, which is based on their defining properties, and extend that modification to the class of boundary-value problems posed by eqn (3.19) and the homogeneous equation

$$\frac{d^4w(x)}{dx^4} = 0, \quad x \in (0, a) \tag{3.22}$$

corresponding to (3.21).

Assume that the boundary-value problem posed by eqns (3.22) and (3.19) has only the trivial solution. This implies that there exists its unique Green's function. Remember that the latter is identified with the influence function of a point force for the associated beam problem.

Let two functions $w_1(x)$ and $w_2(x)$ be not only particular solutions of eqn (3.22), but, in addition, let both of them satisfy the edge conditions imposed at $x = 0$ by the first relation in eqn (3.19). Let also $w_3(x)$ and $w_4(x)$ represent another pair of particular solutions of eqn (3.22) satisfying the edge conditions imposed at $x = a$ by the second relation in eqn (3.19).

Assume also that the set $w_i(x)$, $(i = \overline{1, 4})$, introduced above, is linearly independent on $[0, a]$. This implies that they form a fundamental set of solutions for eqn (3.22). Based on this set, we seek the Green's function to the problem in eqns (3.22) and (3.19) in the form

$$g(x, s) = \begin{cases} a_1(s)w_1(x) + a_2(s)w_2(x), & \text{for } x \le s \\ b_1(s)w_3(x) + b_2(s)w_4(x), & \text{for } s \le x \end{cases} \tag{3.23}$$

From this representation, it follows that the entire set of boundary conditions in eqn (3.19) is satisfied by $g(x, s)$ in this form, regardless of the values of the coefficients $a_i(s)$ and $b_i(s)$, $(i = 1, 2)$. This occurs because the upper branch in $g(x, s)$ is a linear combination of the functions $w_1(x)$ and $w_2(x)$, each of which satisfies the boundary conditions at $x = 0$, while the lower branch is a linear combination of $w_3(x)$ and $w_4(x)$, satisfying respectively the boundary conditions at $x = a$. Hence, $g(x, s)$ in eqn (3.23) meets properties 1 and 4 in the definition of the Green's function.

To compute the coefficients $a_i(s)$ and $b_i(s)$ in eqn (3.23), we take advantage of the remaining defining properties of the Green's function. In compliance with property 2, one recalls that $g(x, s)$ is continuous as $x = s$, that is

$$\lim_{x\to s^+} g(x, s) - \lim_{x\to s^-} g(x, s) = 0$$

This consequently yields

$$w_3(s)b_1(s) + w_4(s)b_2(s) - w_1(s)a_1(s) - w_2(s)a_2(s) = 0 \tag{3.24}$$

According to property 2, the first derivative of $g(x, s)$ with respect to $x$ is also continuous as $x = s$

$$\lim_{x\to s^+} \frac{\partial g(x, s)}{\partial x} - \lim_{x\to s^-} \frac{\partial g(x, s)}{\partial x} = 0$$

resulting in

$$w_3'(s)b_1(s) + w_4'(s)b_2(s) - w_1'(s)a_1(s) - w_2'(s)a_2(s) = 0 \tag{3.25}$$

The second derivative of $g(x, s)$ with respect to $x$ is also continuous as $x = s$

$$\lim_{x\to s^+} \frac{\partial^2 g(x, s)}{\partial x^2} - \lim_{x\to s^-} \frac{\partial^2 g(x, s)}{\partial x^2} = 0$$

which yields

$$w_3''(s)b_1(s) + w_4''(s)b_2(s) - w_1''(s)a_1(s) - w_2''(s)a_2(s) = 0 \tag{3.26}$$

And finally, in compliance with property 3, the third derivative of $g(x, s)$ with respect to $x$ is discontinuous as $x = s$, providing

$$\lim_{x\to s^+} \frac{\partial^3 g(x, s)}{\partial x^3} - \lim_{x\to s^-} \frac{\partial^3 g(x, s)}{\partial x^3} = -1$$

This yields as follows

$$w_3'''(s)b_1(s) + w_4'''(s)b_2(s) - w_1'''(s)a_1(s) - w_2'''(s)a_2(s) = -1 \tag{3.27}$$

Clearly, the relations in eqns (3.24)–(3.27) constitute a system of linear algebraic equations in $a_i(s)$ and $b_i(s)$. It is well-posed, because the determinant of its coefficient matrix represents the Wronskian for a set of linearly independent functions. Thus, upon solving this system and substituting the values of $a_i(s)$ and $b_i(s)$ into eqn (3.23), we complete the construction procedure for the influence function of a point force $g(x, s)$ of the beam whose equilibrium is modeled by the boundary-value problem in eqns (3.22) and (3.19).

In the series of instructive examples that follow we will highlight some key points of the algorithm for the construction of influence functions for beams with various types of edge conditions imposed.

**Example 1.4:** Consider a beam of length $a$ with both edges clamped. To construct the influence function of a point force for such a beam, let us formulate the boundary-value problem

$$w(0) = \frac{dw(0)}{dx} = 0, \quad w(a) = \frac{dw(a)}{dx} = 0 \tag{3.28}$$

for the governing equation in (3.22).

It can easily be shown that the homogeneous boundary-value problem posed by eqns (3.22) and (3.28) has only the trivial solution. Indeed, applying the conditions in eqn (3.28) to the general solution

$$w(x) = C_1 + C_2x + C_3x^2 + C_4x^3 \tag{3.29}$$

we come up with the homogeneous system of linear algebraic equations

$$\begin{pmatrix} 1 & 0 & 0 & 0 \\ 0 & 1 & 0 & 0 \\ 1 & a & a^2 & a^3 \\ 0 & 1 & 2a & 3a^2 \end{pmatrix} \times \begin{pmatrix} C_1 \\ C_2 \\ C_3 \\ C_4 \end{pmatrix} = \begin{pmatrix} 0 \\ 0 \\ 0 \\ 0 \end{pmatrix}$$

whose coefficient matrix is non-singular, justifying the existence of the unique Green's function of the problem in eqns (3.22) and (3.28), which actually represents the influence function of a point force for the clamped-clamped beam shown in Figure 3.6.

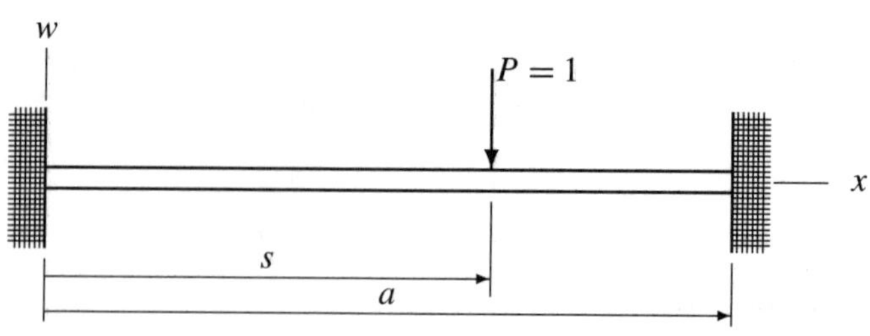


Figure 3.6: Clamped-clamped beam loaded with a point force

Notice that for problems allowing natural physical interpretation, one can draw on intuition to decide whether the corresponding boundary-value problem has a unique Green's function. Speaking of the current problem, for example, it is intuitively clear that the clamped beam uniquely responds to a concentrated force regardless of the point of its application. This observation indirectly justifies the existence and uniqueness of the Green's function of the problem under consideration.

As follows from eqn (3.29), the set of functions

$$w_1(x) \equiv 1, \quad w_2(x) \equiv x, \quad w_3(x) \equiv x^2, \quad w_4(x) \equiv x^3 \tag{3.30}$$

represents a fundamental set of solutions to eqn (3.22). In light of the foregoing discussion, however, we would rather take advantage of a different set of functions

$$\begin{gathered} w_1(x) \equiv x^2, \quad w_2(x) \equiv x^3 \\ w_3(x) \equiv (x-a)^2, \quad w_4(x) \equiv (x-a)^3 \end{gathered} \tag{3.31}$$

which can also be utilized as the fundamental set of solutions to eqn (3.22). Indeed, each of the functions in eqn (3.31), being a polynomial of degree less than or equal to three, represents a particular solution to eqn (3.22). Moreover, the Wronskian of this set

$$\begin{vmatrix} x^2 & x^3 & (x-a)^2 & (x-a)^3 \\ 2x & 3x^2 & 2(x-a) & 3(x-a)^2 \\ 2 & 6x & 2 & 6(x-a) \\ 0 & 6 & 0 & 6 \end{vmatrix} \equiv 12a^4$$

is not identically zero. This consequently implies that the functions in eqn (3.31) are indeed linearly independent on any interval.

It is important to realize the principal distinction between the two fundamental sets of solutions presented in eqns (3.30) and (3.31). What makes the choice of the second of them more promissory in view of the construction procedure just suggested for the influence function? The answer can be given in conjunction with the boundary conditions imposed by eqn (3.28).

The point is that, whereas the components of the system in eqn (3.30) are not at all directly associated with the boundary conditions in eqn (3.28), both $w_1(x)$ and $w_2(x)$ in eqn (3.31) satisfy the conditions imposed at $x = 0$ in eqn (3.28), while both $w_3(x)$ and $w_4(x)$ satisfy the boundary conditions at $x = a$. As a result, the actual computing of the influence function promises to be more compact, because there will be no need to treat the boundary conditions when going through the construction procedure.

We now seek the Green's function $g(x, s)$ of the boundary-value problem in eqns (3.22) and (3.28), which represents the influence function for the clamped-clamped beam, in the following form

$$g(x, s) = \begin{cases} a_1(s)x^2 + a_2(s)x^3, & \text{for } x \le s \\ b_1(s)(x-a)^2 + b_2(s)(x-a)^3, & \text{for } s \le x \end{cases} \tag{3.32}$$

From this representation, it follows that the boundary conditions in eqn (3.28) are satisfied by $g(x, s)$ in this form, because the upper branch in $g(x, s)$ is a linear combination of $w_1(x)$ and $w_2(x)$, each of which satisfies the boundary conditions at $x = 0$, while the lower branch is a linear combination of $w_3(x)$ and $w_4(x)$, satisfying the boundary conditions at $x = a$.

Hence, $g(x, s)$ in eqn (3.32) meets the defining properties 1 and 4 in the definition of the Green's function.

To compute the values of $a_i(s)$ and $b_i(s)$, we use the remaining defining properties of the Green's function. In compliance with property 2, we obtain

$$(s-a)^2 b_1(s) + (s-a)^3 b_2(s) - s^2 a_1(s) - s^3 a_2(s) = 0 \tag{3.33}$$

According to property 2, the first order derivative of $g(x, s)$ with respect to $x$ is also continuous as $x = s$, providing

$$2(s-a)b_1(s) + 3(s-a)^2 b_2(s) - 2s a_1(s) - 3s^2 a_2(s) = 0 \tag{3.34}$$

The second order derivative of $g(x, s)$ is also continuous as $x = s$. This implies

$$2b_1(s) + 6(s-a)b_2(s) - 2a_1(s) - 6s a_2(s) = 0 \tag{3.35}$$

And finally, in compliance with property 3, the third order derivative of $g(x, s)$ is discontinuous as $x = s$, yielding

$$6b_2(s) - 6a_2(s) = -1 \tag{3.36}$$

Equations (3.33)–(3.36) form a well-posed system of linear algebraic equations

$$\begin{pmatrix} (s-a)^2 & (s-a)^3 & -s^2 & -s^3 \\ 2(s-a) & 3(s-a)^2 & -2s & -3s^2 \\ 2 & 6(s-a) & -2 & -6s \\ 0 & 6 & 0 & -6 \end{pmatrix} \times \begin{pmatrix} b_1(s) \\ b_2(s) \\ a_1(s) \\ a_2(s) \end{pmatrix} = \begin{pmatrix} 0 \\ 0 \\ 0 \\ -1 \end{pmatrix}$$

The solution to this system is found as

$$a_1(s) = -\frac{s(s-a)^2}{2a^2}, \quad a_2(s) = \frac{(s-a)^2(2s+a)}{6a^3}$$

$$b_1(s) = \frac{s^2(s-a)}{2a^2}, \quad b_2(s) = \frac{s^2(2s-3a)}{6a^3}$$

Substituting these values in eqn (3.32), one finally obtains the influence function for the clamped-clamped beam in the following form

$$g(x, s) = \frac{1}{6a^3} \begin{cases} x^2(s-a)^2[2s(x-a) + a(x-s)], & \text{for } x \le s \\ s^2(x-a)^2[2x(s-a) + a(s-x)], & \text{for } s \le x \end{cases} \tag{3.37}$$

Thus, $g(x, s)$ represents the deflection of the clamped-clamped beam at an observation point $x$, associated with the concentrated unit force applied to a source point $s$. This implies that $g(x, s)$ determines the response of the clamped-clamped beam to a unit force concentrated at $s$. ▲

The representation that appeared in eqn (3.37) is symmetric in the sense introduced in Chapter 2, that is $g(x, s) = g(s, x)$. As we learned from Chapter 2, this reflects the self-adjointness of the boundary-value problem in eqns (3.22) and (3.28) for which $g(x, s)$ is the Green's function. In some sources in mechanics (see, for example, [32]), this property is called the *Maxwell's reciprocity*. From the mechanics standpoint, it reads as: *the response of the beam at x due to a transverse concentrated force at s, is the same as the response at s due to a force applied at x*.

In order to run any procedure for the construction of Green's function, a set of $n$ linearly independent particular solutions is required for a governing differential equation of order $n$. Recalling the modification of the classical method, which has just been used, let us focus on one specific point in it.

The reader has already grasped the significance of finding pairs of linearly independent particular solutions for a governing equation, which *a priori* satisfy appropriate boundary conditions. Of course, it was not hard to find such solutions for the example setting recently completed, because the boundary conditions of a clamped edge can be easily visualized and a set of appropriate functions could readily be guessed.

This aspect is not trivial, however, in other cases. The question is, how appropriate particular solutions can be obtained in nontrivial situations where boundary conditions do not allow an easy guess for required particular solutions to the governing equation. Addressing this issue, we consider the next example where a universal approach is proposed for finding appropriate fundamental sets of solutions for governing equations.

**Example 1.5:** Let the influence function of a point force be constructed for a beam of length $a$, one edge of which is elastically supported while the other is clamped as shown in Figure 3.7.

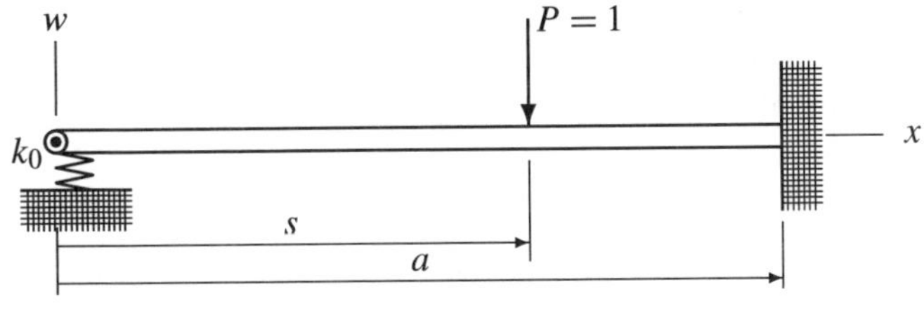


Figure 3.7: Elastically supported–clamped beam

It is intuitive that there exists a unique influence (Green's) function in this case. In other words, it is evident, from a physics viewpoint, that the beam should uniquely respond to a transverse concentrated force regardless of the point of its application. The reader can, of course, easily check it out by directly solving the corresponding homogeneous boundary-value problem for eqn (3.22), as it has, for instance, been done in Example 1.4.

From the beam theory [21, 27, 55], it follows that the edge conditions of elastic support for the beam can be imposed (see eqn (3.9)) in the form

$$\frac{d^2w(0)}{dx^2}=0, \quad \frac{d^3w(0)}{dx^3}+kw(0)=0 \tag{3.38}$$

where $k=k_0/EI$ and $k_0$ represents the elastic support coefficient. Clearly, as the parameter $k$ approaches zero, the conditions in eqn (3.38) reduce to the classical free edge conditions, whereas with $k$ going to infinity, we have the simply supported edge. These particular cases of the boundary conditions will be revisited later when the influence function is constructed.

Conditions at the clamped edge $x=a$, in turn, are evident

$$w(a)=0, \quad \frac{dw(a)}{dx}=0 \tag{3.39}$$

Thus, the influence function of a point force that we are looking for, represents the Green's function of the boundary-value problem posed by eqns (3.22), (3.38), and (3.39).

Clearly, a pair of linearly independent particular solutions $w_3(x)$ and $w_4(x)$ to equation (3.22), satisfying the edge conditions in eqn (3.39) can, as we suggested in Example 1.4, be taken in the form

$$w_3(x)\equiv(x-a)^2, \quad w_4(x)\equiv(x-a)^3 \tag{3.40}$$

It turns out, however, that the choice of two other particular solutions $w_1(x)$ and $w_2(x)$ for eqn (3.22), which satisfy the edge conditions that appeared in eqn (3.38), is not that clear. To show how such solutions can be found in this and other cases, we advocate a common approach.

Namely, to obtain any one of these functions, say $w_1(x)$, we formulate the following problem

$$\frac{d^2w_1(0)}{dx^2}=0, \quad \frac{d^3w_1(0)}{dx^3}+kw_1(0)=0 \tag{3.41}$$

$$\frac{dw_1(0)}{dx}=1, \quad w_1(0)=0 \tag{3.42}$$

for the governing equation (3.22).

Note that all the four conditions of uniqueness are imposed, in this case, at a single point, $x=0$. That is why we refer to the setting in eqns (3.22), (3.41) and (3.42) as the *initial-value problem* (in contrast to most of the settings that we have dealt with so far in this text, which are *boundary-value problems*).

Applying the conditions in eqns (3.41) and (3.42) to the general solution of eqn (3.22) shown in eqn (3.29), we derive the following system of linear algebraic equations

$$\begin{pmatrix} 0 & 0 & 2 & 0 \\ k & 0 & 0 & 6 \\ 0 & 1 & 0 & 0 \\ 1 & 0 & 0 & 0 \end{pmatrix} \times \begin{pmatrix} C_1 \\ C_2 \\ C_3 \\ C_4 \end{pmatrix} = \begin{pmatrix} 0 \\ 0 \\ 1 \\ 0 \end{pmatrix}$$

whose solution can readily be obtained as

$$C_1 = C_3 = C_4 = 0, \quad \text{and} \quad C_2 = 1$$

Thus, upon substituting these in (3.29), we finally have

$$w_1(x) = x \tag{3.43}$$

Clearly, the function $w_1(x)$ represents a particular solution of eqn (3.22) that satisfies the edge conditions of elastic support at the left-hand edge of the beam that are imposed by eqn (3.41). This is what we require from $w_1(x)$ and $w_2(x)$ components of the fundamental set of solutions to make our procedure for the construction of the Green's function work. The conditions in eqn (3.42) have nothing directly in common with the Green's function that we are looking for. They are chosen to just give us a well-posed initial-value problem.

To obtain the second particular solution, $w_2(x)$ to eqn (3.22), which satisfies the boundary conditions in eqn (3.38), we formulate another initial-value problem for eqn (3.22) as follows

$$\frac{d^2w_2(0)}{dx^2} = 0, \quad \frac{d^3w_2(0)}{dx^3} + kw_2(0) = 0$$

$$\frac{dw_2(0)}{dx} = 0, \quad w_2(0) = 1$$

Applying the above conditions to the general solution of eqn (3.22), we derive the following system of linear algebraic equations

$$\begin{pmatrix} 0 & 0 & 2 & 0 \\ k & 0 & 0 & 6 \\ 0 & 1 & 0 & 0 \\ 1 & 0 & 0 & 0 \end{pmatrix} \times \begin{pmatrix} C_1 \\ C_2 \\ C_3 \\ C_4 \end{pmatrix} = \begin{pmatrix} 0 \\ 0 \\ 0 \\ 1 \end{pmatrix}$$

in $C_i$, $(i = \overline{1,4})$. The solution to the above system is found as

$$C_1 = 1, \quad C_2 = C_3 = 0, \quad \text{and} \quad C_4 = -\frac{k}{6}$$

Thus, we finally have for $w_2(x)$

$$w_2(x) = 1 - kx^3/6 \tag{3.44}$$

Note that the set of particular solutions of eqn (3.22) presented in (3.40), (3.43), and (3.44) is linearly independent on $[0, a]$. Indeed, the Wronskian of this set of functions

$$\begin{vmatrix} x & 1-kx^3/6 & (x-a)^2 & (x-a)^3 \\ 1 & -kx^2/2 & 2(x-a) & 3(x-a)^2 \\ 0 & -kx & 2 & 6(x-a) \\ 0 & -k & 0 & 6 \end{vmatrix} = -4(3+ka^3)$$

is a non-zero constant.

This allows the Green's function to the boundary-value problem in eqns (3.22), (3.38), and (3.39) (which is the influence function of a point force for the beam whose left-hand edge is elastically supported and the right-hand edge is clamped) to be sought in the form

$$g(x,s) = \begin{cases} a_1(s)x + a_2(s)(1-kx^3/6), & \text{for } x \le s \\ b_1(s)(x-a)^2 + b_2(s)(x-a)^3, & \text{for } s \le x \end{cases} \tag{3.45}$$

Since the above representation meets the defining properties 1 and 4 of the definition of the Green's function, we satisfy the continuity and discontinuity conditions at $x = s$ (see properties 2 and 3 of the definition). This yields the following well-posed system of linear algebraic equations

$$\begin{pmatrix} (s-a)^2 & (s-a)^3 & -s & ks^3/6-1 \\ 2(s-a) & 3(s-a)^2 & -1 & ks^2/2 \\ 2 & 6(s-a) & 0 & ks \\ 0 & 6 & 0 & k \end{pmatrix} \times \begin{pmatrix} b_1(s) \\ b_2(s) \\ a_1(s) \\ a_2(s) \end{pmatrix} = \begin{pmatrix} 0 \\ 0 \\ 0 \\ -1 \end{pmatrix}$$

in the coefficients $a_1(s)$, $a_2(s)$, $b_1(s)$ and $b_2(s)$ of the representation in eqn (3.45).

The solution of the above system is found as

$$a_1(s) = -\frac{(s-a)^2(ka^2s-6)}{4(3+ka^3)}, \quad b_1(s) = \frac{(s-a)[kas(s+a)+6]}{4(3+ka^3)}$$

$$a_2(s) = \frac{(s-a)^2(s+2a)}{2(3+ka^3)}, \quad b_2(s) = \frac{2(3+ka^3)-3ks(s^2-a^2)}{12(3+ka^3)}$$

Upon substituting these values in eqn (3.45) and performing some routine algebraic transformations, one finally obtains the influence function of a unit force for the beam under consideration in the form

$$g(x,s) = \frac{1}{\Delta}\begin{cases} (s-a)^2\{6[(x-s)+2(x-a)] \\ \qquad + kx[s(x^2-a^2)+2a(x^2-as)]\}, & x \le s \\ (x-a)^2\{6[(s-x)+2(s-a)] \\ \qquad + ks[x(s^2-a^2)+2a(s^2-ax)]\}, & s \le x \end{cases} \tag{3.46}$$

where $\Delta = 12(3+ka^3)$.

The reader should notice that this representation is symmetric in the sense that $g(x, s) = g(s, x)$. This fact reflects the self-adjointness of the boundary-value problem in eqns (3.22), (3.38), and (3.39), for which the above is the Green's function.

As we have earlier mentioned, two particular cases should follow from eqn (3.46). Namely, if the coefficient $k_0$ of elastic support in the second of the boundary conditions in eqn (3.38) approaches zero, then they reduce to the free edge conditions

$$\frac{d^2w(0)}{dx^2} = 0, \qquad \frac{d^3w(0)}{dx^3} = 0$$

and the influence function in eqn (3.46) transforms, consequently, as $k = 0$ or $k_0 = 0$ into

$$g(x, s) = \frac{1}{6}\begin{cases} (s-a)^2[2(x-a)+(x-s)], & x \le s \\ (x-a)^2[2(s-a)+(s-x)], & s \le x \end{cases} \tag{3.47}$$

the influence function for a cantilever beam whose right-hand edge is fixed.

Another particular case of the influence function in eqn (3.46) can readily be derived. That is, if $k_0$ approaches infinity, then one obtains

$$g(x, s) = \frac{1}{12a^3}\begin{cases} x(s-a)^2[s(x^2-a^2)+2a(x^2-as)], & x \le s \\ s(x-a)^2[x(s^2-a^2)+2a(s^2-ax)], & s \le x \end{cases} \tag{3.48}$$

the influence function for the beam whose left-hand edge is simply supported while the right-hand edge is clamped. ▲

For our last example in this section, we will use a different approach for the construction of influence functions. Namely, the alternative technique based on the method of variation of parameters will be employed.

**Example 1.6:** Consider the following boundary-value problem

$$\frac{d^4w(x)}{dx^4} = -q(x), \qquad x \in (0, a) \tag{3.49}$$

$$w(0) = \frac{dw(0)}{dx} = 0, \qquad \frac{d^2w(a)}{dx^2} = \frac{d^3w(a)}{dx^3} = 0 \tag{3.50}$$

which models the bending of a cantilever beam of length $a$, subject to a transverse load that is directly proportional to $q(x)$.

It is intuitive that there exists a unique influence function of a point force for this beam. That is, the beam ought to uniquely respond to a transverse concentrated force (see Figure 3.8). Mathematics readily supports this conclusion. We leave the justification of this issue to the reader.

The influence function of a unit transverse point force for the beam shown in Figure 3.8 represents the Green's function for the homogeneous boundary-value problem corresponding to that posed by eqns (3.49) and (3.50). To construct this function in compliance with the procedure of Lagrange method of variation of

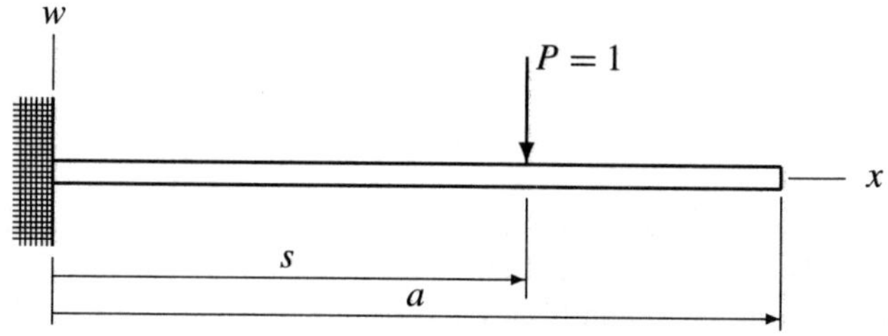


Figure 3.8: Cantilever beam subjected to a point force

parameters described in Chapter 2, we seek the general solution to eqn (3.49) in the form

$$w(x) = C_1(x) + C_2(x)x + C_3(x)x^2 + C_4(x)x^3 \tag{3.51}$$

The system of linear algebraic equations in the derivatives of $C_i(x)$, $(i = \overline{1,4})$, which results from the standard procedure of Lagrange method, appears, in this case, in the form

$$\begin{pmatrix} 1 & x & x^2 & x^3 \\ 0 & 1 & 2x & 3x^2 \\ 0 & 0 & 2 & 6x \\ 0 & 0 & 0 & 6 \end{pmatrix} \times \begin{pmatrix} C_1'(x) \\ C_2'(x) \\ C_3'(x) \\ C_4'(x) \end{pmatrix} = \begin{pmatrix} 0 \\ 0 \\ 0 \\ -q(x) \end{pmatrix}$$

A triangular structure of the coefficient matrix makes the solution of this system as simple as the backward substitution. Thus, starting with the last equation in the system and going up, we obtain

$$C_4'(x) = -\frac{1}{6}q(x), \quad C_3'(x) = \frac{x}{2}q(x)$$

$$C_2'(x) = -\frac{x^2}{2}q(x), \quad C_1'(x) = \frac{x^3}{6}q(x)$$

Hence, values of the coefficients $C_i(x)$ themselves can be obtained by integrating $C_i'(x)$. This yields

$$C_1(x) = \int_0^x \frac{s^3}{6}q(s)\,ds + H_1, \quad C_2(x) = -\int_0^x \frac{s^2}{2}q(s)\,ds + H_2$$

$$C_3(x) = \int_0^x \frac{s}{2}q(s)\,ds + H_3, \quad C_4(x) = -\int_0^x \frac{1}{6}q(s)\,ds + H_4$$

Upon substituting these values in (3.51) and combining then the integral terms, one comes up with the general solution to eqn (3.49) as

$$w(x) = \int_0^x \frac{(s-x)^3}{6}q(s)\,ds + H_1 + H_2x + H_3x^2 + H_4x^3$$

To compute the values of the coefficients $H_i$, we take advantage of the boundary conditions in eqn (3.50). In doing so, the condition $w(0)=0$ yields $H_1=0$, while $w'(0)=0$ results in $H_2=0$. Satisfying the boundary conditions at the right-hand edge $x=a$, one obtains

$$H_3 = -\int_0^a \frac{s}{2} q(s)\,ds, \quad H_4 = \int_0^a \frac{1}{6} q(s)\,ds$$

Thus, the final expression for the solution to the problem in eqns (3.49) and (3.50) is found in the form

$$w(x) = \int_0^x \frac{(s-x)^3}{6} q(s)\,ds + \int_0^a \frac{x^2(x-3s)}{6} q(s)\,ds$$

Following the special transformation explained in Section 1.4, the above representation can now be rewritten in the form of a single integral

$$w(x) = \int_0^a g(x,s) q(s)\,ds, \quad x \in [0,a]$$

where the kernel function $g(x,s)$ is defined in two pieces as

$$g(x,s) = \frac{1}{6}\begin{cases} x^2(x-3s), & \text{for } x \le s \\ s^2(s-3x), & \text{for } s \le x \end{cases} \tag{3.52}$$

By virtue of Theorem 2.4 in Chapter 2, we conclude that $g(x,s)$ does in fact represent the Green's function of the homogeneous boundary-value problem corresponding to that in eqns (3.49) and (3.50). That is, $g(x,s)$ is nothing but the influence function of a transverse concentrated unit force for the cantilever beam clamped at the left-hand edge (see Figure 3.8). ▲

It is worth noting that both the functions exhibited in eqns (3.47) and (3.52) represent influence functions for cantilever beams. The difference is that, in the first case, the beam is clamped at the right-hand edge, whereas, in the second case, the left-hand edge is clamped. The function in eqn (3.52) can formally be obtained from that of eqn (3.47) by a linear transformation. That is, if in (3.47) the expression

$$a - x$$

is interpreted as the force application point $s$, while the expression

$$a - s$$

– as the observation point $x$, then the influence function in (3.47) converts to that of eqn (3.52).

The same transformation can be used to convert other influence functions in cases when the edge conditions are interchanged similarly to the cases related to eqns (3.47) and (3.52). Say, eqn (3.48) represents the influence function for the beam whose left-hand edge is simply supported while the right-hand edge is clamped. The above transformation converts the expression in (3.48) into the influence function

$$g(x, s) = \frac{1}{12a^3} \begin{cases} x^2(s-a)[s(3a-x)(s-2a)+2a^2x], & x \le s \\ s^2(x-a)[x(3a-s)(x-2a)+2a^2s], & x \ge s \end{cases}$$

of the beam whose left-hand edge is clamped while the right-hand edge is simply supported.

This section was designed to assist the reader in developing a close familiarity with the concept of influence function. The material herein is helpful for the actual construction of such functions. By the way, the reader finds here and in the exercises related to this section in the End of Chapter Exercises explicit expressions for influence functions of a point force for a number of single-span Kirchhoff beams. And the collection of influence functions presented in Section 3.5 is, probably, the most comprehensive of all available in literature.

The only thing the reader will still be missing after reading this section is the experience that is needed to be fluent with the developments in the later sections of this text. Such an experience can be gained by going through a set of informative exercises. Therefore, the reader is advised to work through each of the End of Chapter Exercises that are related to the material of the present section. They represent a carefully chosen set of beam problems which are designed to highlight all specific fragments of the construction procedure. This will definitely create the necessary basis for comprehending the material of the remaining topics in this text.

In the next section, we present solutions to a number of problems for Kirchhoff beams with a variety of edge conditions imposed, undergoing various combinations of loads. The discussion will be based on Theorem 2.4 and will utilize the influence functions of a point force which are available in Section 3.1.

## 3.2 Bending of Beams of Uniform Rigidity

A number of standard Kirchhoff beam problems will be discussed in this section. Since those settings can be simulated with boundary-value problems for a linear ordinary differential equation allowing analytic integration, the exact solutions to these problems are well-known and can be found in many standard texts and handbooks in the mechanical engineering curriculum.

The purpose of including such problems in our text is just to show how their classical solutions can be obtained by means of the *influence function method*, which was never before suggested as a standard tool. Another reason for considering these standard problems is to maintain a background for the reader to tackle later on more complex problems for which this method looks to be the most reasonable alternative.

Consider a beam made of an elastic material, having a uniform flexural rigidity $EI$ and undergoing a transverse continuously distributed load $q(x)$. Recall the displacement formulation of this problem

$$\frac{d^4w(x)}{dx^4}=\frac{q(x)}{EI}, \quad x\in(0,a) \tag{3.53}$$

$$B_{0,i}[w(0)]=0, \quad B_{a,i}[w(a)]=0, \quad (i=1,2) \tag{3.54}$$

where the relations in eqn (3.54) deliver a combination of edge conditions that are supposed to be feasible. Remember that saying *feasible* we imply that the above problem has a unique solution.

### 3.2.1 Deflection function

Let $g(x,s)$ represent the deflection of the beam at $x$, caused by a transverse unit force concentrated at $s$. In other words, we assume that $g(x,s)$ is the Green's function of the homogeneous boundary-value problem associated with that in eqns (3.53) and (3.54).

In compliance with Theorem 2.4 from Chapter 2, the deflection $w_q(x)$ caused by the transverse load $q(x)$, distributed over the entire beam length, can be expressed as follows

$$w_q(x)=\frac{1}{EI}\int_0^a g(x,s)q(s)\,ds, \quad x\in[0,a] \tag{3.55}$$

Since the deflection is written as a definite integral containing a parameter $x$, it represents a function of that parameter. Recall that the influence function is given in two pieces by definition. Therefore, in computing the above integral, we break it down onto two integrals as shown

$$w_q(x)=\frac{1}{EI}\left\{\int_0^x g^-(x,s)q(s)\,ds+\int_x^a g^+(x,s)q(s)\,ds\right\}, \quad x\in[0,a] \tag{3.56}$$

where $g^-(x,s)$ and $g^+(x,s)$ represent the branches of the influence function defined for $s\le x$ and $x\le s$, respectively.

Let us now analyze various kinds of load that may occur in practice (such as concentrated forces and moments and distributed forces). We need to find out how the integral representation in (3.56) can be utilized to handle any reasonable combination of such loads.

Clearly, if the load $q(x)$ is applied to only a certain sub-segment $[\alpha,\beta]$ of $(0,a)$, then, breaking down the interval $[0,a]$ onto three segments, we obtain

$$\begin{aligned}w_q(x)&=\frac{1}{EI}\left\{\int_0^\alpha g(x,s)\cdot 0\,ds+\int_\alpha^\beta g(x,s)q(s)\,ds+\int_\beta^a g(x,s)\cdot 0\,ds\right\}\\&=\frac{1}{EI}\int_\alpha^\beta g(x,s)q(s)\,ds, \quad x\in[0,a]\end{aligned} \tag{3.57}$$

So, the deflection at any point in the beam is obtained in this case by the integration over the loaded segment $[\alpha,\beta]$. This looks simple. However, in computing the above integral, there is an issue that requires a close attention. That is, one ought to discern three different options for the location of the observation point $x$. Namely, for $x \leq \alpha$ (to the left of the loaded segment), eqn (3.57) transforms into

$$w_q(x) = \frac{1}{EI} \int_{\alpha}^{\beta} g^{+}(x, s) q(s) \, ds, \qquad x \in [0, a]$$

because the variable of integration $s$, ranging from $\alpha$ to $\beta$, remains greater than or equal to $x$. That is why the branch $g^{+}(x, s)$ of the influence function ought to be employed.

For the values of $x$ determined by $\alpha \leq x \leq \beta$ (the observation point is inside the loaded segment), the integral in eqn (3.57) has to be broken as shown

$$w_q(x) = \frac{1}{EI} \left\{ \int_{\alpha}^{x} g^{-}(x, s) q(s) \, ds + \int_{x}^{\beta} g^{+}(x, s) q(s) \, ds \right\}, \qquad x \in [\alpha, \beta]$$

And for $x \geq \beta$ (to the right of the loaded interval), we obtain

$$w_q(x) = \frac{1}{EI} \int_{\alpha}^{\beta} g^{-}(x, s) q(s) \, ds, \qquad x \in [\beta, a]$$

because the variable of integration $s$, ranging from $\alpha$ to $\beta$, remains, in this case, less than or equal to $x$, and $g^{-}(x, s)$ is defined just for $s \leq x$.

Let us consider some other possible types of loading. If, for example, the beam of length $a$ subject to a single transverse force of magnitude $P_0$, concentrated at $x = s_0 \in (0, a)$, then (as we have mentioned earlier) the right-hand side of eqn (3.53) can be thought of as a scalar multiple of the Dirac delta function

$$q(x) = P_0 \delta(x - s_0)$$

yielding as follows

$$w_{P_0}(x) = \frac{1}{EI} \int_{0}^{a} g(x, s) P_0 \delta(s - s_0) \, ds = \frac{P_0}{EI} g(x, s_0) \tag{3.58}$$

Notice that the above transformation has been carried out by virtue of the sifting property of the Dirac function.

Let us return to the relation in (3.55), which provides the response of a beam to a load $q(x)$ continuously distributed over the entire length of the beam. This relation was derived while proving Theorem 2.4 in Chapter 2. At this point, we will present a different derivation of the relation in (3.55). Namely, we will take advantage of eqn (3.58) for the deflection caused by a concentrated force.

In doing so, we consider a beam of length $a$ subject to a continuously distributed load $q(x)$ and let $g(x, s)$ represent the influence function for the beam. We partition the interval $[0, a]$ onto $n$ sub-intervals by choosing a set of distinct mesh points $0 = x_0 < x_1 < x_2 < \cdots < x_n = a$. Then we pick up an arbitrary point $s_k$ inside of the $k$-th sub-interval $[x_{k-1}, x_k]$ of the partition, and replace the load function $q(s)$ within each of the sub-intervals with a concentrated force whose magnitude $P_k$ is equal to

$$P_k = q(s_k)\Delta s_k, \quad (k = \overline{1, n})$$

with $\Delta s_k$ representing the length of $[x_{k-1}, x_k]$.

Assume that $P_k$ is applied to $s_k$. In compliance with eqn (3.58), the beam's response to each of the forces $P_k$ (the deflection $w_{p_k}(x)$ caused by $P_k$) can be written in the form

$$w_{p_k}(x) = \frac{1}{EI} g(x, s_k) q(s_k) \Delta s_k, \quad (k = \overline{1, n})$$

Summing up the responses to each force $P_k$, we obtain the resultant deflection $w_n(x)$ in the form

$$w_n(x) = \frac{1}{EI} \sum_{k=1}^{n} g(x, s_k) q(s_k) \Delta s_k$$

Clearly, the limit of $w_n(x)$ as $n$ goes to infinity

$$\lim_{n\to\infty} w_n(x) = \frac{1}{EI} \lim_{n\to\infty} \sum_{k=1}^{n} g(x, s_k) q(s_k) \Delta s_k$$

represents the deflection $w_q(x)$ caused by the distributed load $q(x)$. But, the above limit represents the definite integral of $g(x, s)q(s)$ taken from 0 to $a$

$$w_q(x) = \frac{1}{EI} \int_0^a g(x, s) q(s)\, ds$$

In other words, we have shown that $w_q(x)$ can indeed be written as the integral in eqn (3.55).

We now go to other types of loading that could be treated by means of the influence function method. If, for instance, the beam is subject not to a single concentrated transverse force but to a finite number of such forces with magnitudes $P_i$, $(i = \overline{1, k})$ and respective locations at $s_i$, then the deflection of the beam, caused by such a loading, can be obtained in the form

$$\begin{aligned} w(x) &= \frac{1}{EI} \int_0^a g(x, s) \sum_{i=1}^{k} P_i \delta(s - s_i)\, ds \\ &= \frac{1}{EI} \sum_{i=1}^{k} P_i \int_0^a g(x, s) \delta(s - s_i)\, ds = \frac{1}{EI} \sum_{i=1}^{k} P_i g(x, s_i) \qquad (3.59) \end{aligned}$$

Thus, within the scope of the geometrically linear statement of the problem, the response of the beam to a finite number of concentrated forces can be obtained as a sum of the responses to each of those forces.

Let us now consider a problem where a beam is subject to a bending couple of magnitude $m_0$, which is applied to a point $s_0$. It can readily be shown that this case can also be treated in terms of the influence function of a single transverse concentrated force.

To show how it can be realized, we consider two equal but opposing transverse forces of magnitude $P_0$, which are concentrated at two next to each other points $s_0$ and $s_0 + h$. In addition, we assume that the following relation $m_0 = P_0 h$ holds for the quantities $m_0$, $P_0$, and $h$.

Thus, in compliance with eqn (3.58), the deflection $w_h(x)$ of the beam, caused by the two forces recently introduced can consequently be written as

$$w_h(x) = \frac{P_0}{EI}[g(x, s_0 + h) - g(x, s_0)]$$

As we replace $P_0$ with the quotient $m_0/h$ that follows from the assumption recently made, the above equation reads

$$w_h(x) = \frac{m_0}{EIh}[g(x, s_0 + h) - g(x, s_0)]$$

From what we assumed earlier, it follows that the limit of $w_h$, as $h$ approaches zero, equals the value of the deflection $w_{m_0}(x)$ of the beam, caused by the concentrated bending couple $m_0$, that is

$$w_{m_0}(x) = \lim_{h \to 0} w_h(x)$$

Hence, taking into account the expression for $w_h$, one obtains

$$w_{m_0}(x) = \frac{m_0}{EI} \lim_{h \to 0} \frac{g(x, s_0 + h) - g(x, s_0)}{h}$$

It turns out that the above limit represents the value of the partial derivative of $g(x, s)$ with respect to $s$ at $s = s_0$. Hence, for the deflection of the beam, caused by the concentrated bending moment $m_0$, we finally have

$$w_{m_0}(x) = \frac{m_0}{EI} \frac{\partial g(x, s_0)}{\partial s}, \quad x \in [0, a] \tag{3.60}$$

Thus, to obtain the value of the deflection function at any point $x$ of the beam, caused by the bending moment concentrated at a certain point $s_0$, we need to know the value of the partial derivative

$$\frac{\partial g(x, s_0)}{\partial s}$$

of the influence function at $s = s_0$. In other words, the derivative

$$\frac{\partial g(x, s)}{\partial s}$$

being a function of $x$ and $s$, represents the deflection (output) of the beam at a point $x$ caused by a unit concentrated bending moment (input) applied to a point $s$.

In the discussion that follows, it will be referred to as the *influence function of a point force of the second order.*

Similarly to the case of a finite number of concentrated forces, one can readily derive the following expression

$$w(x) = \frac{1}{EI}\sum_{i=1}^{k} m_i \frac{\partial g(x, s_i)}{\partial s}, \quad x \in [0, a] \tag{3.61}$$

for the deflection of the beam subject to a finite number of concentrated bending moments $m_i$, $(i = \overline{1, k})$ applied to the points $s_i$, respectively.

So, from what we have recently found, it follows that the influence function $g(x, s)$ of a transverse concentrated unit force can successfully be employed for computing analytic expressions for the deflection of a beam, caused by a simultaneous action of several concentrated and distributed transverse loads and bending moments. The resultant deflection of a beam subject to a combination of such loads can be obtained by a superposition of the deflections caused by each individual input.

In other words, the resultant deflection is a proper combination of those in eqns (3.55) and (3.58)–(3.61). A word of caution is appropriate at this moment. The above conclusion is true, if the combined result of the individual loads does not cause either physical nonlinearity (violation of the Hooke's law for the material of which the beam is made, resulting in the nonlinear stress-strain relationship) or geometric nonlinearity (large deflection). In either case, the aforementioned boundary-value problem in eqns (3.53) and (3.54) is no longer adequately applicable to the physical problem.

### 3.2.2 Stress-related components

Thus, the influence function method provides a universal technique for computing the deflection function in an analytic form, for all of the physically feasible combinations of transverse and bending loads applied to a beam. This makes it possible to obtain also in analytic form the bending moment $M(x)$ and the shear force $Q(x)$

$$M(x) = EI\frac{d^2w(x)}{dx^2}, \quad Q(x) = EI\frac{d^3w(x)}{dx^3}$$

occurring in any cross-section of the beam. This becomes possible, because in doing so one can analytically differentiate the corresponding expression for the deflection function. Indeed, for a beam undergoing a transverse load $q(x)$ distributed over the interval $(\alpha, \beta)$, for example, upon differentiating $w_q(x)$ in equation (3.57), one obtains

$$M(x) = \int_{\alpha}^{\beta} \frac{\partial^2 g(x, s)}{\partial x^2} q(s)\, ds, \quad x \in [0, a] \tag{3.62}$$

$$Q(x) = \int_{\alpha}^{\beta} \frac{\partial^3 g(x, s)}{\partial x^3} q(s)\, ds, \quad x \in [0, a] \tag{3.63}$$

As we have learned from the foregoing discussion, which touched upon the integral in eqn (3.57), to practically compute $M(x)$ and $Q(x)$, one ought to account for three different locations of the field point $x$ with respect to the interval of integration in eqns (3.62) and (3.63). That is, the branch $g^{+}(x, s)$ represents the influence function when $x \leq s$, whereas $g^{-}(x, s)$ does so when $x \geq s$.

The relations in eqns (3.62) and (3.63) are valid for computing bending moments and shear forces for a beam subject to a continuously distributed transverse load, regardless of the edge conditions prescribed. This highlights the nature of the influence function method in which the edge conditions are treated at the stage of the method when the influence function is constructed. And when that $g(x, s)$ is used in integrals like those in eqns (3.55) or (3.57), for example, the entire integral representations satisfy prescribed edge conditions.

Note that eqns (3.62) and (3.63) suggest that computing of bending moments and shear forces within the influence function method does not require numerical differentiation and can be done analytically regardless of the complexity of the right-hand side function $q(x)$ in eqn (3.53). This aspect makes the influence function method radically different from pure numerical approaches like the finite difference or the finite element method. The distinction becomes crucial when $q(x)$ does not allow for the boundary-value problem in eqns (3.53) and (3.54) to be solved exactly. This implies that approximate differentiation of the deflection function (which badly deteriorates the accuracy level) is unavoidable in the numerical methods. Clearly, the influence function method, for such cases, is significantly superior compared to pure numerical techniques. Later in this section, we will present some data illustrating the last point.

We turn now to other types of elementary loads (different to a transverse point concentrated force) that can possibly be applied to the beam. Using the corresponding differentiation of the expression for $w(x)$ in eqn (3.58), one readily obtains the bending moment

$$M(x) = P_0 \frac{\partial^2 g(x, s_0)}{\partial x^2}, \quad x \in [0, a] \tag{3.64}$$

and the shear force

$$Q(x) = P_0 \frac{\partial^3 g(x, s_0)}{\partial x^3}, \quad x \in [0, a] \tag{3.65}$$

at any cross-section $x$ of the beam subject to a single transverse force of magnitude $P_0$, concentrated at $s_0$.

For a finite number of concentrated forces $P_i$, $(i = \overline{1, k})$ applied to points $s_i$, the bending moment

$$M(x) = \sum_{i=1}^{k} P_i \frac{\partial^2 g(x, s_i)}{\partial x^2}, \quad x \in [0, a] \tag{3.66}$$

and the shear force

$$Q(x) = \sum_{i=1}^{k} P_i \frac{\partial^3 g(x, s_i)}{\partial x^3}, \quad x \in [0, a] \tag{3.67}$$

are to be computed at any cross-section of the beam by the corresponding differentiation of $w(x)$ in eqn (3.59).

For the input in the form of a single concentrated bending moment of magnitude $m_0$, the bending moment and the shear force at any cross-section can be obtained by an appropriate differentiation of the influence function of the second order. This yields

$$M(x) = m_0 \frac{\partial^3 g(x, s_0)}{\partial s\, \partial x^2}, \quad x \in [0, a] \tag{3.68}$$

and

$$Q(x) = m_0 \frac{\partial^4 g(x, s_0)}{\partial s\, \partial x^3}, \quad x \in [0, a] \tag{3.69}$$

by properly differentiating $w_{m_0}(x)$ in eqn (3.60).

Given the input in the form of a finite sum of concentrated bending moments $m_i$, $(i = \overline{1, k})$, the values of the bending moment $M(x)$

$$M(x) = \sum_{i=1}^{k} m_i \frac{\partial^3 g(x, s_i)}{\partial s\, \partial x^2}, \quad x \in [0, a] \tag{3.70}$$

and the shear force $Q(x)$

$$Q(x) = \sum_{i=1}^{k} m_i \frac{\partial^4 g(x, s_i)}{\partial s\, \partial x^3}, \quad x \in [0, a] \tag{3.71}$$

at any cross-section are derived from eqn (3.61).

Remember again that the formulae in eqns (3.62)–(3.71) are valid in computing bending moments and shear forces generated in a beam by corresponding loads, with any edge conditions imposed in a particular problem. The edge conditions, in turn, are taken care of by an appropriate choice of the influence function.

### 3.2.3 Illustrative examples

A number of instructive examples are presented below. They show how the classical solutions from Kirchhoff beam theory can be obtained within the scope of the influence function method.

**Example 2.1:** Consider a cantilever beam of length $a$, undergoing a transverse load $q(x)$ given as

$$q(x) = q_0 x(a - x), \quad q_0 = \text{const} \tag{3.72}$$

and distributed throughout the entire beam span as shown in Figure 3.9.

As we have shown earlier in this section, the deflection function $w(x)$ of a beam undergoing a continuously distributed load $q(x)$ and subject to standard edge conditions can be expressed in terms of the corresponding influence function

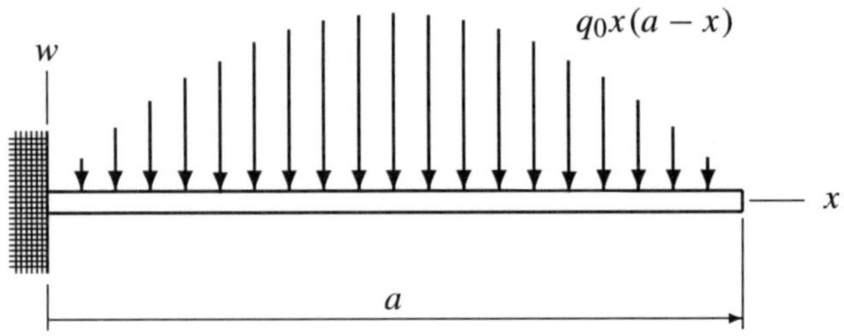


Figure 3.9: Cantilever beam subject to a distributed load

$g(x, s)$ as

$$w(x) = \frac{1}{EI}\left\{\int_0^x g^-(x, s)q(s)\,ds + \int_x^a g^+(x, s)q(s)\,ds\right\}, \quad x \in [0, a]$$

where $g^-(x, s)$, with $x \geq s$ and $g^+(x, s)$, with $x \leq s$ are the branches of the influence function derived earlier in eqn (3.52). Substituting the load function $q(x)$ from eqn (3.72) and the branches of the influence function from eqn (3.52) in the above equation, one obtains the deflection function (of the cantilever beam depicted in Figure 3.9) in the form

$$w(x) = \frac{q_0}{6EI}\left\{\int_0^x s^2(s-3x)s(a-s)\,ds + \int_x^a x^2(x-3s)s(a-s)\,ds\right\}, \quad x \in [0, a]$$

which results in

$$w(x) = \frac{q_0}{360EI}x^2(x^4 - 3ax^3 + 10a^3x - 15a^4), \quad x \in [0, a]$$

One can compute the bending moment $M(x)$ and the shear force $Q(x)$ by properly differentiating the deflection function just obtained. This yields the following expressions

$$M(x) = \frac{q_0}{12}(x^4 - 2ax^3 + 2a^3x - a^4), \quad x \in [0, a]$$

for the bending moment

$$Q(x) = \frac{q_0}{6}(2x^3 - 3ax^2 + a^3), \quad x \in [0, a]$$

and for the shear force, respectively.

If the same cantilever beam is loaded with a uniformly distributed load $q_0$ over a portion $[b, a]$ of its span as shown in Figure 3.10, then the same influence function

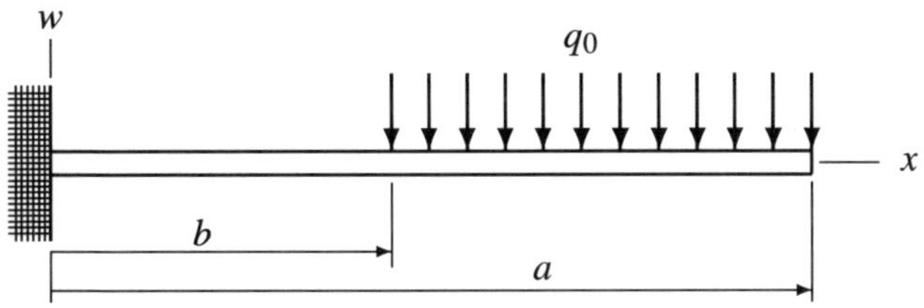


Figure 3.10: Cantilever beam loaded over a portion of its span

from eqn (3.52) can be used to determine the stress-strain state. The deflection at any field point $x$ belonging to the unloaded portion $[0, b]$ of the beam's span is found in this case as

$$\begin{aligned} w(x) &= \frac{1}{EI}\int_b^a g^+(x,s)q_0\,ds = \frac{q_0}{6EI}\int_b^a x^2(x-3s)\,ds \\ &= \frac{q_0x^2(a-b)}{72EI}[2x-3(a+b)], \quad x\in[0,b] \end{aligned}$$

If the field point is located on the loaded sub-segment $[b, a]$, then the deflection function is computed as

$$\begin{aligned} w(x) &= \frac{1}{EI}\int_b^x g^-(x,s)q_0\,ds + \frac{1}{EI}\int_x^a g^+(x,s)q_0\,ds \\ &= \frac{q_0}{6EI}\left[\int_b^x s^2(s-3x)\,ds + \int_x^a x^2(x-3s)\,ds\right] \\ &= \frac{q_0}{24EI}\left\{2x^2(a-b)[2x-3(a+b)]-(x-b)^4\right\}, \quad x\in[b,a] \end{aligned}$$

As to the stress components, they can readily be obtained by corresponding differentiation of the above expressions for the deflection function. ▲

**Example 2.2:** A beam of length $a$, with the left-hand edge elastically supported ($k$ represents elastic constant of the supporting spring) and the right-hand edge clamped, is loaded with a single bending moment $m_0$ applied to a point $x = s_0$ (see Figure 3.11).

Remember that the edge conditions are written in this case as

$$\frac{d^2w(0)}{dx^2}=0, \quad EI\frac{d^3w(0)}{dx^3}+kw(0)=0$$

$$w(a)=0, \quad \frac{dw(a)}{dx}=0$$

and the influence function $g(x, s)$ of a unit transverse concentrated force for the beam

$$g(x,s)=\frac{1}{\Delta}\begin{cases}(s-a)^2\{6[(x-s)+2(x-a)]\\ \qquad +kx[s(x^2-a^2)+2a(x^2-as)]\}, & x\le s\\ (x-a)^2\{6[(s-x)+2(s-a)]\\ \qquad +ks[x(s^2-a^2)+2a(s^2-ax)]\}, & s\le x\end{cases}$$

where $\Delta = 12(3+ka^3)$, was obtained in Section 3.1 (see eqn (3.46)).

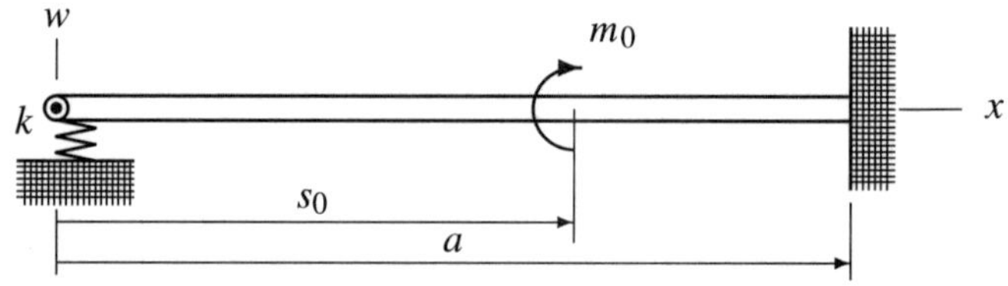

Figure 3.11: A beam loaded with a concentrated moment

Earlier in this section we have shown that for a beam subject to any set of edge conditions and undergoing a single bending moment of magnitude $m_0$ applied at a point $s_0$, the deflection function $w_{m_0}(x)$ can be expressed in terms of the corresponding influence function of the second order (see eqn (3.60)) as

$$w_{m_0}(x)=\frac{m_0}{EI}\frac{\partial g(x,s_0)}{\partial s},\qquad x\in[0,a]$$

Upon differentiating $g(x, s)$ with respect to $s$ and substituting the resultant expression for the derivative in the above equation, one obtains the deflection of the beam under consideration in the form

$$w_{m_0}(x)=\frac{3m_0}{\Delta EI}\begin{cases}(s_0-a)[(kx^3-1)(s_0+a)+(ka^3-3ka^2s_0+2)x], & x\le s_0\\ (x-a)^2[k(x+2a)s_0^2+(1-ka^2x)], & s_0\le x\end{cases}$$

Distribution of the bending moment $M_{m_0}(x)$

$$M_{m_0}(x)=\frac{6m_0}{\Delta}\begin{cases}3k(s_0^2-a^2)x, & x<s_0\\ 3k(s_0^2-a^2)x+\Delta/6, & s_0<x\end{cases}$$

is in this case computed by using eqn (3.68). It is clearly seen that, in agreement with physics, $M_{m_0}(x)$ makes a jump of discontinuity of magnitude $m_0$ at $s_0$.

The shear force $Q_{m_0}(x)$ caused in the beam by $m_0$ is found as

$$Q_{m_0}(x)=\frac{18m_0}{\Delta}k(s_0^2-a^2)$$

which is uniform throughout the beam's length. ▲

In the next example, we show that the influence function technique is easily applicable to a beam undergoing a combination of elementary loads.

**Example 2.3:** A beam of length $a$, with both edges simply supported, undergoes a combination of loads (a uniform load $q_0$ distributed over the interval $[s_1, s_2]$, a single bending moment $m_0$ concentrated at $s_3$, and a transverse force of magnitude $P_0$ concentrated at $s_4$) applied as shown in Figure 3.12.

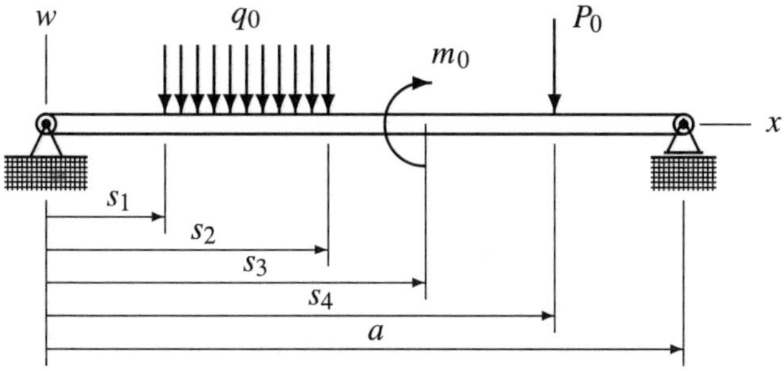


Figure 3.12: A beam subject to a combination of loads

The influence function

$$g(x, s) = \frac{1}{6a}\begin{cases} x(a-s)(x^2+s^2-2as), & x \le s \\ s(a-x)(s^2+x^2-2ax), & s \le x \end{cases}$$

of a point force for the simply supported beam is to be obtained by the reader in Section 3.1 (see the End of Chapter Exercises where influence functions are to be constructed for dozens of other single-span Kirchhoff beams).

The output (deflection, bending moment, and shear force) caused by each single input specified in a problem statement can readily be found in terms of the influence function $g(x, s)$. Indeed, in accordance with the relation in eqn (3.57), the deflection $w_{q_0}(x)$ caused by the load $q_0$ uniformly distributed over the interval $[s_1, s_2]$, can be obtained as

$$w_{q_0}(x) = \frac{1}{EI}\int_{s_1}^{s_2} g(x, s) q_0 \, ds, \quad x \in [0, a]$$

Due to the piecewise appearance of the influence function $g(x, s)$, different expressions result from the above integral for different locations of the observation point $x$ with respect to $s_1$ and $s_2$. For $x \le s_1$, for instance, we obtain

$$\begin{aligned} w_{q_0}(x) &= \frac{1}{EI}\int_{s_1}^{s_2} g^{+}(x, s) q_0 \, ds \\ &= \frac{q_0}{6aEI}\int_{s_1}^{s_2} x(a-s)(x^2+s^2-2as) \, ds, \quad x \in [0, s_1] \end{aligned}$$

If the observation point $x$ is located within the loaded interval (that is, $s_1 < x < s_2$), then the deflection function is expressed as a sum of the two integrals

$$w_{q_0}(x) = \frac{q_0}{6aEI}\left\{\int_{s_1}^{x} s(a-x)(s^2+x^2-2ax)\,ds + \int_{x}^{s_2} x(a-s)(x^2+s^2-2as)\,ds\right\}, \quad x \in [s_1, s_2]$$

For an observation point located to the right of $s_2$ ($x \geq s_2$), the deflection function is found as

$$w_{q_0}(x) = \frac{q_0}{6aEI}\int_{s_1}^{s_2} s(a-x)(s^2+x^2-2ax)\,ds$$

The above integrals representing the deflection function caused by the uniformly distributed load $q_0$ can be routinely found.

Deflection caused by the concentrated moment $m_0$ can be computed by means of the influence function of the second order as

$$w_{m_0}(x) = \frac{m_0}{6aEI}\begin{cases} x[2as - s_3^2 - x^2 - 2(a-s_3)^2], & x \leq s_3 \\ (a-x)(x^2+3s_3^2-2ax), & x \geq s_3 \end{cases}$$

Using eqn (3.58), we write down the deflection, caused by the single point force $P_0$ applied at $s_4$, in the form

$$w_{p_0}(x) = \frac{P_0}{6aEI}\begin{cases} x(a-s_4)(x^2+s_4^2-2as_4), & x \leq s_4 \\ s_4(a-x)(s_4^2+x^2-2ax), & s_4 \leq x \end{cases}$$

Due to the linear nature of the problem, the sum of the deflection components $w_{q_0}(x)$, $w_{m_0}(x)$, and $w_{p_0}(x)$ just found represents the resultant deflection function $w(x)$ caused by the combination of the loads specified in the statement of the problem.

Stress-related components generated in the beam by the given combination of loads can also be expressed analytically. In compliance with the relations in eqns (3.62), (3.64), and (3.68), the resultant bending moment $M(x)$, for example, is computed as

$$M(x) = \frac{\partial^2}{\partial x^2}\left(P_0 g(x, s_4) + m_0\frac{\partial g(x, s_3)}{\partial s} + \int_{s_1}^{s_2} g(x, s)q_0\,ds\right) \qquad \blacktriangle$$

The material in this section suggests that, if a boundary-value problem of the type in eqns (3.53) and (3.54) allows an exact solution, then the influence function method can be considered as a valuable option to practically obtain that solution. The method presents a standard approach for the treatment of Kirchhoff beams subject to both distributed and concentrated loads. The next section brings an important extension of the method to beam problems that do not allow analytic solution.

### 3.2.4 Numerical implementations

Note that beneficiary features of the influence function method become more significant when it is applied to more complicated problems. If, for instance, the loading function $q(x)$ is complex in the sense that it does not allow for the boundary-value problem in eqns (3.53) and (3.54) to be solved analytically and, consequently, only a numerical solution is possible, then it is hard to find a viable alternative to the influence function method in terms of the computational efficiency. To illustrate this point, we consider the clamped-clamped beam of length $a$, having a uniform flexural rigidity $EI$ and subject to a transverse load given as

$$q(x) = q_0 \sin \frac{\pi x^2}{a^2}, \qquad q_0 = \text{const}$$

distributed over the entire beam's length as shown in Figure 3.12. Note that the choice of a clamped-clamped beam is conditional in this case and a single-span beam with any feasible set of edge conditions can be considered instead.

To determine the deflection function $w(x)$, bending moment $M(x)$, and shear force $Q(x)$, the statement in eqns (3.53) and (3.54) appears in this case as

$$\frac{d^4w(x)}{dx^4} = -\frac{q_0}{EI} \sin \frac{\pi x^2}{a^2}, \qquad x \in (0, a) \tag{3.73}$$

$$w(0) = \frac{dw(0)}{dx} = 0, \qquad w(a) = \frac{dw(a)}{dx} = 0 \tag{3.74}$$

Remember that the general solution of eqn (3.73) is a sum of the general solution of the corresponding homogeneous equation, the obtaining of which is not an issue at all, with a particular solution of eqn (3.73), computing of which is definitely an issue, because a particular solution to eqn (3.73) cannot, unfortunately, be found analytically. Indeed, two standard analytic approaches (the method of undetermined coefficients and the method of variation of parameters) are potentially available for this purpose. But none of them works in the case of eqn (3.73). Indeed, the method of undetermined coefficients is not simply applicable, because the right-hand side function in the statement is too cumbersome for it, while the method of variation of parameters results in integrals that cannot be computed analytically. This makes the setting in eqns (3.73) and (3.74) unsolvable analytically.

Since the stress-strain state of the beam depicted in Figure 3.13 cannot be determined analytically, the only option that we are left with, is to numerically tackle the boundary-value problem in eqns (3.73) and (3.74) in order to obtain approximate values of the components of the beam's stress-strain state. And once a certain numerical method is applied and an approximation of the deflection function $w(x)$ is found, the stress related components (the bending moment $M(x)$ and the shear force $Q(x)$) ought to be computed by *repeatedly* differentiating the deflection function (again numerically). But, from numerical analysis, it is well known that a significant effort needs to be put forth to attain a high accuracy level of outputs in a numerical differentiation.

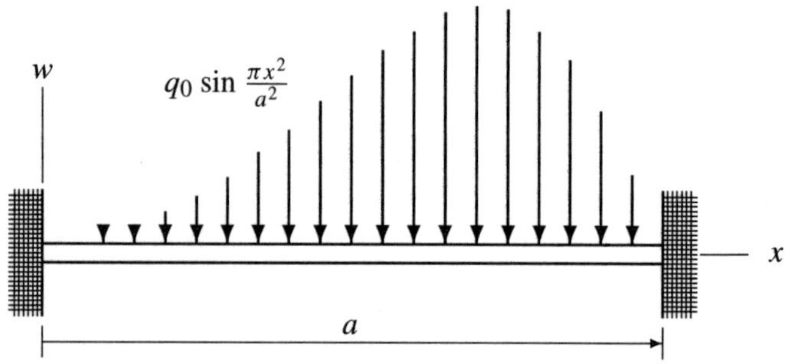


Figure 3.13: Clamped-clamped beam subject to a distributed load

So, it does not take an advanced degree in mathematics or mechanics to figure out that to attain a high accuracy level in a numerical algorithm, the use of numerical differentiation should be as minimal as possible if not avoided at all. That is why the influence function method is a fortunate option whenever it is applicable, because it does, in fact, represent a technique that completely avoids numerical differentiation. All the derivatives, required to compute the stress related components in beam problems, can be obtained *analytically* within the scope of this method. Indeed, the solution to the boundary-value problem in eqns (3.73) and (3.74) is expressible, in accordance with Theorem 2.4 in Chapter 2, in the form

$$w(x) = \frac{q_0}{EI} \int_0^a g(x, s) \sin \frac{\pi s^2}{a^2} \, ds, \quad x \in [0, a] \tag{3.75}$$

where $g(x, s)$ represents the influence function of a unit point force for a clamped-clamped beam. That influence function was derived earlier (see eqn (3.37) of Section 3.1).

Having the deflection function expressed by eqn (3.75), the bending moment $M(x)$ and the shear force $Q(x)$ in the beam can be found in compliance with the relations in eqns (3.3) and (3.4) as

$$M(x) = q_0 \int_0^a \frac{\partial^2 g(x, s)}{\partial x^2} \sin \frac{\pi s^2}{a^2} \, ds, \quad x \in [0, a] \tag{3.76}$$

and

$$Q(x) = q_0 \int_0^a \frac{\partial^3 g(x, s)}{\partial x^3} \sin \frac{\pi s^2}{a^2} \, ds, \quad x \in [0, a] \tag{3.77}$$

Thus, all the components of the stress-strain state for the beam depicted in Figure 3.13 are written in terms of the influence function and its repeated derivatives taken with respect to the observation variable $x$. Note that if a beam with another feasible set of edge conditions is considered, then the corresponding influence function replaces $g(x, s)$ in eqns (3.75)–(3.77).

Since eqn (3.37) delivers an analytic expression for $g(x, s)$, the derivatives in eqns (3.76) and (3.77) can be taken analytically. This clearly illustrates the issue that we have earlier raised concerning the complete elimination of numerical differentiation within the influence function method. So, no numerical differentiation is required, but the reader may bring another reasonable concern. The integrals in eqns (3.75)–(3.77) cannot be taken analytically or, in other words, cannot be obtained in a closed form. This is true, because in any attempt to analytically compute such integrals, we face the so-called Fresnel integrals

$$\int \sin(x^2)\, dx \quad \text{or} \quad \int \cos(x^2)\, dx$$

which, as we learned from calculus [60], cannot be taken in a closed form. This means that the influence function method does not bring a pure analytical solution to the problem. But we have never declared that it does so and never pretended to analytically solve the problem in eqns (3.73) and (3.74). This is simply impossible because the mother-nature cannot be deceived. What we do actually state is that our approach lays a different stress compared to pure numerical methods. It is free of a numerical differentiation, but the integrals in eqns (3.75)–(3.77) are still to be computed numerically.

As it follows from what we have just declared, the influence function method represents a semi-analytic approach in a sense that all the output data can be expressed analytically but some numerical effort is still required to obtain quantitative information. In the discussion that follows, we will show that the influence function method-based numerical procedures are substantially more efficient compared to classical numerical alternatives when solving the boundary-value problem in eqns (3.73) and (3.74).

To highlight the effectiveness of the influence function method in solving beam problems, we are going to conduct a numerical experiment where this method will be compared against the standard finite difference method. That is, before going any further with the statement in eqns (3.73) and (3.74), we pose a test example allowing an exact solution, and apply to it both the numerical procedures.

**Example 2.4:** Consider the boundary-value problem in eqn (3.74) stated for the following equation

$$\frac{d^4 w(x)}{dx^4} = x \sin x, \quad x \in (0, \pi) \tag{3.78}$$

Clearly, this is a relatively trivial set up. Either the method of undetermined coefficients or the method of variation of parameters can be used or, given a specific form of the differential operator in eqn (3.78), one can also obtain the solution by simply integrating eqn (3.78) four times successively and then satisfying the boundary conditions in eqn (3.74). Note though that whatever method is used, a quite tedious and time consuming algebra is required to finally obtain the exact solution to the setting in eqns (3.78) and (3.74) as

$$w(x) = -4(1 - \cos x) + x \sin x + \frac{x^2}{\pi^3}[(\pi^2 - 16)x - \pi(\pi^2 - 24)] \tag{3.79}$$

On the other hand, the exact solution of the problem in eqns (3.78) and (3.74) can be written down by means of the influence function method which yields

$$w(x) = -\int_0^{\pi} g(x, s)s \sin s \, ds \tag{3.80}$$

where $g(x, s)$ is the influence function of eqn (3.37), with $a = \pi$. Substituting $g(x, s)$ in (3.80) and breaking the integral onto two, we obtain

$$\begin{aligned} w(x) &= -\int_0^{\pi} g(x, s)s \sin s \, ds \\ &= -\left\{\int_0^{x} g^{-}(x, s)s \sin s ds + \int_x^{\pi} g^{+}(x, s)s \sin s \, ds\right\} \\ &= -\frac{1}{6\pi^3}\left\{\int_0^{x} s^2(x - s)^2[2\pi(s - \pi) + \pi(s - x)]s \sin s \, ds \right. \\ &\quad \left. + \int_x^{\pi} x^2(x - s)^2[2\pi(x - \pi) + \pi(x - s)]s \sin s \, ds\right\} \end{aligned} \tag{3.81}$$

It takes again a trivial but quite cumbersome algebra to show that the above integrals lead to the exact solution (3.79) of the problem. But as long as in our numerical experiment we are going to check out approximate results computed by two different numerical methods, approximate values of the function $w(x)$ will be obtained by a numerical computation of the integrals in eqn (3.81). These results will be compared against an approximate solution obtained for the problem in eqns (3.78) and (3.74) with a classical finite difference approach.

The same uniform discretization $0 = x_0, x_1, \ldots, x_n = \pi$, where

$$x_k = kh, \quad \text{with } h = \frac{\pi}{n}, \quad 0 \le k \le n \tag{3.82}$$

of the interval $[0, \pi]$ is utilized in both the influence function method-based (IFM) and the finite difference method-based (FDM) procedures. We will call $x_k$, $(k = \overline{0, n})$ the mesh-points.

Both procedures that we are going to apply, are supposed to provide the same order of accuracy $O(h^m)$. This notation is conventional in mathematics and is especially widely used in numerical analysis [5, 6, 60]. It implies that as $h \to 0$, the difference between the exact and approximate solutions approaches zero at the same rate that $h^m$ does. In other words, the error of computation is directly proportional to $h^m$. So, the greater is the exponent $m$ in $O(h^m)$, the more accurate is the method.

To compute the integrals in eqn (3.81), we write them for exact values $w(x_k)$ of the function $w(x)$ obtained at the mesh-points $x_k$, defined by eqn (3.82), as

$$w(x_k) = -\left\{\int_0^{x_k} g^-(x_k, s)s \sin s\, ds + \int_{x_k}^{\pi} g^+(x_k, s)s \sin s\, ds\right\}, \quad k = \overline{1, n-1}$$

and use then the standard trapezoid rule [1, 2, 3] based on the uniform partition (3.82) to obtain approximate values $w_k$ for $w(x_k)$ as

$$w_k \simeq -\frac{\pi}{2n}\left\{\sum_{i=1}^{k}[g^-(x_k, x_{i-1})x_{i-1} \sin x_{i-1} + g^-(x_k, x_i)x_i \sin x_i]\right.$$
$$\left.+\sum_{i=k}^{n-1}[g^+(x_k, x_i)x_i \sin x_i + g^+(x_k, x_{i+1})x_{i+1} \sin x_{i+1}]\right\}, \quad k = \overline{1, n-1} \tag{3.83}$$

As it has been proven in numerical analysis, the order of accuracy of this standard trapezoid rule procedure is $O(h^2)$. That is, the absolute value of the difference $w(x_k) - w_k$, for $k = \overline{1, n-1}$, approaches zero at the rate that $h^2$ does. Since the exact solution (3.79) of the problem posed by eqns (3.74) and (3.78) is available, it is clear that upon carrying out an actual computation of values $w_k$ with various discretization parameter $n$, one can test the practical convergence of the described influence function-based numerical procedure and control the level of accuracy that is attained with it.

Alternatively to the numerical influence function approach, we have computed an approximate solution to the setting in eqns (3.78) and (3.74) by a standard finite difference method [32]. The simplest version of this method has been used, the one based on the uniform partition introduced by eqn (3.82) and which approximates the derivatives (as shown in Chapter 1) in the governing differential equation with central differences as

$$\frac{dw_k(x)}{dx} \simeq \frac{1}{2h}(w_{k+1} - w_{k-1}), \qquad \frac{d^2w_k(x)}{dx^2} \simeq \frac{1}{h^2}(w_{k+1} - 2w_k + w_{k-1}),$$

$$\frac{d^3w_k(x)}{dx^3} \simeq \frac{1}{2h^3}(w_{k+2} - 2w_{k+1} + 2w_{k-1} - w_{k-2})$$

and

$$\frac{d^4w_k(x)}{dx^4} \simeq \frac{1}{h^4}(w_{k+2} - 4w_{k+1} + 6w_k - 4w_{k-1} + w_{k-2})$$

This reduces the boundary-value problem in eqns (3.74) and (3.78) to the following system of $n-3$ linear algebraic equations

$$w_{k+2} - 4w_{k+1} + 6w_k - 4w_{k-1} + w_{k-2} = \frac{\pi^4 x_k}{n^4}\sin(x_k), \quad k = \overline{2, n-2} \tag{3.84}$$

in $n+1$ unknowns $w_k$, $(k = \overline{0, n})$. So, the number of equations in the above system is fewer than the number of unknowns, meaning that the system is formally ill-posed. But note that four of those unknowns can be obtained separately by means of the boundary conditions. Indeed, the end-values $w_0$ and $w_n$ as well as the next to them values $w_1$ and $w_{n-1}$ are zero in compliance with eqn (3.74). This makes the system in eqn (3.84) consistent and well-posed, and its solution provides us with the remaining approximate values $w_2, w_3, \ldots, w_{n-2}$ of $w(x)$.

As it is shown in numerical analysis, the order of accuracy of the described finite difference method approximation is $O(h^2)$, which is the same as for the influence function method-based algorithm presented in eqn (3.83). This allows a fair comparison of the results obtained by the two approaches.

Hence, the output of the finite difference method (FDM) and the influence function method (IFM) ought to be equivalently accurate. At least, this is what the *a priori* (obtained up-front) estimation suggests. Practice shows, however, differently. But this does not contradict the theory, it simply illustrates the conditionality of *a priori* estimations. The point is that although the error of approximation, in both numerical procedures, is directly proportional to the value of $h^2$, coefficients of proportionality in $O(h^2)$ appear to be notably different for each of the methods. This yields a significant difference in their results.

**Table 3.1:** Approximate values $w_k$ of the function $w(x)$ computed for the problem in eqns (3.78) and (3.74)

| Mesh points, $x_k/\pi$ | Discretization parameter, $n$ | | | | | | Exact values, $w(x)$ |
|---|---|---|---|---|---|---|---|
| | FDM | | | IFM | | | |
| | 10 | 50 | 100 | 10 | 50 | 100 | |
| 0.1 | 0.00000 | 0.02870 | 0.03255 | 0.03649 | 0.03648 | 0.03648 | 0.03648 |
| 0.2 | 0.05522 | 0.10732 | 0.11437 | 0.12158 | 0.12156 | 0.12156 | 0.12156 |
| 0.3 | 0.13088 | 0.20062 | 0.21013 | 0.21987 | 0.21984 | 0.21984 | 0.21984 |
| 0.4 | 0.19578 | 0.27716 | 0.28832 | 0.29977 | 0.29972 | 0.29972 | 0.29972 |
| 0.5 | 0.22615 | 0.31295 | 0.32489 | 0.33717 | 0.33711 | 0.33710 | 0.33710 |
| 0.6 | 0.20986 | 0.29560 | 0.30737 | 0.31949 | 0.31942 | 0.31941 | 0.31941 |
| 0.7 | 0.15009 | 0.22774 | 0.23833 | 0.24923 | 0.24915 | 0.24915 | 0.24915 |
| 0.8 | 0.06784 | 0.12912 | 0.13742 | 0.14596 | 0.14590 | 0.14590 | 0.14590 |
| 0.9 | 0.00000 | 0.03635 | 0.04117 | 0.04611 | 0.04606 | 0.04606 | 0.04606 |

The data in Table 3.1 reveal a notable difference in the accuracy level attained by the two algorithms. Namely, the IFM is substantially more accurate in computing approximate values of the beam's deflection function (the data have been obtained for a set of uniformly spaced interior mesh-points $x_k$). Indeed, the accuracy of the results obtained by the IFM with the discretization parameter as small as

$n = 10$ is notably higher compared to the FDM results obtained with much finer discretization $n = 100$.

The superiority of the influence function method against the finite difference approach becomes even more evident when computing values of the stress-related components such as the bending moment and the shear force, which require repeated differentiation of the deflection function $w(x)$. In order to address this issue in more detail, we present exact and approximate values of the second order derivative of $w(x)$ (Table 3.2) and of the third order derivative of $w(x)$ (Table 3.3) for the boundary-value problem posed by eqns (3.74) and (3.78). Data for two different values of the discretization parameter $n$ are shown to illustrate the convergence issue for both procedures.

**Table 3.2:** Approximate values of $d^2w(x)/dx^2$ computed by the IFM and FDM

| Mesh point, $x_k/\pi$ | $n = 10$ | | $n = 50$ | | Exact values |
|---|---|---|---|---|---|
| | IFM | FDM | IFM | FDM | |
| 0.1 | 0.490871 | 0.552253 | 0.491513 | 0.505041 | 0.491540 |
| 0.2 | 0.128746 | 0.207030 | 0.130697 | 0.145830 | 0.130701 |
| 0.3 | −0.198931 | −0.109039 | −0.192721 | −0.175964 | −0.192684 |
| 0.4 | −0.448035 | −0.349854 | −0.440567 | −0.422187 | −0.440483 |
| 0.5 | −0.583820 | −0.472714 | −0.570918 | −0.550906 | −0.570796 |
| 0.6 | −0.560256 | −0.440543 | −0.547690 | −0.525811 | −0.547349 |
| 0.7 | −0.353760 | −0.231439 | −0.349217 | −0.325657 | −0.348918 |
| 0.8 | 0.013279 | 0.153256 | 0.022382 | 0.047797 | 0.022718 |
| 0.9 | 0.530217 | 0.678437 | 0.536957 | 0.564651 | 0.537654 |

From the data presented in Tables 3.2 and 3.3, we learn that in computing the bending moment and the shear force, the numerical influence function method could practically be as accurate as for the deflection function. This feature of the method is not surprising, because following the IFM procedure, we compute values of $M(x)$ and $Q(x)$ by a numerical integration (see eqns (3.62) and (3.63)). In other words, numerical differentiation is completely avoided, which, in fact, predetermines high accuracy level of IFM procedures.

On the contrary, the finite difference schemes of the type that was used in our experiment can hardly be recommended for computing either bending moments or shear forces, unless some radical adjustments are undertaken. For example, being within the scope of the finite difference method, the simplest way to improve the solution accuracy would be to radically increase the discretization parameter $n$. Because of the five-diagonal structure of the coefficient matrix of the system in eqn (3.84), such an increase does not generate an obstacle that cannot be overcome.

**Table 3.3:** Approximate values of $d^3w(x)/dx^3$ computed by the IFM and FDM

| Observation point, $x/a$ | $n = 10$ | | $n = 50$ | | Exact values |
|---|---|---|---|---|---|
| | IFM | FDM | IFM | FDM | |
| 0.1 | −1.171116 | – | −1.175871 | −1.170383 | −1.176054 |
| 0.2 | −1.103854 | −1.064091 | −1.106217 | −1.100971 | −1.106823 |
| 0.3 | −0.927510 | −0.886308 | −0.931014 | −0.925363 | −0.931246 |
| 0.4 | −0.618571 | −0.578807 | −0.622478 | −0.617642 | −0.623554 |
| 0.5 | −0.181700 | −0.144336 | −0.185902 | −0.180486 | −0.186288 |
| 0.6 | 0.346023 | 0.384002 | 0.347164 | 0.352846 | 0.347252 |
| 0.7 | 0.914813 | 0.945061 | 0.915281 | 0.920648 | 0.915336 |
| 0.8 | 1.425281 | 1.456573 | 1.434518 | 1.441774 | 1.434778 |
| 0.9 | 1.797210 | – | 1.811613 | 1.826466 | 1.811777 |

Another natural way of getting more accurate in the finite difference approach is the use of more accurate approximations for derivatives.

Let us now turn to the original formulation in eqns (3.73) and (3.74). Table 3.4 shows some results of solving that problem by the influence function procedure where we assumed $q_0/EI = 1$. The deflection, bending moment and shear force were computed throughout the entire segment $[0, \pi]$ for three different values of the discretization parameter $n$. A set of five uniformly spaced interior mesh-points has been examined. In this case, we cannot compare approximate data against an exact solution since the latter is not available. But the rapid convergence of the results obtained at each mesh-point is a reliable indication of high accuracy of the approximate solution.

From what we observe in Table 3.4, it follows that for the problem under consideration, the numerical procedure of the influence function method is rapidly converging for all of the solution components $w(x)$, $M(x)$ and $Q(x)$ required in applications. Our experience in the use of this method for a broad variety of beam problems has always been productive showing higher computational potential compared to other numerical approaches. Based on this, use of the IFM procedures can definitely be encouraged. ▲

Later in this text, the reader will find out that static equilibrium problems for single-span Kirchhoff beams with various feasible combinations of edge conditions do not represent the only class of beam problems that can successfully be dealt with by means of the influence function method. In the next section, for example, we will show that beams resting on elastic foundation also allow a productive analysis with the aid of this method.

**Table 3.4:** Approximate solution of the problem in eqns (3.73) and (3.74)

| Solution values | Discretiz. param., $n$ | Interior mesh-points, $x_k/\pi$ | | | | |
|---|---|---|---|---|---|---|
| | | 0.1 | 0.3 | 0.5 | 0.7 | 0.9 |
| $w(x_k)$ | 10 | 0.016827 | 0.103205 | 0.162465 | 0.123897 | 0.023586 |
| | 50 | 0.016824 | 0.103186 | 0.162422 | 0.123844 | 0.023556 |
| | 100 | 0.016824 | 0.103186 | 0.162422 | 0.123844 | 0.023556 |
| $M(x_k)$ | 10 | 0.231002 | −0.080374 | −0.276761 | −0.196721 | 0.259219 |
| | 50 | 0.231186 | −0.078197 | −0.271142 | −0.189692 | 0.263548 |
| | 100 | 0.231193 | −0.078149 | −0.270987 | −0.189445 | 0.263784 |
| $Q(x_k)$ | 10 | −0.520190 | −0.432058 | −0.125815 | 0.426520 | 0.956073 |
| | 50 | −0.521718 | −0.436563 | −0.131375 | 0.426185 | 0.967917 |
| | 100 | −0.521767 | −0.436706 | −0.131549 | 0.426174 | 0.968285 |

## 3.3 Beams on Elastic Foundation

Our attention turns, in this section, to Poisson–Kirchhoff beams of uniform flexural rigidity resting on a *single-parameter elastic foundation*, where we presume that the shear force generated in the beam is directly proportional to the beam's deflection (the Hooke's law). Influence functions of a point force for such beams are constructed by the customary procedure and boundary-value problems simulating bending of beams that undergo different loading are discussed.

### 3.3.1 Bending of infinite beam

Let a beam of infinite length undergo a transverse load $q(x)$ applied to a certain part of it. Let $EI$ and $\beta$ represent the flexural rigidity of the beam and the elastic coefficient of the foundation, respectively. Infinite length of the beam notably simplifies the solution procedure and makes it more compact. Such a simplifying assumption is often appropriate for a really long beam when the edge conditions do not, according to the Saint Venant principle, practically affect a local stress-strain state. If, however, an immediate neighborhood of the edges is of interest, then beams of a finite length ought to be considered.

A differential equation that governs the equilibrium state of the beam can be presented (see [27, 31]) in the form

$$\frac{d^4w(x)}{dx^4} + \frac{\beta}{EI}w(x) = -\frac{q(x)}{EI}, \quad x \in (-\infty, \infty)$$

For the sake of compactness in the development that follows, we introduce a parameter

$$k = \sqrt[4]{\frac{\beta}{4EI}}$$

that transforms the governing equation into

$$\frac{d^4 w(x)}{dx^4} + 4k^4 w(x) = -\frac{q(x)}{EI}, \quad x \in (-\infty, \infty) \tag{3.85}$$

The right-hand side function $q(x)$ in eqn (3.85) is supposed to be defined over a finite interval $[a, b]$, beyond of which it is identically zero. Clearly, this feature of $q(x)$ is quite realistic, because in reality a load is always applied to a finite portion of a beam.

Instead of formally imposing edge conditions for the infinite beam, we assume that all the components of the stress-strain state vanish when $x$ approaches positive and negative infinity. In what follows, these will be referred to as the *conditions at infinity*.

Let, for the beam under consideration, $g(x, s)$ be the influence function of a transverse unit force concentrated at $s$. In other words, let $g(x, s)$ represent the Green's function to the homogeneous equation corresponding to (3.85) and subject to the conditions at positive and negative infinity. In compliance with Theorem 2.4 in Chapter 2, the influence function allows the deflection function of the beam to be written as

$$w(x) = \frac{1}{EI} \int_a^b g(x, s) q(s)\, ds, \quad x \in (-\infty, \infty) \tag{3.86}$$

With the aid of this representation, one can compute the bending moment $M(x)$ and the shear force $Q(x)$ caused in the beam by the load $q(x)$ as

$$M(x) = \int_a^b \frac{\partial^2 g(x, s)}{\partial x^2} q(s)\, ds, \quad x \in (-\infty, \infty) \tag{3.87}$$

and

$$Q(x) = \int_a^b \frac{\partial^3 g(x, s)}{\partial x^3} q(s)\, ds, \quad x \in (-\infty, \infty) \tag{3.88}$$

So, the influence function $g(x, s)$ and its partial derivatives with respect to $x$ are required for obtaining all the components of the beam's stress-strain state. Therefore, we focus first on the construction of the influence function.

As we mentioned earlier, $g(x, s)$ represents the Green's function for the homogeneous equation

$$\frac{d^4 w(x)}{dx^4} + 4k^4 w(x) = 0, \quad x \in (-\infty, \infty) \tag{3.89}$$

subject to the conditions at infinity.

To come up with a fundamental set of solutions to eqn (3.89), we take a look at its characteristic equation

$$m^4 + 4k^4 = 0$$

for which the following two pairs of complex conjugates

$$(1 \pm i)k \quad \text{and} \quad (-1 \pm i)k$$

are found as distinct complex roots. This yields a fundamental set of solutions to eqn (3.89) in the following form

$$\begin{aligned} w_1(x) &= e^{kx} \cos kx, \quad w_2(x) = e^{kx} \sin kx \\ w_3(x) &= e^{-kx} \cos kx, \quad w_4(x) = e^{-kx} \sin kx \end{aligned} \tag{3.90}$$

Clearly, the first two components, $w_1(x)$ and $w_2(x)$ in the above set approach zero as $x$ approaches negative infinity, whereas $w_3(x)$ and $w_4(x)$ vanish as $x$ goes to positive infinity. This special feature of the components of the fundamental set of solutions enable us to trace out the technique proposed in Section 3.1 and to express the Green's function $g(x, s)$ that we are looking for as

$$g(x, s) = \begin{cases} a_1(s)e^{kx} \cos kx + a_2(s)e^{kx} \sin kx, & x \leq s \\ b_1(s)e^{-kx} \cos kx + b_2(s)e^{-kx} \sin kx, & s \leq x \end{cases} \tag{3.91}$$

It is evident that this representation satisfies defining property 1 of a Green's function, because both branches of $g(x, s)$ are linear combinations of particular solutions to eqn (3.89) and are, therefore, also its solutions. Defining property 4, dealing with boundary conditions, is also satisfied. Indeed, the upper branch of $g(x, s)$ that is valid for $x \leq s$, and all its partial derivatives with respect to $x$ have a common factor of $e^{kx}$ and thus vanish as $x$ goes to negative infinity, regardless of $a_1(s)$ and $a_2(s)$. The conditions at positive infinity are satisfied by the lower branch of $g(x, s)$, regardless of $b_1(s)$ and $b_2(s)$, because both terms in it have a common factor of $e^{-kx}$.

To compute the coefficients $a_i(s)$ and $b_i(s)$, we take advantage of the two remaining defining properties of a Green's function. By virtue of property 2, which claims the continuity of $g(x, s)$ as $x = s$, one obtains

$$b_1(s)e^{-ks} \cos ks + b_2(s)e^{-ks} \sin ks - a_1(s)e^{ks} \cos ks - a_2(s)e^{ks} \sin ks = 0 \tag{3.92}$$

The continuity of the first order partial derivative of $g(x, s)$ with respect to $x$ as $x = s$, required in property 2, yields

$$\begin{aligned} &- b_1(s)e^{-ks}(\cos ks + \sin ks) + b_2(s)e^{-ks}(\cos ks - \sin ks) \\ &\quad - a_1(s)e^{ks}(\cos ks - \sin ks) - a_2(s)e^{ks}(\cos ks + \sin ks) = 0 \end{aligned} \tag{3.93}$$

When the continuity is stated for the second order partial derivative of $g(x, s)$ with respect to $x$ as $x = s$, the following relation holds

$$b_1(s)e^{-ks} \sin ks - b_2(s)e^{-ks} \cos ks + a_1(s)e^{ks} \sin ks - a_2(s)e^{ks} \cos ks = 0 \tag{3.94}$$

By virtue of defining property 3, which claims the discontinuity of the third order partial derivative of $g(x, s)$ with respect to $x$ as $x = s$, we obtain

$$b_1(s)e^{-ks}(\cos ks - \sin ks) + b_2(s)e^{-ks}(\cos ks + \sin ks)$$
$$+ a_1(s)e^{ks}(\cos ks + \sin ks - a_2(s)e^{ks}(\cos ks - \sin ks) = -1/2k^3 \quad (3.95)$$

The relations in eqns (3.92)–(3.95) constitute a system of linear algebraic equations in $a_1(s)$, $a_2(s)$, $b_1(s)$, and $b_2(s)$. It is evident that the system is well-posed, because its coefficient matrix represents the Wronskian of the fundamental set of solutions in eqn (3.90). Upon solving the system, one obtains

$$a_1(s) = -\frac{e^{-ks}}{8k^3}(\cos ks + \sin ks), \quad a_2(s) = \frac{e^{-ks}}{8k^3}(\cos ks - \sin ks)$$
$$b_1(s) = -\frac{e^{ks}}{8k^3}(\cos(ks) - \sin(ks)), \quad b_2(s) = -\frac{e^{ks}}{8k^3}(\cos(ks) + \sin(ks))$$

Substituting the above in eqn (3.91) and going through a rather straightforward algebra, we ultimately obtain the influence function of a transverse unit force concentrated at $s$ for the infinite beam resting on a simple elastic foundation in the form

$$g(x, s) = -\frac{1}{8k^3}\begin{cases} e^{k(x-s)}[\cos k(x - s) - \sin k(x - s)], & x \le s \\ e^{k(s-x)}[\cos k(x - s) + \sin k(x - s)], & s \le x \end{cases} \quad (3.96)$$

Analogously to the development in Section 3.2, one can apply the influence function formalism to analytically compute the deflection function, bending moment, and shear force for the beam resting on an elastic foundation and undergoing a diversity of transverse loads.

Compact expression for the influence function in eqn (3.96) allows one to easily account for any load applied to the beam in the form of either concentrated transverse forces and bending moments or continuously distributed loads as well as in the form of a reasonable collection of those (given that the linearity of the problem is not violated).

**Example 3.1:** If, for instance, a transverse distributed load $q(x)$ is applied to a finite interval $[a, b]$ as depicted in Figure 3.14, then we obtain the deflection at any point located to the left of $a$ by the integral

$$w(x) = \frac{1}{8k^3EI}\int_a^b e^{k(x-s)}[\cos k(x - s) - \sin k(x - s)]q(s)\,ds, \quad x \le a \quad (3.97)$$

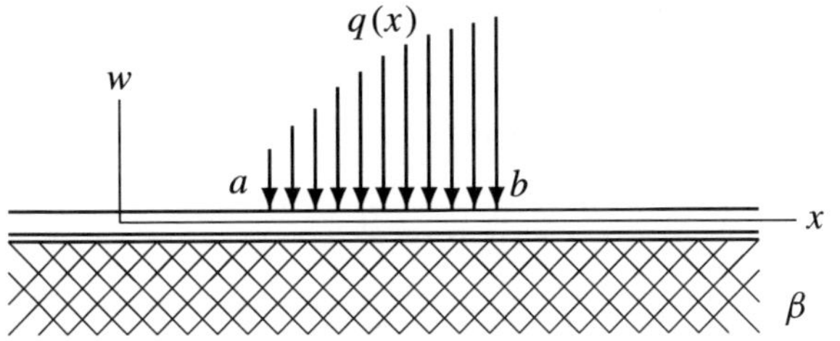


Figure 3.14: A load $q(x)$ applied to a finite interval

For the observation point $x$ within the loaded interval, the deflection can be computed as

$$w(x) = \frac{1}{8k^3EI}\left\{\int_a^x e^{k(s-x)}[\cos k(x-s) + \sin k(x-s)]q(s)\,ds \right.$$
$$\left. + \int_x^b e^{k(x-s)}[\cos k(x-s) - \sin k(x-s)]q(s)\,ds\right\}, \quad a \le x \le b \tag{3.98}$$

When the deflection is to be computed to the right of the loaded interval (if $x \ge b$), then we obtain

$$w(x) = \frac{1}{8k^3EI}\int_a^b e^{k(s-x)}[\cos k(x-s) + \sin k(x-s)]q(s)\,ds, \quad x \ge b \tag{3.99}$$

If the loading function $q(x)$ is simple enough (polynomial, exponential, trigonometric, or in the form of a product of those), then the integrals in eqns (3.97)–(3.99) can be computed analytically. If, however, the analytic integration is either impossible or results in a too cumbersome algebra, then approximate results can be obtained by using numerical integration. In the latter case, as we have shown in Section 3.2, the accuracy attained by the IFM approach is at a much higher level than for the finite difference method.

Stress-related components of the stress-strain state of an infinite beam resting on an elastic foundation can be found in terms of the deflection function $w(x)$ as

$$M(x) = EI\frac{d^2w(x)}{dx^2}, \quad Q(x) = EI\frac{d^3w(x)}{dx^3} \tag{3.100}$$

In finding these derivatives, we repeatedly differentiate the kernel functions of the integrals in eqns (3.97)–(3.99). Hence, to the left of the loaded interval, the bending moment and the shear force caused by $q(x)$ are obtained as

$$M(x) = \frac{1}{4k}\int_a^b e^{k(x-s)}[\cos k(x-s) + \sin k(x-s)]q(s)\,ds, \qquad x \le a$$

and

$$Q(x) = \frac{1}{2}\int_a^b e^{k(x-s)}\cos k(x-s)q(s)\,ds, \qquad x \le a$$

If the observation point is within the loaded interval, then one obtains

$$\begin{aligned} M(x) = &\frac{1}{4k}\int_a^x e^{k(s-x)}[\cos k(x-s) - \sin k(x-s)]q(s)\,ds \\ &+ \frac{1}{4k}\int_x^b e^{k(x-s)}[\cos k(x-s) + \sin k(x-s)]q(s)\,ds, \qquad a \le x \le b \end{aligned}$$

and

$$\begin{aligned} Q(x) = &-\frac{1}{2}\int_a^x e^{k(s-x)}\cos k(x-s)q(s)\,ds \\ &+ \frac{1}{2}\int_x^b e^{k(x-s)}\cos k(x-s)q(s)\,ds, \qquad a \le x \le b \end{aligned}$$

For $x \ge b$, we have

$$M(x) = \frac{1}{4k}\int_a^b e^{k(s-x)}[\cos k(x-s) - \sin k(x-s)]q(s)\,ds, \qquad x \ge b$$

and

$$Q(x) = -\frac{1}{2}\int_a^b e^{k(s-x)}\cos k(x-s)q(s)\,ds, \qquad x \ge b$$

The above integrals are readily computable regardless of the complexity of the loading function $q(x)$. Other transverse either point concentrated or distributed loads can also be readily accounted for. ▲

### 3.3.2 Semi-infinite beam under transverse load

Later in this section, we will formulate and solve some settings on the bending of a semi-infinite beam resting on the elastic foundation, with different edge conditions imposed at $x = 0$ and subject to different types of load.

But before going any further with this, let us construct the influence function of a point force for a beam whose edge $x = 0$ is clamped, which can be identified as the Green's function $g(x, s)$ of a boundary-value problem posed by the equation

$$\frac{d^4w(x)}{dx^4} + 4k^4w(x) = 0, \quad x \in (0, \infty) \tag{3.101}$$

subject to the following set of boundary conditions

$$w(0) = 0, \quad \frac{dw(0)}{dx} = 0, \quad |w(\infty)| < \infty, \quad \left|\frac{dw(\infty)}{dx}\right| < \infty \tag{3.102}$$

where the last two are the *conditions at infinity* that have earlier been introduced in this section.

As we have already learned, a fundamental set of solutions for eqn (3.101) is required to construct the Green's function by using either the defining properties-based procedure or the procedure that uses the variation of parameters method. Let us follow the version of the variation of parameters procedure used in Example 1.5 of Section 3.1. A specifically designed fundamental set of solutions is required in this case. Namely, we claim that each of two of the four components in that set is supposed to satisfy those boundary conditions in eqn (3.102) that are imposed at $x = 0$, while the other two components ought to satisfy the conditions at infinity.

Clearly, the functions $w_3(x)$ and $w_4(x)$ in the fundamental set of solutions exposed in eqn (3.90) satisfy the conditions at infinity. Hence, their linear combination

$$b_1(s)e^{-kx}\cos kx + b_2(s)e^{-kx}\sin kx$$

could be used as the branch of the influence function $g(x, s)$ that is valid for $x \geq s$. However, none of the remaining two components in eqn (3.90) satisfies both conditions imposed at $x = 0$. To obtain one of such functions, say $W_1(x)$, we set up the initial-value problem

$$w(0) = 0, \quad \frac{dw(0)}{dx} = 0, \quad \frac{d^2w(0)}{dx^2} = 0, \quad \frac{d^3w(0)}{dx^3} = 1 \tag{3.103}$$

for eqn (3.101). It is evident that not only the solution of this problem but also any its scalar multiple could be taken as $W_1(x)$.

Based on the fundamental set of solutions from eqn (3.90), the general solution to eqn (3.101) can be written as

$$w(x) = C_1e^{kx}\cos kx + C_2e^{kx}\sin kx + C_3e^{-kx}\cos kx + C_4e^{-kx}\sin kx \tag{3.104}$$

where $C_i, i = \overline{1, 4}$ are arbitrary constants.

Substituting (3.104) in the relations of eqn (3.103), we obtain the following well-posed system of linear algebraic equations

$$\begin{pmatrix} 1 & 0 & 1 & 0 \\ 1 & 1 & -1 & 1 \\ 0 & 1 & 0 & -1 \\ -1 & 1 & 1 & 1 \end{pmatrix} \times \begin{pmatrix} C_1 \\ C_2 \\ C_3 \\ C_4 \end{pmatrix} = \begin{pmatrix} 0 \\ 0 \\ 0 \\ 1/2k^3 \end{pmatrix}$$

in $C_i$, whose solution is found as

$$C_1 = -\frac{1}{8k^3}, \quad C_2 = \frac{1}{8k^3}, \quad C_3 = \frac{1}{8k^3}, \quad C_4 = \frac{1}{8k^3}$$

Upon substituting these in eqn (3.104), we obtain the following expression for the solution

$$w(x) = \frac{1}{8k^3}[-e^{kx}\cos kx + e^{kx}\sin kx + e^{-kx}\cos kx + e^{-kx}\sin kx] \quad (3.105)$$

of the initial-value problem stated in eqns (3.101) and (3.103).

Since any scalar multiple of $w(x)$ in eqn (3.105) can be considered as $W_1(x)$, we present the latter in the form

$$W_1(x) = -e^{kx}\cos kx + e^{kx}\sin kx + e^{-kx}\cos kx + e^{-kx}\sin kx$$

The same exact approach can be used to obtain another particular solution $W_2(x)$ of eqn (3.101), which is linearly independent on $W_1(x)$ and satisfies both boundary conditions in eqn (3.102) imposed at $x = 0$. In doing so, let us set up another initial-value problem

$$w(0) = 0, \qquad \frac{dw(0)}{dx} = 0, \qquad \frac{d^2w(0)}{dx^2} = 1, \qquad \frac{d^3w(0)}{dx^3} = 0$$

to eqn (3.101). The solution to this problem as well as any its scalar multiple could be taken as $W_2(x)$.

Proceeding analogously to the case with $W_1(x)$, one obtains $W_2(x)$ in the form

$$W_2(x) = -e^{kx}\sin kx + e^{-kx}\sin kx$$

The set of four functions ($W_1(x)$ and $W_2(x)$ just found, along with $w_3(x) = e^{-kx}\cos ks$ and $w_4(x) = e^{-kx}\sin kx$ taken from eqn (3.90)), is linearly independent on any finite interval. The proof of this statement is rather cumbersome, but with the aid of a computer algebra, the reader can readily form the Wronskian

$$Wr[W_1(x), W_2(x), w_3(x), w_4(x)] = \begin{pmatrix} W_1(x) & W_2(x) & w_3(x) & w_4(x) \\ W_1'(x) & W_2'(x) & w_3'(x) & w_4'(x) \\ W_1''(x) & W_2''(x) & w_3''(x) & w_4''(x) \\ W_1'''(x) & W_2'''(x) & w_3'''(x) & w_4'''(x) \end{pmatrix}$$

for these four functions and find then its determinant

$$\det(Wr[W_1(x), W_2(x), w_3(x), w_4(x)]) = 32 \neq 0 \tag{3.106}$$

Thus, the Green's function for the boundary-value problem in eqns (3.101) and (3.102) or, in other words, the influence function of a point force for the semi-infinite beam shown in Figure 3.15, can be written as

$$g(x, s) = \begin{cases} a_1(s)W_1(x) + a_2(s)W_2(x), & x \le s \\ b_1(s)w_3(x) + b_2(s)w_4(x), & s \le x \end{cases} \tag{3.107}$$

Clearly, the upper branch of $g(x, s)$ satisfies the boundary conditions at $x = 0$ as of eqn (3.102), since $W_1(x)$ and $W_2(x)$ do so. The lower branch of $g(x, s)$, in turn, satisfies the boundedness conditions at infinity, since both $w_3(x)$ and $w_4(x)$ as well as their derivatives approach zero as $x$ goes to infinity. Hence, $g(x, s)$ in (3.107) meets defining properties 1 and 4 for the Green's function.

Satisfying then properties 2 and 3, one derives a system of four linear algebraic equations

$$\begin{pmatrix} W_1(s) & W_2(s) & -w_3(s) & -w_4(s) \\ W_1'(s) & W_2'(s) & -w_3'(s) & -w_4'(s) \\ W_1''(s) & W_2''(s) & -w_3''(s) & -w_4''(s) \\ W_1'''(s) & W_2'''(s) & -w_3'''(s) & -w_4'''(s) \end{pmatrix} \times \begin{pmatrix} a_1 \\ a_2 \\ b_1 \\ b_2 \end{pmatrix} = \begin{pmatrix} 0 \\ 0 \\ 0 \\ -1 \end{pmatrix}$$

in $a_1(s)$, $a_2(s)$, $b_1(s)$, and $b_2(s)$. This system is well-posed, given that the determinant of its coefficient matrix, representing the Wronskian for the set of functions $W_1(x)$, $W_2(x)$, $-w_3(x)$ and $-w_4(x)$, is the same as in eqn (3.106). This is true because from linear algebra [67], we learn that multiplying a single column of a matrix by negative one changes the sign of its determinant to the opposite.

Hence, the above system does have a unique solution. Upon obtaining it, substituting then the found expressions for $a_1(s)$, $a_2(s)$, $b_1(s)$, and $b_2(s)$ in eqn (3.107) and going through rather simple but quite unwieldy algebra, one ultimately obtains the influence function of a point force $g(x, s)$ for the semi-infinite beam resting on a simple foundation if the edge $x = 0$ is clamped. The branch $g^+(x, s)$ of $g(x, s)$ that is valid for $x \leq s$ is found as

$$g^+(x, s) = \frac{1}{8k^3}\{e^{k(x-s)}[\sin k(x-s) - \cos k(x-s)] + e^{-k(s+x)}[\sin k(x+s) + 2\sin kx \sin ks + \cos k(x-s)]\} \quad (3.108)$$

while for $x \geq s$, we have

$$g^-(x, s) = \frac{1}{8k^3}\{e^{k(s-x)}[\sin k(s-x) - \cos k(x-s)] + e^{-k(s+x)}[\sin k(x+s) + 2\sin kx \sin ks + \cos k(x-s)]\} \quad (3.109)$$

The influence function, whose compact expressions are just found, enables us to compute components of the stress-strain state of the beam subject to any conventional load. In the case of a transverse load $q(x)$ continuously distributed over a finite interval $[0, a]$, the deflection function at any point within the loaded interval can be written as

$$w(x) = -\frac{1}{EI}\left\{\int_0^x g^-(x, s)q(s)\,ds + \int_x^a g^+(x, s)q(s)\,ds\right\}, \quad x \leq a \quad (3.110)$$

The deflection function cannot, however, be found by the above formula if the observation point is outside the loaded interval ($x \geq a$). If so then the deflection function must be computed as

$$w(x) = -\frac{1}{EI}\int_0^a g^-(x, s)q(s)\,ds, \quad x \geq a \quad (3.111)$$

Other components of the stress-strain state of the beam can be obtained by repeated analytic differentiation of the above expressions for the deflection function. Hence, for the bending moment we have

$$M(x) = -\left\{\int_0^x \frac{\partial^2 g^-(x, s)}{\partial x^2}q(s)\,ds + \int_x^a \frac{\partial^2 g^+(x, s)}{\partial x^2}q(s)\,ds\right\}, \quad x \leq a$$

and

$$M(x) = -\int_0^a \frac{\partial^2 g^-(x, s)}{\partial x^2}q(s)\,ds, \quad x \geq a$$

while the expressions

$$Q(x) = -\left\{\int_0^x \frac{\partial^3 g^-(x, s)}{\partial x^3}q(s)\,ds + \int_x^a \frac{\partial^3 g^+(x, s)}{\partial x^3}q(s)\,ds\right\}, \quad x \leq a$$

and

$$Q(x) = -\int_0^a \frac{\partial^3 g^-(x, s)}{\partial x^3}q(s)\,ds, \quad x \geq a$$

should be implemented to compute distribution of the shear force. Clearly, a manual differentiation would be too time consuming for such a computation and the use of computer algebra is recommended.

**Example 3.2:** Let a semi–infinite beam, whose edge $x = 0$ is clamped, be loaded with a transverse load defined as $q(x) = q_0x^2 + q_1$, where $q_0$ and $q_1$ represent positive constants, and applied to a finite interval $[0, a]$ as shown in Figure 3.15.

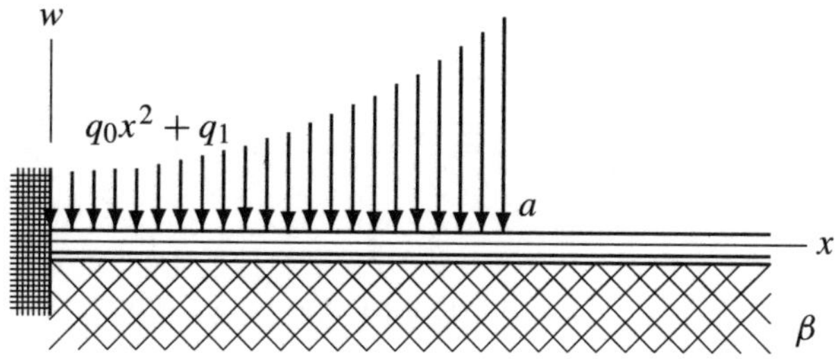


Figure 3.15: Semi-infinite beam subject to a distributed load

Upon substituting the expressions for $g^+(x, s)$ and $g^-(x, s)$ from eqns (3.108) and (3.109) in (3.110) and (3.111), we obtain the deflection function for $x \le a$ as

$$\begin{aligned} w(x) = &-\frac{1}{8k^3EI}\int_0^x \Big\{e^{k(s-x)}[\sin k(s-x) - \cos k(x-s)] \\ &+ e^{-k(s+x)}[\sin k(x+s) + 2\sin kx \sin ks + \cos k(x-s)]\Big\}(q_0s^2 + q_1)\, ds \\ &- \frac{1}{8k^3EI}\int_x^a \Big\{e^{k(x-s)}[\sin k(x-s) - \cos k(x-s)] \\ &+ e^{-k(x+s)}[\sin k(x+s) + 2\sin kx \sin ks + \cos k(x-s)]\Big\}(q_0s^2 + q_1)\, ds \end{aligned}$$

If, however, $x \ge a$, then we obtain

$$\begin{aligned} w(x) = &-\frac{1}{8k^3EI}\int_0^a \Big\{e^{k(s-x)}[\sin k(s-x) - \cos k(x-s)] \\ &+ e^{-k(s+x)}[\sin k(x+s) + 2\sin kx \sin ks + \cos k(x-s)]\Big\}(q_0s^2 + q_1)\, ds \end{aligned}$$

To compute the bending moment $M(x)$ and the shear force $Q(x)$ caused by $q(x)$, one appropriately differentiates the kernel functions in the above representations for the deflection function. Computer algebra is recommended to facilitate this procedure. ▲

We present below another example on the use of the influence function method in computing components of the stress-strain state for a semi-infinite beam resting on an elastic foundation. Different edge conditions at $x = 0$ and different loading will be considered.

**Example 3.3:** Let three point concentrated loads be applied to a semi-infinite beam whose edge $x = 0$ is free of tension. Two transverse forces of magnitude $P_1$

and $P_2$ along with a concentrated moment $m_0$ are acting at $x = a_1$, $x = a_2$ and $x = a_0$, respectively (see Figure 3.16).

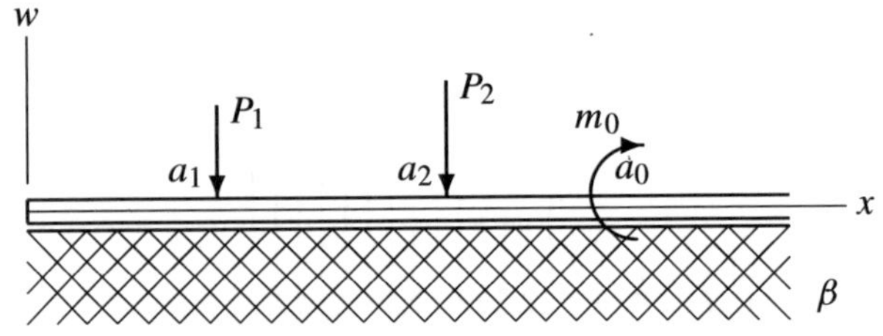


Figure 3.16: Beam freely resting on an elastic foundation

Clearly, the influence function of a point force for this beam represents the Green's function of the following boundary-value problem

$$\frac{d^2w(0)}{dx^2} = 0, \quad \frac{d^3w(0)}{dx^3} = 0, \quad |w(\infty)| < \infty, \quad \left|\frac{dw(\infty)}{dx}\right| < \infty$$

stated for eqn (3.101).

Tracing out the procedure described earlier in this section, one obtains the influence function of a point force for the beam, which for $x \le s$ is found in the form

$$g^+(x, s) = \frac{1}{8k^3}\{e^{k(x-s)}[\sin k(x-s) - \cos k(x-s)] \\ + e^{-k(s+x)}[\sin k(x+s) - 2\cos kx \cos ks - \cos k(x-s)]\},$$

while for $x \ge s$, we have

$$g^-(x, s) = \frac{1}{8k^3}\{e^{k(s-x)}[\sin k(s-x) - \cos k(x-s)] \\ + e^{-k(s+x)}[\sin k(x+s) - 2\cos kx \cos ks - \cos k(x-s)]\}$$

The actual computation of the deflection function $w(x)$ caused by $P_1$, $P_2$ and $m_0$ ought to be performed on the basis of the piecewise format of the influence function $g(x, s)$. For $x < a_1$ (to the left to the application point of $P_1$), for example, the deflection function is expressed as

$$w(x) = -\frac{1}{EI}\left[P_1 g^+(x, a_1) + P_2 g^+(x, a_2) + m_0 \frac{\partial g^+(x, a_0)}{\partial s}\right]$$

For $a_1 \le x \le a_2$, the computation is carried out as

$$w(x) = -\frac{1}{EI}\left[P_1 g^-(x, a_1) + P_2 g^+(x, a_2) + m_0 \frac{\partial g^+(x, a_0)}{\partial s}\right]$$

For $a_2 \le x \le a_0$, we have

$$w(x) = -\frac{1}{EI}\left[P_1 g^-(x, a_1) + P_2 g^-(x, a_2) + m_0 \frac{\partial g^+(x, a_0)}{\partial s}\right]$$

And for $x > a_0$ (to the right to the application point of $m_0$) the beam's deflection is determined as

$$w(x) = -\frac{1}{EI}\left[P_1 g^-(x, a_1) + P_2 g^-(x, a_2) + m_0 \frac{\partial g^-(x, a_0)}{\partial s}\right]$$

Clearly, by using the above expressions for the deflection function, the bending moment $M(x)$ and the shear force $Q(x)$ in any cross-section of the beam can readily be computed by the corresponding differentiation. Indeed, for $x < a_1$, the bending moment is found as

$$M(x) = -\left[P_1 \frac{\partial^2 g^+(x, a_1)}{\partial x^2} + P_2 \frac{\partial^2 g^+(x, a_2)}{\partial x^2} + m_0 \frac{\partial^3 g^+(x, a_0)}{\partial x^2 \partial s}\right]$$

For $a_1 \le x \le a_2$, it is computed as

$$M(x) = -\left[P_1 \frac{\partial^2 g^-(x, a_1)}{\partial x^2} + P_2 \frac{\partial^2 g^+(x, a_2)}{\partial x^2} + m_0 \frac{\partial^3 g^+(x, a_0)}{\partial x^2 \partial s}\right]$$

For $a_2 \le x \le a_0$, we have

$$M(x) = -\left[P_1 \frac{\partial^2 g^-(x, a_1)}{\partial x^2} + P_2 \frac{\partial^2 g^-(x, a_2)}{\partial x^2} + m_0 \frac{\partial^3 g^+(x, a_0)}{\partial x^2 \partial s}\right]$$

And for $x > a_0$, the bending moment is determined as

$$M(x) = -\left[P_1 \frac{\partial^2 g^-(x, a_1)}{\partial x^2} + P_2 \frac{\partial^2 g^-(x, a_2)}{\partial x^2} + m_0 \frac{\partial^3 g^-(x, a_0)}{\partial x^2 \partial s}\right]$$

The shear force, caused by the loads shown in Figure 3.16, can be computed by differentiation, with respect to $x$, of the above representations for the bending moment. Thus, all required components of the stress-strain state of the beam can, in this case and for other conventional loads, be computed analytically. ▲

## 3.4 End Chapter Exercises

**3.1** For a beam of length $a$, both edges of which are simply supported, construct the influence function of a transverse unit force. Use the classical method based on the defining properties of Green's function.

**3.2** For a beam of length $a$, both edges of which are simply supported, construct the influence function by the modification of the classical method proposed in Section 3.1.3. Compare your routine against that of the classical method.

**3.3** Construct the influence function for the beam, whose edge $x = 0$ is simply supported while the edge $x = a$ is clamped. Use two procedures: (a) the classical approach based on the defining properties of Green's function; (b) the modification of the classical approach discussed in Section 3.1.3. Compare the latter routine against that of the classical method.

**3.4** For a beam of length $a$ with both edges elastically supported as shown in Figure 3.17, construct the influence function by the modification of the classical method proposed in Section 3.1.3.

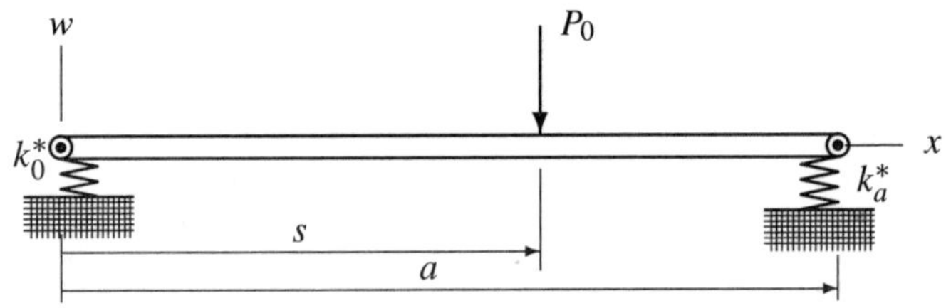


Figure 3.17: An elastically supported beam

**3.5** Consider particular cases of the influence function constructed in Exercise 3.4 that occur when the elastic coefficients $k_0^*$ and $k_a^*$ approach either zero or infinity (as listed below in parts (a) through (d)). Provide physical interpretation of each of these statements and explain why case (d) is meaningless:

**a)** either $k_0^*$ or $k_a^*$ approaches zero;

**b)** both $k_0^*$ and $k_a^*$ approach infinity;

**c)** $k_0^*$ approaches infinity, while $k_a^*$ remains finite.

**d)** both $k_0^*$ and $k_a^*$ approach zero;

**3.6** For a beam, with edge $x = 0$ simply supported, while edge $x = a$ is elastically supported ($k_a^*$), construct the influence function by the method of variation of parameters. What this influence function transforms to as the parameter $k_a^*$ approaches zero (infinity)?

**3.7** Construct the influence function of a point force for a beam, whose edge $x = 0$ is simply supported, while the edge $x = a$ is sliding against a rigid wall (see Figure 3.18).

**3.8** Construct the influence function of a point force for a beam, with edge $x = 0$ being clamped, while edge $x = a$ is sliding.

**3.9** For a beam, with edge $x = 0$ and edge $x = a$ simply supported and clamped, respectively, determine the deflection, the bending moment, and the shear force caused by:

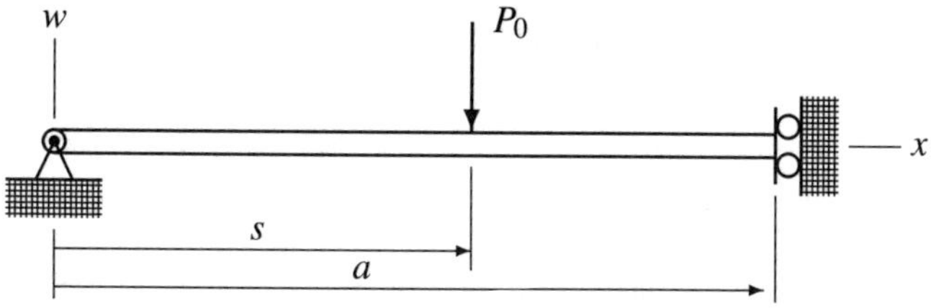


Figure 3.18: A simply supported–sliding beam

**a)** a transverse load $q(x) = q_0(x - a/4)(x - 3a/4)$ applied over the interval $[a/4, 3a/4]$;

**b)** two concentrated forces $P_1$ and $P_2$, spaced at $x = a/3$ and $x = 2a/3$, respectively;

**c)** two concentrated bending moments $M_1$ and $M_2$, spaced at $x = a/3$ and $x = 2a/3$, respectively;

**3.10** For a beam, with edge $x = 0$ and edge $x = a$ simply and elastically supported ($k_a^*$), respectively, determine the deflection, the bending moment, and the shear force caused by:

**a)** a transverse load $q(x) = q_0 x(x - a)$ distributed over the entire beam;

**b)** two concentrated bending moments $M_1$ and $M_2$, spaced at $x = a/4$ and $x = a/2$, respectively;

**c)** two concentrated forces $P_1$ and $P_2$, spaced at $x = a/3$ and $x = 2a/3$, respectively.

**3.11** For a beam, with both edges being elastically supported ($k_0^*$ and $k_a^*$), determine the deflection function caused by:

**a)** a transverse load $q(x) = q_0 x(x - a)$ distributed over the entire beam;

**b)** a combination of the bending moment $M_0$ spaced at $x = a/2$ and a uniform load $q(x) \equiv q_0$ distributed over the entire beam;

**c)** a combination of the concentrated force $P_0$ and the bending moment $M_0$ spaced at $x = a/4$ and $x = a/2$, respectively;

**d)** a transverse load $q(x) = q_0 \exp(-x^2)$ distributed over the entire beam.

**3.12** For the infinite beam resting on elastic foundation, with $EI$ and $k_0$ representing the flexural rigidity of the beam and the elastic constant of the foundation, respectively, determine its deflection function caused by the transverse loads shown below:

**a)** a transverse load $q(x) = q_0(1 + x^2)$ continuously distributed over the interval $[a, b]$;

**b)** two transverse concentrated forces $P_1$ and $P_2$, spaced at $x = a$ and $x = b$, respectively, with $b > a$;

**c)** a combination of the transverse load $q_0$ uniformly distributed over the interval $[a_1, a_2]$ and the concentrated force $P_0$ spaced at $x = b$, where $b > a_2$;

**d)** two concentrated bending moments $M_1$ and $M_2$, spaced at $x = a$ and $x = b$, respectively, with $b > a$.

**3.13** For a semi-infinite beam resting on an elastic foundation and having a free edge, determine its response to the loads shown below:

**a)** two concentrated bending moments, $M_1$ and $M_2$, spaced at $x = a$ and $x = b$, respectively, with $b > a$;

**b)** a combination of the concentrated force $P_0$ and the bending moment $M_0$ spaced at $x = a$ and $x = b$, respectively, with $b > a$.

**3.14** Construct the influence function for a semi-infinite beam $(EI)$ resting on an elastic foundation $(k_0)$, if its edge is simply supported.

**3.15** For the beam in Exercise 3.14, determine its deflection function caused by the loads shown below:

**a)** a combination of the concentrated moment $M_0$ spaced at $x = a$ and the concentrated force $P_0$ spaced at $x = b$, with $b > a$;

**b)** two concentrated bending moments, $M_1$ and $M_2$, spaced at $x = a$ and $x = b$, respectively, with $b > a$.

The next section summarizes all the work in this chapter. It brings an extensive catalogue of influence functions of a point force, constructed for a single-span Kirchhoff beam with a variety of edge conditions imposed. This is supposed to be helpful to users of our approach. Note that only the branch of $g(x, s)$ that is valid for $x \leq s$ is shown, while the other branch can be obtained from that presented by interchanging in it the $x$ and the $s$ variables.

## 3.5 Compendium of Influence Functions for Beams

| No. | Edge conditions | Influence function, for $x \leq s$ |
|---|---|---|
| 1 | $w(0) = w'(0) = w''(a) = w'''(a) = 0$ | $\frac{1}{6}x^2(x - 3s)$ |
| 2 | $w(0) = w'(0) = w(a) = w'(a) = 0$ | $\dfrac{1}{6a^3}x^2(s - a)^2[2s(x - a) + a(x - s)]$ |

*continued on next page*

| No. | Edge conditions | Influence function, for $x \le s$ |
|---|---|---|
| 3 | $w''(0) = w'''(0) + k_0 w(0) = 0$ <br> $w(a) = w'(a) = 0$ | $\frac{1}{\Delta}(s-a)^2\{6[(x-s)+2(x-a)]$ <br> $+ k_0 x[s(x^2-a^2)+2a(x^2-as)]\}$, <br> $\Delta = 12(3+k_0 a^3)$ |
| 4 | $w''(0) = w'''(0) = w(a) = w'(a) = 0$ | $\frac{1}{6}(s-a)^2[2(x-a)+(x-s)]$ |
| 5 | $w(0) = w'(0) = w(a) = w''(a) = 0$ | $\frac{1}{12a^3}x^2(s-a)[s(3a-x)(s-2a)+2a^2x]$ |
| 6 | $w(0) = w''(0) = w(a) = w''(a) = 0$ | $\frac{1}{6a}x(a-s)(x^2+s^2-2as)$ |
| 7 | $w''(0) = w'''(0) + k_0 w(0) = 0$ <br> $w''(a) = w'''(a) - k_a w(a) = 0$ | $\frac{1}{\Delta}\{(a-s)[k_0 k_a a x(x^2+s^2-2as)$ <br> $+ 6k_a(x-a)] - 6k_0 x s\}$, $\quad \Delta = 6a^2 k_0 k_a$ |
| 8 | $w(0) = w''(0) = 0$, <br> $w''(a) = w'''(a) - k_a w(a) = 0$ | $\frac{1}{\Delta}x[k_a a(a-s)(x^2+s^2-2as)-6s]$, <br> $\Delta = 6a^2 k_a$ |
| 9 | $w(0) = w''(0) = w'(a) = w'''(a) = 0$ | $\frac{1}{6}x(x^2+3s^2-6as)$ |
| 10 | $w(0) = w'(0) = w'(a) = w'''(a) = 0$ | $\frac{1}{12a}x^2(3s^2+2ax-6as)$ |
| 11 | $w''(0) = w'''(0) + k_0 w(0) = 0$ <br> $w'(a) = w'''(a) = 0$ | $\frac{1}{6}x(x^2+3s^2-6as) - 1/k_0$ |
| 12 | $w'(0) = w'''(0) + k_0 w(0) = 0$ <br> $w'(a) = w'''(a) = 0$ | $\frac{1}{12a}x^2(3s^2-6as+2ax) - 1/k_0$ |
| 13 | $w(0) = w''(0) - k_0 w'(0) = 0$ <br> $w'(a) = w'''(a) = 0$ | $\frac{1}{\Delta}x[3s(s-2a)(2+k_0 x)+2x^2(1+k_0 a)]$, <br> $\Delta = 12(1+k_0 a)$ |
| 14 | $w''(0) - k_1 w'(0) = 0$ <br> $w'''(0) + k_2 w(0) = 0$ <br> $w'(a) = w'''(a) = 0$ | $\frac{1}{\Delta}x[3s(s-2a)(2+k_1 x)+2x^2(1+k_1 a)] - 1/k_2$ <br> $\Delta = 12(1+k_1 a)$ |
| 15 | $w(0) = w'(0) = 0$ <br> $w''(a) = w'''(a) - k_a w(a) = 0$ | $\frac{1}{\Delta}x^2[k_a s^2(3a-x)(3a-s) - 2(3s-x)$ <br> $\times (6+k_a a^3)]$, <br> $\Delta = 12(3+k_a a^3)$ |

*continued on next page*

| No. | Edge conditions | Influence function, for $x \leq s$ |
|---|---|---|
| 16 | $w(0) = w'(0) = 0$<br>$w'(a) = w'''(a) - k_a w(a) = 0$ | $\frac{1}{\Delta} x^2\{6[3s(s-2a) + 2ax] - k_a a$<br>$\times (a-s)^2[2s(a-x) + a(s-x)]\}$<br>$\Delta = 6a(12 + k_a a^3)$ |
| 17 | $w(0) = w'(0) = 0$<br>$w(a) = w''(a) + k_a w'(a) = 0$ | $\frac{1}{\Delta} x^2(s-a)\{2[3as(2a-s) - x(2a(a+s) - s^2)]$<br>$+ k_a a[3as(a-s) + x(s(2s-a) - a^2)]\}$<br>$\Delta = 6a^3(4 + k_a a)$ |
| 18 | $w(0) = w'(0) = 0$<br>$w'''(a) = w''(a) + k_a w'(a) = 0$ | $\frac{1}{\Delta} x^2[2(x-3s)(1 + k_a a) + 3k_a s^2]$<br>$\Delta = 12(1 + k_a a)$ |
| 19 | $w(0) = w''(0) = 0$<br>$w'(a) = w'''(a) - k_a w(a) = 0$ | $\frac{1}{\Delta} x\{6[3s(s-2a) + x^2] - k_a(a-s)$<br>$\times [x^2(s(s+a) - 2a^2) + 3a^2 s(a-s)]\}$<br>$\Delta = 12(3 + k_a a^3)$ |
| 20 | $w(0) = w''(0) = 0$<br>$w(a) = w''(a) + k_a w'(a) = 0$ | $\frac{1}{\Delta} x(s-a)\{6a(2as - x^2 - s^2) + k_a$<br>$\times [3a^2 s(a-s) + x^2(s(a+s) - 2a^2)]\}$<br>$\Delta = 12a^2(3 + k_a a)$ |
| 21 | $w(0) = w''(0) = 0$<br>$w'''(a) = w''(a) + k_a w'(a) = 0$ | $\frac{1}{6} x[(3s^2 + x^2 - 6as) - 6s/k_a]$ |
| 22 | $w(0) = w''(0) = 0$<br>$w''(a) + k_1 w'(a) = 0$<br>$w'''(a) - k_2 w(a) = 0$ | $-\frac{1}{\Delta} x\{36s + 6k_1[3s(2a-s) - x^2]$<br>$+ 6k_2 a(a-s)[s(2a-s) - x^2]$<br>$+ k_1 k_2 (a-s)^2[3a^2 s - x^2(2a+s)]\}$<br>$\Delta = 12(3k_1 + k_1 k_2 a^3 + 3k_2 a^2)$ |
| 23 | $w'''(0) = w''(0) - k_0 w'(0) = 0$<br>$w(a) = w''(a) + k_a w'(a) = 0$ | $-\frac{1}{\Delta}(a-s)\{12(a-x) + 2k_0[2a(a+s)$<br>$- s^2 - 3x^2] + k_a(a-s)[2(2a+s-3x)$<br>$+ k_0(a(2s+a) - 3x^2)]\}$<br>$\Delta = 12(k_0 + k_0 k_a a + k_a)$ |
| 24 | $w(0) = w''(0) - k_0 w'(0) = 0$<br>$w''(a) = w'''(a) - k_a w(a) = 0$ | $\frac{1}{\Delta} x\{6k_0 x(x-3s) - 36s + k_a(a-s)$<br>$\times [6a(s(s-2a) + x^2) + k_0 x$<br>$\times (3as(s-2a) + x(2a(a+s) - s^2))]\}$<br>$\Delta = 12(3k_0 + k_0 k_a a^3 + 3k_a a^2)$ |

*continued on next page*

| No. | Edge conditions | Influence function, for $x \leq s$ |
|---|---|---|
| 25 | $w'''(0) = w''(0) - k_0 w'(0) = 0$<br>$w'(a) = w'''(a) - k_a w(a) = 0$ | $-\frac{1}{\Delta}\{12(1 + k_0 a) + k_a(a - s)^2[2(2a + s - 3x)$<br>$+ k_0(a(a + 2s) - 3x^2)]\}$<br>$\Delta = 12k_a(1 + k_0 a)$ |
| 26 | $w'''(0) = w''(0) - k_0 w'(0) = 0,$<br>$w''(a) = w'''(a) - k_a w(a) = 0$ | $-\frac{1}{\Delta}\{6k_0 + k_a(a - s)[6(a - x) + k_0(2a(s + a)$<br>$- s^2 - 3x^2)]\}, \ \Delta = 6k_0 k_a$ |
| 27 | $w'(0) = w'''(0) + k_0 w(0) = 0$<br>$w''(a) = w'''(a) - k_a w(a) = 0$ | $-\frac{1}{\Delta}\{36 - 6k_0 x^2(x - 3s) - k_a(a - s)[6(s^2 + 3x^2$<br>$- 2a(a + s)) + k_0(3ax^2 s(s - 2a) + x^3(2a$<br>$\times (a + s) - s^2))]\}$<br>$\Delta = 12(3k_0 + k_0 k_a a^3 + 3k_a)$ |
| 28 | $w'(0) = w'''(0) + k_0 w(0) = 0$<br>$w'(a) = w'''(a) - k_a w(a) = 0$ | $-\frac{1}{\Delta}\{72a - 6k_0 x^2[3s(s - 2a) + 2ax] + k_a(a - s)$<br>$\times [6(a^2(a + s) - 2as^2 - 3x^2(a - s)) + k_0 ax^2$<br>$\times (3as(a - s) + x(2s^2 - a(a + s)))]\}$<br>$\Delta = 6a(12k_0 + 12k_a + k_0 k_a a^3)$ |

# Chapter 4

# Other Beam Problems

A broad discussion is continued in this chapter on possible applications of the influence function method to other beam problems which are treated within the scope of Kirchhoff theory. Bending of beams of variable flexural rigidity $EI(x)$ is covered in Section 4.1 where influence functions of a point force are analytically constructed for some particular cases of the variation of flexural rigidity with the $x$ coordinate and an effective numerical approach is proposed for a general case that does not allow an analytic treatment.

The rest of the material in this chapter touches upon a number of problem settings in the beam theory, to which the influence function method had been applied just recently. In [43, 45] it was shown for the first time how some of such settings can be handled by means of this method. This text aims at the development of influence function-based computational procedures that can be recommended to engineers as an alternative to the classical approaches usually employed in structural mechanics.

In Section 4.2 algorithms are developed and tested for application of the influence function method to the transverse natural vibration beam problems. Some buckling problems for beams are targeted in Section 4.3, where the classical Euler approach is applied to the buckling phenomenon. The last section deals with the bending of multi-span beams undergoing concentrated and distributed transverse loads. Note that before the reader actually goes to Section 4.4, the material of Section 2.4 in Chapter 2 is recommended for a careful review.

## 4.1 Beams of Variable Flexural Rigidity

An extension of the influence function approach is proposed in this section to another class of beam problems. These are problems that simulate static equilibrium of beams with variable flexural rigidity when they undergo point concentrated as well as distributed loads.

We will show that once the influence function of a point force for a beam is available, the computational procedure for obtaining components of the stress-strain state is analogous to that described earlier for a beam of a uniform flexural rigidity. A number of particular variations (linear, quadratic, exponential, etc.) of the flexural rigidity, for which the analytic form of the influence function is easily attainable, are considered in detail. A general procedure is also developed for cases which do not allow an analytic solution.

Bending of a beam of variable rigidity $EI(x)$ is treated here within the scope of Kirchhoff theory. This implies that the beam's deflection function $w(x)$ satisfies a boundary-value problem for the Euler–Bernoulli equation

$$\frac{d^2}{dx^2}\left(EI(x)\frac{d^2w(x)}{dx^2}\right) = -q(x), \quad x \in (0, a) \tag{4.1}$$

$$B_{0,i}[w(0)] = 0, \quad B_{a,i}[w(a)] = 0, \quad i = 1, 2 \tag{4.2}$$

One can find in this section a number of particular statements of this type. Our approach makes the technique for finding the solution to the above problem fairly standard. We focus first on the construction of the influence function of a point force related to the statement and then actually solve the problem by the influence function method.

Before going any further with actual solutions of particular problems, let us reveal the relations for components of the stress-strain state in terms of the influence function of a unit point force for a beam having a variable flexural rigidity $EI(x)$. Those relations are slightly different from the corresponding ones presented in Chapter 3 for a beam of uniform flexural rigidity.

Let $g(x, s)$ be the influence function of a point force for the beam under consideration. That is, $g(x, s)$ represents the Green's function for the homogeneous boundary-value problem associated with that in eqns (4.1) and (4.2).

In compliance with Theorem 2.4 of Chapter 2, for a beam having a variable flexural rigidity, the deflection caused by the transverse load $q(x)$ continuously distributed over the interval $[\alpha, \beta]$ can be computed, for $x \leq \alpha$ (to the left of the loaded segment), as

$$w(x) = \int_{\alpha}^{\beta} g^{+}(x, s)q(s)\, ds, \quad x \in [0, \alpha]$$

For $\alpha \leq x \leq \beta$, in turn, it can be obtained as

$$w(x) = \int_{\alpha}^{x} g^{-}(x, s)q(s)\, ds + \int_{x}^{\beta} g^{+}(x, s)q(s)\, ds, \quad x \in [\alpha, \beta] \tag{4.3}$$

For $x \geq \beta$ (to the right of the loaded interval) we have

$$w(x) = \int_{\alpha}^{\beta} g^{-}(x, s)q(s)\, ds, \quad x \in [\beta, a]$$

where $g^{-}(x, s)$ and $g^{+}(x, s)$ represent the branch of the influence function $g(x, s)$ defined for $s \leq x$ and $x \leq s$, respectively.

Following the standard technique described earlier, one can readily derive the formula

$$w(x) = \sum_{i=1}^{k} P_i g(x, s_i), \qquad x \in [, 0, a]$$

for the deflection of a beam caused by a set of transverse point forces of magnitudes $P_i$, $(i = \overline{1, k})$, located at $s_i$ respectively.

For a beam having a variable flexural rigidity, the deflection caused by a set of concentrated bending moments of magnitudes $M_i$, $(i = \overline{1, k})$ acting at $s_i$ can be computed by means of the influence function of the second order as

$$w(x) = \sum_{i=1}^{k} M_i \frac{\partial g(x, s_i)}{\partial s}, \qquad x \in [0, a] \tag{4.4}$$

Note that, as we emphasized earlier, actual use of the above equations implies that corresponding branches of the influence function are implemented for different locations of the observation point. And analogously to the situation with beams of uniform flexural rigidity, the response to a reasonable combination of elementary loads can be computed based on the superposition principle. Saying *reasonable* we again mean that the simultaneous action of individual loads must not cause either geometrical or physical nonlinearity.

For a beam having variable flexural rigidity, the bending moment $M(x)$ and the shear force $Q(x)$ are expressed in terms of the deflection function $w(x)$ as

$$M(x) = EI(x)\frac{d^2w(x)}{dx^2}, \qquad Q(x) = \frac{d}{dx}\left(EI(x)\frac{d^2w(x)}{dx^2}\right) \tag{4.5}$$

Hence, the expressions for $M(x)$ and $Q(x)$ in any particular problem for a beam of variable flexural rigidity can be obtained by a proper differentiation of the deflection function. For example, the bending moment $M(x)$ caused by a set of concentrated bending moments of magnitudes $M_i$, $(i = \overline{1, k})$ acting at $s_i$ can be computed by formally taking the second order derivative of $w(x)$ with respect to $x$ in eqn (4.4) and substituting it in the first relation in (4.5). This yields

$$M(x) = EI(x) \sum_{i=1}^{k} M_i \frac{\partial^3 g(x, s_i)}{\partial s \partial x^2}, \qquad x \in [0, a]$$

In what follows, one finds a set of particular examples where we demonstrate the practical solvability of problems for beams of variable flexural rigidity by means of the influence function method.

### 4.1.1 Linear rigidity

Consider a cantilever beam of length $a$ whose edge $x = 0$ is clamped, if its flexural rigidity is represented by a linear function $EI(x) = px + r$ of the observation variable $x$. Since the flexural rigidity ought to be non-negative by physical nature and could take on zero value only at $x = a$, some constraints must be imposed on the parameters $p$ and $r$. That is, we assume that $r$ is necessarily positive and if $p$ is negative then the relation $|p|a \leq r$ holds.

It is intuitive that a transverse point force applied to an arbitrary point of such a cantilever beam causes its unique response. In other words, there exists a unique influence function of a point force for this beam, which must be identified with the Green's function of the homogeneous equation

$$\frac{d^2}{dx^2}\left((px+r)\frac{d^2w(x)}{dx^2}\right) = 0 \tag{4.6}$$

subject to the following set of boundary conditions

$$w(0) = 0, \qquad \frac{dw(0)}{dx} = 0, \qquad \frac{d^2w(a)}{dx^2} = 0, \qquad \frac{d^3w(a)}{dx^3} = 0 \tag{4.7}$$

The method of variation of parameters will be used here for the construction of the Green's function. To obtain a fundamental set of solutions for eqn (4.6), we first rewrite it in the form

$$(px+r)\frac{d^4w(x)}{dx^4} + 2p\frac{d^3w(x)}{dx^3} = 0 \tag{4.8}$$

by simply performing the outer differentiation in (4.6).

Because of the specific form of this equation (it does not contain derivatives of $w(x)$ of up to the second order included), the first three components

$$w_1(x) \equiv 1, \quad w_2(x) \equiv x, \quad w_3(x) \equiv x^2$$

of its fundamental set of solutions are evident. To determine the fourth component $w_4(x)$, we introduce a new function $u(x)$ as

$$u(x) = \frac{d^3w(x)}{dx^3} \tag{4.9}$$

which reduces eqn (4.8) to the first order separable equation

$$(px+r)\frac{du(x)}{dx} + 2pu(x) = 0$$

A particular solution of this equation

$$u(x) = \frac{1}{(px+r)^2}$$

can readily be obtained by the straightforward integration. Upon substituting $u(x)$ in eqn (4.9) and integrating it successively three times, we eventually obtain $w_4(x)$

in the form

$$w_4(x) \equiv (px + r)\ln(px + r)$$

Based on the fundamental set of solutions consisting of $w_1(x)$,$w_2(x)$, $w_3(x)$, along with $w_4(x)$ just obtained, we start the actual construction of the influence function. In doing so, we seek the general solution to eqn (4.6) in the form

$$w(x) = C_1(x) + C_2(x)x + C_3(x)x^2 + C_4(x)(px + r)\ln(px + r) \tag{4.10}$$

In compliance with the customary procedure of the method of variation of parameters, one obtains the following system

$$\begin{pmatrix} 1 & x & x^2 & (px+r)\ln(px+r) \\ 0 & 1 & 2x & p\ln(px+r)+p \\ 0 & 0 & 2 & p^2/(px+r) \\ 0 & 0 & 0 & -p^3/(px+r)^2 \end{pmatrix} \times \begin{pmatrix} C_1'(x) \\ C_2'(x) \\ C_3'(x) \\ C_4'(x) \end{pmatrix} = \begin{pmatrix} 0 \\ 0 \\ 0 \\ q^*(x) \end{pmatrix}$$

of linear algebraic equations in $C_i'(x)$, $(i = \overline{1,4})$. Here $q^*(x) = -q(x)/(px + r)$. Since the coefficient matrix of this system has an upper triangular form, its determinant $-2p^3/(px + r)^2$, as a product of the diagonal entries, is not zero. The well-posedness of this system reflects the linear independence of the components $w_i(x)$, $(i = \overline{1,4})$ in the fundamental set of solutions.

Solving the above system, we obtain

$$C_1'(x) = \frac{1}{2p^3}[px(px + 2r) - r(px + r)\ln(px + r)]q(x)$$

$$C_2'(x) = -\frac{1}{p^2}[r + (px + r)\ln(px + r)]q(x)$$

and

$$C_3'(x) = -\frac{1}{2p}q(x), \quad C_4'(x) = \frac{px + r}{p^3}q(x)$$

Integration of these relations yields

$$C_1(x) = \int_0^x \frac{1}{2p^3}[ps(ps + 2r) - r(ps + r)\ln(ps + r)]q(s)\,ds + H_1$$

$$C_2(x) = -\int_0^x \frac{1}{p^2}[r + (ps + r)\ln(ps + r)]q(s)\,ds + H_2$$

and

$$C_3(x) = -\int_0^x \frac{1}{2p}q(s)\,ds + H_3, \quad C_4(x) = \int_0^x \frac{(ps + r)}{p^3}q(s)\,ds + H_4$$

Upon substituting these in eqn (4.10) and performing some routine algebra, one finally obtains the general solution of eqn (4.6) as

$$w(x) = H_1 + H_2 x + H_3 x^2 + H_4(px+r)\ln(px+r) + \int_0^x \frac{1}{2p^3}\left\{2(px+r)(ps+r)\ln\frac{px+r}{ps+r} - p(x-s)[p(x+s)+2r]\right\}q(s)\,ds$$

Completing the construction procedure for the influence function, we ought to compute the coefficients $H_i$ upon satisfying the boundary conditions imposed by eqn (4.7). Substituting then $H_i$ into the above equation, we express the deflection $w(x)$ in the form of a single integral over the interval $[0, a]$. Recalling then Theorem 2.4 from Chapter 2, one finally obtains the explicit expression for the influence function that we are looking for. Completing this development, the branch $g^+(x, s)$ of the influence function, which is valid for $x \leq s$, is finally presented in the form

$$g^+(x,s) = \frac{1}{2p^3}\left\{px[px + 2(ps+r)] + 2(px+r)(ps+r)\ln\frac{r}{px+r}\right\} \tag{4.11}$$

while, for the branch $g^-(x, s)$ valid for $x \geq s$, we obtain

$$g^-(x,s) = \frac{1}{2p^3}\left\{ps[ps + 2(px+r)] + 2(px+r)(ps+r)\ln\frac{r}{ps+r}\right\} \tag{4.12}$$

Based on the influence function just derived, one can compute components of the stress–strain state of the cantilever beam, caused by a combination of transverse loads, as it is shown, for instance, in the example that follows.

**Example 1.1:** Let the beam under consideration be subject to a combination of two loads. Suppose a concentrated transverse force of magnitude $P$ is applied at $x = a_1$ and a concentrated bending moment of magnitude $m_0$ is applied at $x = a_2$, with $a_1 < a_2$ (see Figure 4.1).

The resultant deflection $w(x)$ of this beam can be expressed as

$$w(x) = Pg(x, a_1) + m_0\frac{\partial g(x, a_2)}{\partial s}$$

In computing this function, one should account for a piecewise format of $g(x, s)$ determined by eqns (4.11) and (4.12). That is, to the left of $a_1$, the deflection

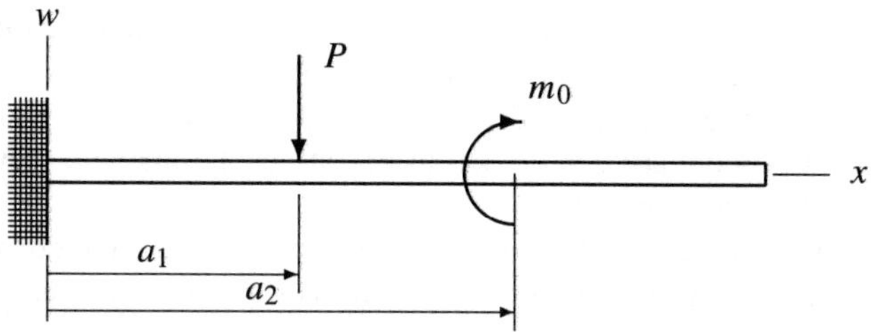


Figure 4.1: Cantilever beam of variable rigidity, $EI(x) = px + r$

function is defined as

$$w(x) = \frac{P}{2p^3}\left\{px[px + 2(pa_1 + r)] + 2(px + r)(pa_1 + r)\ln\frac{r}{px + r}\right\} + \frac{m_0}{p^2}\left[px + (px + r)\ln\frac{r}{px + r}\right], \quad x \le a_1$$

At any point located between $a_1$ and $a_2$, the beam's deflection function $w(x)$ is expressed as

$$w(x) = \frac{P}{2p^3}\left\{pa_1[pa_1 + 2(px + r)] + 2(px + r)(pa_1 + r)\ln\frac{r}{pa_1 + r}\right\} + \frac{m_0}{p^2}\left[px + (px + r)\ln\frac{r}{px + r}\right], \quad a_1 < x < a_2$$

Whereas, to the right of $a_2$, for $x \ge a_2$, we have

$$w(x) = \frac{P}{2p^3}\left\{pa_1[pa_1 + 2(px + r)] + 2(px + r)(pa_1 + r)\ln\frac{r}{pa_1 + r}\right\} + \frac{m_0}{p^2}\left[pa_2 + (px + r)\ln\frac{r}{pa_2 + r}\right], \quad x \ge a_2$$

The above expressions for the deflection function can be used to analytically compute the stress-related components of the beam. To obtain either the bending moment $M(x)$ or the shear force $Q(x)$ generated at any cross-section of the beam by the load depicted in Figure 4.1, one is required to analytically differentiate the expressions for the deflection function just obtained in compliance with the standard relations in eqn (4.5). So, this part of the analysis is absolutely routine and we leave it as an exercise for the reader. ▲

### 4.1.2 Exponential rigidity

In the previous subsection, it has been shown that, if the flexural rigidity of a beam represents a linear function of $x$, then the influence function of a point force for that beam can analytically be expressed in terms of elementary functions (see eqns (4.11) and (4.12)). An analytic expression was found there for the influence function of a cantilever beam. But it can also be found for any other physically feasible set of boundary conditions imposed at the end-points of the beam. This is so because boundary conditions do not affect the analytic solvability of the governing differential equation.

In the example that follows, we consider a beam with another form of flexural rigidity and show that the influence function also appears in that case in an analytic form.

**Example 1.2:** Use the influence function method to determine the deflection $w(x)$, the bending moment $M(x)$, and the shear force $Q(x)$ of a simply-supported beam of length $a$, with the flexural rigidity $EI(x)$ being an exponential function $pe^{\lambda x}$. The beam is loaded with a transverse load $q^*(x)$ continuously distributed over a portion $[\alpha, \beta]$ of the beam's span as shown in Figure 4.2.

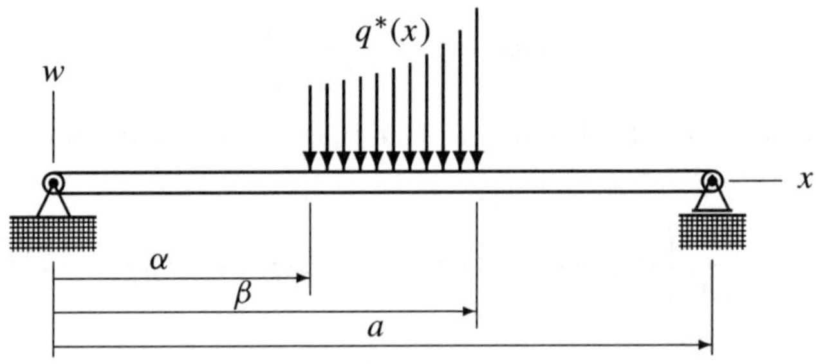


Figure 4.2: A beam of a variable rigidity $EI(x) = pe^{\lambda x}$

Clearly, the boundary-value problem modeling the equilibrium state of the beam in this setting can be formulated as follows

$$\frac{d^2}{dx^2}\left(pe^{\lambda x}\frac{d^2w(x)}{dx^2}\right) = -q(x), \quad x \in (0, a) \tag{4.13}$$

$$w(0) = \frac{d^2w(0)}{dx^2} = 0, \quad w(a) = \frac{d^2w(a)}{dx^2} = 0 \tag{4.14}$$

where

$$q(x) = \begin{cases} 0, & x < \alpha \\ q^*(x), & \alpha \leq x \leq \beta \\ 0, & x > \beta \end{cases}$$

The construction procedure for an influence function (which represents in this case the Green's function to the homogeneous boundary-value problem corresponding to that posed by eqns (4.13) and (4.14)) can be developed on the standard basis. We are not going to provide its detailed description. One issue in this procedure is, however, worth focusing on. That is, how to obtain a fundamental set of solutions for the homogeneous equation associated with that in (4.13). To address this issue, we accomplish the outer differentiation in eqn (4.13) by using the product rule. This yields

$$\frac{d^4w(x)}{dx^4} + 2\lambda\frac{d^3w(x)}{dx^3} + \lambda^2\frac{d^2w(x)}{dx^2} = -\frac{1}{p}q(x)e^{-\lambda x}$$

Hence, eqn (4.13) reduces to the one with constant coefficients. Since its characteristic equation has two roots ($k = 0$ and $k = -\lambda$, each of multiplicity two), a fundamental set of solutions for the homogeneous equation corresponding to (4.13) can be represented by the following set of functions

$$w_1(x) \equiv 1, \quad w_2(x) \equiv x, \quad w_3(x) \equiv e^{-\lambda x}, \quad w_4(x) \equiv xe^{-\lambda x}$$

Upon using this set, one readily obtains the influence function $g(x, s)$ for the simply supported beam having an exponential flexural rigidity. In doing so, we follow the standard procedure. The branch of this function, which is defined for $x \leq s$ is finally found as

$$g^+(x, s) = \frac{1}{p\lambda^3a^2}\{2xse^{-\lambda a} - 2(a-x)(a-s) + xa[\lambda(a-s) - 2]e^{-\lambda s} + a(a-s)(2+\lambda x)e^{-\lambda x}\}$$

while for $x \geq s$, we obtain

$$g^-(x, s) = \frac{1}{p\lambda^3a^2}\{2xse^{-\lambda a} - 2(a-s)(a-x) + sa[\lambda(a-x) - 2]e^{-\lambda x} + a(a-x)(2+\lambda s)e^{-\lambda s}\}$$

Since the analytic expression for the influence function of a point force is available, we can now turn to the original statement of the problem posed by eqns (4.13) and (4.14). Upon utilizing the influence function just obtained, one determines the solution of this problem (that is the deflection, caused by the transverse load $q(x)$ applied to the interval $[\alpha, \beta]$). In this case, for $x \leq \alpha$ (to the left of the loaded interval), we have

$$w(x) = \frac{1}{p\lambda^3a^2}\int_\alpha^\beta \{2xse^{-\lambda a} - 2(a-x)(a-s) + xa[\lambda(a-s) - 2]e^{-\lambda s} + a(a-s)(2+\lambda x)e^{-\lambda x}\}\overline{q}(s)\,ds, \quad x \in [0, \alpha]$$

where

$$\overline{q}(s) = \frac{1}{p}q^*(s)e^{-\lambda s}$$

For $\alpha \leq x \leq \beta$, in turn, $w(x)$ is obtained as

$$w(x) = \frac{1}{p\lambda^3a^2}\int_\alpha^x \{2xse^{-\lambda a} - 2(a-s)(a-x) + sa[\lambda(a-x) - 2]e^{-\lambda x} + a(a-x)(2+\lambda s)e^{-\lambda s}\}\overline{q}(s)\,ds$$

$$+\frac{1}{p\lambda^3 a^2}\int_x^\beta \{2xse^{-\lambda a} - 2(a-x)(a-s)$$
$$+xa[\lambda(a-s)-2]e^{-\lambda s} + a(a-s)(2+\lambda x)e^{-\lambda x}\}\overline{q}(s)\,ds, \quad x\in[\alpha,\beta]$$

For $x \geq \beta$ (to the right of the loaded interval), we finally obtain

$$w(x) = \frac{1}{p\lambda^3 a^2}\int_\alpha^\beta \{2xse^{-\lambda a} - 2(a-s)(a-x)$$
$$+sa[\lambda(a-x)-2]e^{-\lambda x} + a(a-x)(2+\lambda s)e^{-\lambda s}\}\overline{q}(s)\,ds, \quad x\in[\beta,a]$$

If the loading function $q(x)$ in eqn (4.13) has a simple form, the integrals in the expressions for $w(x)$ just obtained can be computed analytically. When $q(x)$ is too complicated to practically obtain an analytic solution, one should numerically integrate by choosing an appropriate quadrature formula. The choice of such a formula is determined by the accuracy level that is required.

In computing either the bending moments $M(x)$ or the shear forces $Q(x)$, caused by $q(x)$, one is required to analytically differentiate the expression for $w(x)$ in compliance with the relations in eqn (4.5). ▲

Note that for both cases of variable flexural rigidity considered so far in this section, fundamental sets of solutions for governing equations (see eqns (4.6) and (4.13)) are expressed in elementary functions. This makes it possible to construct corresponding influence functions of a point force in analytic form.

There also exist some other variations of flexural rigidity for which analytic construction of influence functions is potentially possible. These include, for example, quadratic or rational polynomial functions of some type. It is worth noting, however, that the analytic form of the influence function in such cases becomes too cumbersome and inconvenient to operate with.

### 4.1.3 General case

In what follows in this section, we will sketch out a part analytic–part numeric procedure that enables the obtaining of influence functions of a point force for a beam of a variable rigidity $EI(x)$ in the case for which the exact solution of the homogeneous Euler–Bernoulli equation

$$\frac{d^2}{dx^2}\left(EI(x)\frac{d^2w(x)}{dx^2}\right) = 0, \quad x\in(0,a) \tag{4.15}$$

is either impossible at all or results in too cumbersome analytic development.

Whichever standard procedure is utilized for the construction of the Green's function to the homogeneous boundary-value problem in eqns (4.15) and (4.2), a fundamental set of solutions of eqn (4.15) is required. Since, in light of our assumption, this equation does not allow exact solution, the components of its fundamental set of solutions ought to be computed numerically.

Since any four linearly independent on $[0, a]$ particular solutions $\{w_i(x)\}$, $(i = \overline{1, 4})$ of eqn (4.15) could constitute a fundamental set of solutions (FSS) of (4.15), the following strategy is proposed to practically obtain its components.

The first component $w_1(x)$ of the FSS is suggested to be looked for as the solution of the initial-value problem

$$w_1(0) = 1, \quad \frac{dw_1(0)}{dx} = \frac{d^2w_2(0)}{dx^2} = \frac{d^3w_1(0)}{dx^3} = 0 \tag{4.16}$$

for the governing eqn (4.15).

The second component $w_2(x)$ of the FSS will be found as the solution to another initial-value problem

$$\frac{dw_2(0)}{dx} = 1, \quad w_2(0) = \frac{d^2w_2(0)}{dx^2} = \frac{d^3w_2(0)}{dx^3} = 0 \tag{4.17}$$

posed for the same governing equation.

The third component $w_3(x)$ of the FSS will represent the solution to the initial-value problem

$$\frac{d^2w_3(0)}{dx^2} = 1, \quad \frac{dw_3(0)}{dx} = w_3(0) = \frac{d^3w_3(0)}{dx^3} = 0 \tag{4.18}$$

posed for eqn (4.15).

And finally the last component $w_4(x)$ of the FSS will be determined as the solution to the problem

$$\frac{d^3w_4(0)}{dx^3} = 1, \quad \frac{dw_4(0)}{dx} = \frac{d^2w_4(0)}{dx^2} = w_4(0) = 0 \tag{4.19}$$

for eqn (4.15).

Numerical solution of such initial-value problems, with any feasible flexural rigidity $EI(x)$, could not be an issue. It can be obtained with a high accuracy level by employing standard numerical routines. The Runge–Kutta method of the fourth order [5, 6, 19], for example, can be recommended in this regard, since it provides an extremely high accuracy and subroutines based on this method are widely available in existing computers software.

It can easily be shown that the solutions of the four initial-value problems posed by eqns (4.15) and (4.16); (4.15) and (4.17); (4.15) and (4.18); and (4.15) and (4.19) represent a set of linearly independent functions on $[0, a]$. Indeed, their linear combination

$$W(x) = C_1w_1(x) + C_2w_2(x) + C_3w_3(x) + C_4w_4(x)$$

with arbitrary coefficients $C_i$, $(i = \overline{1, 4})$ represents a solution of the initial-value problem written as

$$\frac{d^2}{dx^2}\left(EI(x)\frac{d^2W(x)}{dx^2}\right) = 0, \quad x \geq 0$$

$$W(0) = C_1, \quad \frac{dW(0)}{dx} = C_2, \quad \frac{d^2W(0)}{dx^2} = C_3, \quad \frac{d^3W(0)}{dx^3} = C_4$$

To grasp the point, the reader is recommended to satisfy these initial conditions with the above expression for $W(x)$ keeping in mind that each component $w_i(x)$ of the fundamental set of solutions that we are dealing with, satisfies the corresponding set of initial conditions imposed by eqns (4.16)–(4.19).

However, it is evident that the above initial-value problem has a nontrivial solution if at least one of the four constants $C_i$ is non-zero. And the only case for which $W(x)$ is identical zero on $[0, a]$ is that with all $C_i$ equal zero. Hence, the functions $w_i(x)$, $(i = \overline{1, 4})$, which represent solutions of the initial-value problems posed by eqns (4.15) and (4.16); (4.15) and (4.17); (4.15) and (4.18); and (4.15) and (4.19) are really linearly independent on $[0, a]$. They could therefore constitute a fundamental set of solutions for eqn (4.15), based on which the Green's function to the boundary-value problem in eqns (4.15) and (4.2) can be routinely obtained and used then in computing required components of the stress-strain state of the beam undergoing a given combination of loads.

A special sample problem in the Example 1.3 that follows was chosen to illustrate the productivity of the part analytic–part numeric influence function approach just sketched. The point is that the exact solution of the sample problem can be obtained analytically. This means that solving it with our approach, we are able to check out the accuracy in this case and to make some observations concerning the accuracy level that can potentially be attained.

**Example 1.3:** We formulate a boundary-value problem where the solution of the nonhomogeneous equation

$$\frac{d^2}{dx^2}\left(\frac{B}{x+b}\frac{d^2w(x)}{dx^2}\right) = -q(x) \tag{4.20}$$

is subject to the boundary conditions

$$w(0) = \frac{dw(0)}{dx} = 0, \quad w(a) = \frac{d^2w(a)}{dx^2} = 0 \tag{4.21}$$

where $B$ and $b$ represent positive constants. The loading function $q(x)$ is assumed to be continuous on $(0, a)$.

The above problem simulates the bending of a beam of length $a$ subject to a distributed transverse load, with one edge clamped while the other is simply supported. The beam's flexural rigidity is a rational function of $x$ which is decreasing towards the right-hand edge.

It can be shown that if the loading function $q(x)$ in (4.20) is either polynomial, or trigonometric (either of the sine or the cosine type), or exponential, then the exact solution to the problem in eqns (4.20) and (4.21) can be found as an elementary function. Indeed, as to the homogeneous equation

$$\frac{d^2}{dx^2}\left(\frac{B}{x+b}\frac{d^2w(x)}{dx^2}\right) = 0 \tag{4.22}$$

corresponding to (4.20), its first two linearly independent particular solutions

$$w_1(x) \equiv 1 \quad \text{and} \quad w_2(x) \equiv x \tag{4.23}$$

are evident. To obtain another two linearly independent particular solutions of that equation, we make a substitution

$$u(x) = \frac{d^2 w(x)}{dx^2}$$

that reduces (4.22) to the second order equation

$$\frac{d^2}{dx^2}\left(\frac{B}{x+b}u(x)\right) = 0$$

in $u(x)$. Two linearly independent particular solutions to this equation

$$u_1(x) \equiv (x+b) \quad \text{and} \quad u_2(x) \equiv (x+b)^2$$

follow just from observation. And then by two successive integrations, one comes up with another two linearly independent particular solutions

$$w_3(x) \equiv (x+b)^3 \quad \text{and} \quad w_4(x) \equiv (x+b)^4 \tag{4.24}$$

for eqn (4.22).

Thus, the functions $w_i(x)$, $(i = \overline{1,4})$, just obtained (see eqns (4.23) and (4.24)), represent a fundamental set of solutions to equation (4.22), allowing its general solution in the form

$$w(x) = C_1 + C_2 x + C_3(x+b)^3 + C_4(x+b)^4 \tag{4.25}$$

To obtain the general solution to eqn (4.20), we first specify its right-hand side function $q(x)$. Let it, for simplicity, be a constant, that is $q(x) \equiv q_0$. If so then a particular solution to eqn (4.20) can be found by the method of variation of parameters in the form

$$w_p(x) = -\frac{q_0}{40B}(x+b)^5$$

So, the general solution to eqn (4.20) can be written as the sum of the general solution of equation (4.22) that we recently obtained (see (4.25)) and the particular solution $w_p(x)$. This results in

$$w(x) = C_1 + C_2 x + C_3(x+b)^3 + C_4(x+b)^4 - \frac{q_0}{40B}(x+b)^5 \tag{4.26}$$

Based on (4.26), the solution to the boundary-value problem in (4.20) and (4.21) can be found by satisfying the boundary conditions in (4.21). This yields the well-posed system of linear algebraic equations

$$\begin{pmatrix} 1 & 0 & b^3 & b^4 \\ 0 & 1 & 3b^2 & 4b^3 \\ 1 & a & (a+b)^3 & (a+b)^4 \\ 0 & 0 & 6(a+b) & 12((a+b)^2 \end{pmatrix} \times \begin{pmatrix} C_1 \\ C_2 \\ C_3 \\ C_4 \end{pmatrix} = \frac{q_0}{40B}\begin{pmatrix} b^5 \\ 5b^4 \\ (a+b)^5 \\ 20(a+b)^3 \end{pmatrix}$$

in $C_i$, $(i = \overline{1,4})$, from which we have

$$C_1 = \frac{q_0 b^3}{120B(a+4b)}(4a^3 + 17a^2b + 28ab^2 + 12b^3)$$

$$C_2 = \frac{q_0 b^2}{120B(a+4b)}(12a^3 + 44a^2b + 55ab^2 + 20b^3)$$

$$C_3 = -\frac{q_0}{30B(a+4b)}(a^3 + 6a^2b + 15ab^2 + 10b^3)$$

and

$$C_4 = \frac{q_0}{120B(a+4b)}(7a^2 + 35ab + 40b^2)$$

Thus, upon substituting the above values of $C_i$ in (4.26), one obtains the exact solution to the boundary-value problem in eqns (4.20) and (4.21), with the right-hand side term being a constant $q(x) = q_0$. This solution will later be used as a sample in checking out the numerical influence function method-based procedure. Once the Green's function $g(x, s)$ is obtained for the homogeneous setting in eqns (4.22) and (4.21), the solution of the problem posed by eqns (4.20) and (4.21) can be found in terms of $g(x, s)$ and the right-hand side term of eqn (4.20), in compliance with Theorem 2.4 of Chapter 2.

Before going any further with our numerical experiment, let us make some important comments as to the fundamental set of solutions to equation (4.22). This set is required for the construction of the Green's function $g(x, s)$. As it has been shown in Chapter 1, the fundamental set of solutions to a linear homogeneous differential equation is not unique (Example 1.6 in Chapter 1 could refresh the reader's mind on this point). Eqns (4.23) and (4.24), for instance, present one of the fundamental sets of solutions to equation (4.22). The solutions of the four initial-value problems stated by eqns (4.16)–(4.19) that we recommended earlier, could produce another fundamental set of solutions to that equation. As it follows from eqn (4.26), the solution of eqn (4.22) satisfying the initial-value problem in eqn (4.16) appears as

$$w_1(x) \equiv 1$$

while the setting in eqn (4.17) yields

$$w_2(x) \equiv x$$

These two components of the fundamental set of solutions to eqn (4.22) appear to be the same as those in eqn (4.23).

Satisfying the initial conditions in eqns (4.18) and (4.19) by the expression from eqn (4.26), we obtain another two components which are different of those in eqn (4.24). Indeed, for the third one we obtained

$$w_3(x) \equiv \frac{x^2}{12b^2}(6b^2 - x^2)$$

while the fourth component is found as

$$w_4(x) \equiv \frac{x^3}{12b}(x + 2b)$$

So, we came up with exact solutions to each of the initial-value problems in eqns (4.16)–(4.19) formulated for the governing equation (4.22). These could be used to just check out the accuracy of approximately obtained components of the fundamental set of solutions to eqn (4.22), when the initial-value problems in eqns (4.16)–(4.19) are solved numerically.

In Table 4.1 some results are presented on the comparison of the analytic solution and the numerical influence function treatment for the problem in eqns (4.20) and (4.21), where we assumed: $a = 1$, $b = 1$, $B = 1$ and $q_0 = 10$. Values of the deflection function $w(x)$ as well as of the bending moment $M(x)$ are exhibited. Once the influence function of a point force is numerically obtained with the aid of the Runge–Kutta method, the values of $w(x)$ and $M(x)$ in the integrals of eqns (4.3) and (4.5) are computed using the standard trapezoid rule with a uniform partition of the interval $(0, a)$. The number of partitions (the partition parameter) is denoted in Table 4.1 with $n$.

Recall that the numerical version of the influence function approach is nearly identical to the analytical version of this method, except for the manner in which the components $w_i(x)$ of the fundamental set of solutions and their derivatives required for the influence function itself are obtained. The relatively rapid convergence of the numerical version is evident from the data of Table 4.1. This brings a confidence in high efficiency of the influence function method applied to problems that are related to beams having variable flexural rigidity, if the governing differential equation cannot be solved analytically.

Notice that the numerical version of the influence function method does not look computationally expensive unless the partition parameter $n$ exceeds the level of 100. CPU time for the numerical version for $n = 10$, for instance, is about the same as that required for the analytical version. For $n = 100$, however, the numerical version becomes ten to fifteen times as computationally expensive as the analytical version. But the accuracy level attained with $n = 10$ is relatively high to satisfy most practical needs. ▲

It is worth noting that in the analytical version of the influence function method the accuracy level for the bending moment values does not practically differ from that for the deflection function. In the numerical version of this method, as it follows from the data in Table 4.5, the bending moment values are still computed with a high accuracy which, however, notably deteriorates compared to the accuracy of the deflection function. The cause of this phenomenon is in the

**Table 4.1:** Effectiveness of the numerical version of the influence function method

| Field point, $x$ | Analytic $w(x)$ | Approximate $w(x)$ $n=10$ | Approximate $w(x)$ $n=100$ | Analytic $M(x)$ | Approximate $M(x)$ $n=10$ | Approximate $M(x)$ $n=100$ |
|---|---|---|---|---|---|---|
| 0.1 | 0.061575 | 0.061437 | 0.061572 | 8.1 | 8.082513 | 8.100376 |
| 0.2 | 0.212267 | 0.212208 | 0.212265 | 2.2 | 2.176287 | 2.201028 |
| 0.3 | 0.401625 | 0.401547 | 0.401621 | −0.7 | −0.681765 | −0.700794 |
| 0.4 | 0.582400 | 0.582324 | 0.582397 | −2.6 | −2.578453 | −2.601082 |
| 0.5 | 0.713542 | 0.713419 | 0.713540 | −5.5 | −5.476754 | −5.501276 |
| 0.6 | 0.763198 | 0.763103 | 0.763195 | −6.4 | −6.381965 | −6.400786 |
| 0.7 | 0.711724 | 0.711637 | 0.711720 | −6.3 | −6.283465 | −6.300653 |
| 0.8 | 0.554667 | 0.554576 | 0.554664 | −5.2 | −5.176541 | −5.201186 |
| 0.9 | 0.305774 | 0.305708 | 0.305770 | −2.1 | −2.082765 | −2.100843 |
| Error, % | | 0.01 | 0.0001 | – | 0.2 | 0.01 |

approximate computation of the components of the fundamental set of solutions for the governing equation.

So far in this chapter we have been involved with the ways of possible utilization of the influence (Green's) function method in solving some elementary beam problems. In the following sections, the reader will find the extension of this method to some other problem classes that occur in Kirchhoff beam theory. Some of those problems have traditionally been formulated in structural mechanics and their solutions are usually described in detail in standard texts. But they have never been tackled with the aid of the influence function method. We will apply this method to the determination of natural frequencies of transverse vibrations of beams, to some classical buckling beam problems and to the bending of multi-span elastic beams.

The reader can readily figure out that each of the problem classes in the beam theory is usually modeled with a certain boundary-value problem for a differential equation. Each of those classes is usually tackled, in standard texts, by a specific individual approach. In other words, there is no a universal method that could be equally effective for different problem classes. One of the objectives in this text is to bring convincing evidence that the influence function method could be considered as such a universal approach.

## 4.2 Transverse Natural Vibrations

We included this and the two following sections in this chapter to show how the influence function method works for some beam problems for which it has not been traditionally suggested in literature as a possible approach.

Another purpose for presenting these sections is to offer some non-traditional topics for undergraduate research projects within the existing curricula in mechanical engineering or in other related fields. The author's objective here is to provoke the reader's interest in the implementation of the influence function method for a number of linear (and potentially some nonlinear) formulations which could be even more complicated compared to those which are actually considered within this text.

Earlier in this chapter, implementations of influence functions were considered to problems for which these functions have actually been constructed. We have analyzed, for example, stress-strain states caused by a variety of loads applied to a beam, by means of the beam's response to a transverse concentrated unit force applied to an arbitrary point in the beam. In other words, direct applications of influence functions have been considered so far.

We intend to advance a little further in this and the following sections. That is, some indirect applications of influence functions will be explored. We will utilize influence functions, constructed for problems of the linear bending of beams, for the solution of some more complicated beam problems. Expanding the range of productive applications of this method, we will recall, in this section, another problem class from Kirchhoff beam theory. This method happens to be fairly efficient in computing frequencies and mode shapes of free transverse vibrations of a beam with any physically feasible set of edge conditions imposed.

In standard texts (see, for example, [13, 63]), this problem class for beams of variable flexural rigidity is associated with a special kind of boundary-value problems which are called *eigenvalue problems* and are written as

$$\frac{d^2}{dx^2}\left(EI(x)\frac{d^2w(x)}{dx^2}\right) - p^2 m(x)A(x)w(x) = 0, \quad x \in [0, a] \tag{4.27}$$

$$B_{0,k}[w(0)] = 0, \quad B_{a,k}[w(a)] = 0, \quad k = 1, 2 \tag{4.28}$$

where, in addition to our customary notations for beam problems, $m(x)$ and $A(x)$ represent the mass density of the material and the cross-sectional area of the beam, respectively. The parameter $p$ is to be determined. The boundary conditions in (4.28) are presented in a general form, since we do not need them to be specified at the moment. However, as we have earlier mentioned, any certain physically feasible set of edge conditions can be viewed as a particular case of the relations in (4.28).

From the standard undergraduate course of differential equations, the reader learns that those values of the parameter $p$, for which the homogeneous boundary-value problem in (4.27) and (4.28) has nontrivial (non-zero) solutions, are referred to as the *eigenvalues* of this problem, while the corresponding nontrivial solutions themselves are called the *eigenfunctions* of this problem. Physical interpretation of eigenvalues and eigenfunctions directly leads to the natural frequencies and modes of transverse vibration for the beam under consideration.

It is evident that the homogeneity of eqn (4.27) implies that, if a certain function $w(x)$ represents its particular solution, then the function $Cw(x)$, where $C$ is an arbitrary constant, is also a solution. In other words, an eigenfunction of the problem in (4.27) and (4.28) is defined up to a scalar multiple.

### 4.2.1 Influence function algorithm

To instruct the reader regarding the way of using the influence function method in solving the eigenvalue problem posed in (4.27) and (4.28), we reduce the latter to a regular integral equation whose approximate solution can easily be computed by standard numerical methods.

In doing so, let $g(x, s)$ be the influence function of a transverse unit point force for the beam under consideration. In other words, we assume that $g(x, s)$ represents the Green's function for the boundary-value problem posed by the homogeneous equation

$$\frac{d^2}{dx^2}\left(EI(x)\frac{d^2w(x)}{dx^2}\right)=0, \quad x \in [0, a] \tag{4.29}$$

subject to the boundary conditions imposed by eqn (4.28).

Taking the second term in eqn (4.27) to the right-hand side

$$\frac{d^2}{dx^2}\left(EI(x)\frac{d^2w(x)}{dx^2}\right)=p^2m(x)A(x)w(x), \quad x \in [0, a] \tag{4.30}$$

we interpret it formally as the right-hand side function in the differential equation. In compliance with Theorem 2.4 of Chapter 2, one can use the Green's function $g(x, s)$ and express then the solution of the problem in (4.30) and (4.28) in the form

$$w(x) = -\int_0^a g(x, s)[p^2m(s)A(s)w(s)]\, ds$$

from which by factoring out the parameter $p^2$, we obtain

$$w(x) = -p^2\int_0^a g(x, s)m(s)A(s)w(s)\, ds \tag{4.31}$$

Note that this is not an explicit form for $w(x)$, because the latter is expressed in (4.31) in terms of itself. Such relations are called *integral equations*. The one in (4.31) represents the so-called *homogeneous Fredholm integral equation of the second kind*. Basic concepts of integral equations and their qualitative theory can be found in [60]. In this text (see Section 1.6), however, the reader is just briefly instructed on the classification of integral equations and on numerical approaches to their approximate solution.

The equation in (4.31) poses an eigenvalue problem equivalent to that in eqns (4.27) and (4.28). Thus, as a result of the development just completed, the original differential eigenvalue problem reduces to the eigenvalue problem for the integral equation in (4.31), where we are looking for those values of $p$ that yield nontrivial solutions.

Generally speaking, (4.27), as an equation with variable coefficients, does not allow an analytical solution. Neither does, of course, eqn (4.31). Therefore, in the developing of a computational procedure for this equation, one ought to rely only on approximate methods. And this is where a superiority of integral equations (compared to differential equations) is very notable. In what follows, the last statement will be supported with a number of examples.

Approximate eigenvalues and eigenfunctions to an integral equation of the type in (4.31) can successfully be obtained by a variety of traditional numerical methods. Let, for example, the quadrature formulae method [60] be utilized. In this method we approximate the definite integral with a finite sum (see Section 1.6). Let $x_k$ and $B_k$, $(k = \overline{1, n})$ represent mesh-points and the quadrature coefficients, respectively. Let in addition $w_k$ represent the approximate values of $w(x)$ at $x = x_k$. In accordance with the standard scheme of the quadrature formulae method, eqn (4.31) is approximated, at every mesh-point $x_j$, with the following homogeneous linear algebraic equation

$$w_j = -p^2 \sum_{k=1}^{n} B_k g(x_j, s_k) m(s_k) A(s_k) w_k, \quad (j = \overline{1, n})$$

in $n$ unknowns $w_k$, so that the entire set of the above relations (as the parameter $j$ goes from 1 to $n$) constitutes a standard eigenvalue problem of linear algebra and can therefore be solved in a standard way. This ultimately yields approximate values of the $n$ lowest components of the eigenvalue spectrum of eqn (4.31) along with approximate values of the corresponding eigenfunctions computed at the mesh-points.

In the discussion that follows, the reader will find some data indicating a high accuracy level attained when the approach described here is used in practice. The results appear to be relatively accurate even if one uses such a primitive quadrature technique as the trapezoid rule with an equally spaced set of a limited number $n$ of mesh-points.

We will also conduct a computational experiment on the comparison of the results obtained by the finite difference method directly applied to the differential formulation in eqns (4.27) and (4.28) against those obtained by the direct tackling of eqn (4.31) by the quadrature formulae method.

**Example 2.1:** This will be a validation example that represents a classical formulation with a well-known solution. Namely, let us seek natural frequencies and mode shapes of transverse vibrations of a single-span simply supported beam of length $a$, having a uniform flexural rigidity $EI = const$.

This problem results in the standard eigenvalue formulation

$$\frac{d^4 w(x)}{dx^4} - \lambda^4 w(x) = 0, \quad x \in [0, a] \tag{4.32}$$

$$w(0) = \frac{d^2 w(0)}{dx^2} = 0, \quad w(a) = \frac{d^2 w(a)}{dx^2} = 0 \tag{4.33}$$

where a specific notation $\lambda^4$ for the parameter in eqn (4.32) is used just for a notational convenience.

Physical interpretation of the eigenvalues and eigenfunctions of this problem directly leads to the natural frequencies and modes of transverse vibration of the beam under consideration. Namely, the circular natural frequency $f$ (in hertz) can be found for this beam in terms of $\lambda$ as

$$f = \frac{\lambda^2}{2\pi}\sqrt{\frac{EI}{mA}}$$

where the constants $m$ and $A$ represent mass density of the material and cross-sectional area of the beam, respectively. The eigenfunctions of the problem in eqns (4.32) and (4.33) represent, in turn, the mode shapes.

From observation, it follows that the following functions

$$w_l(x) = C \sin\frac{l\pi x}{a}, \quad l = 1, 2, 3, \ldots \tag{4.34}$$

where $C$ is an arbitrary constant, represent solutions to eqn (4.32) if the parameter $\lambda$ takes on the values

$$\lambda_l = \frac{l\pi}{a} \tag{4.35}$$

Moreover, it can easily be seen that each of the functions in (4.34) satisfies the boundary conditions imposed by eqn (4.33). Hence, there exist infinitely many eigenvalues $\lambda_l$ for the problem posed by eqns (4.32) and (4.33). They are given by eqn (4.35) and each of them, in turn, is associated with an infinite set of eigenfunctions defined by eqn (4.34).

Thus, equations (4.34) and (4.35) present the exact solution to the problem. In what follows, while computing the approximate solution to this problem, we will take advantage of the fact that its exact solution is available. This makes it convenient to test computational algorithms to be developed and to estimate the accuracy level attained.

For developing a numerical procedure based on the influence function method, let us reduce the eigenvalue problem posed by eqns (4.32) and (4.33) to the corresponding integral equation. This reduction can be accomplished upon implementing the approach introduced earlier in this section. In doing so, let $g(x, s)$ represent the Green's function for a boundary-value problem posed with the following homogeneous equation

$$\frac{d^4w(x)}{dx^4} = 0, \quad x \in [0, a] \tag{4.36}$$

subject to the boundary conditions of eqn (4.33). That is, $g(x, s)$ represents the influence function of a unit transverse force concentrated at a point $s$ for a simply supported beam of length $a$, with a uniform flexural rigidity. This influence function is found in Chapter 3 as

$$g(x, s) = \frac{1}{6a} \begin{cases} x(a-s)(s^2+x^2-2as), & x \le s \\ s(a-x)(x^2+s^2-2ax), & s \le x \end{cases}$$

Based on this compact representation of the influence function, the integral equation that brings an alternative formulation to the eigenvalue problem posed by eqns (4.36) and (4.33), in the sense introduced earlier, is written as

$$w(x) = -\lambda^4 \int_0^a g(x, s) w(s)\, ds \tag{4.37}$$

In Table 4.2 we exhibit the first six components of the eigenvalue spectrum for the integral equation in (4.37) with $a = 1$. The trapezoid rule has been used with a uniform partition of the interval $[0, a]$ into $n = 10$ subintervals where mesh-points are defined as

$$x_k = \frac{ak}{n}, \qquad (k = \overline{0, n})$$

It is clearly seen from the exhibited data that although a very coarse partition ($n = 10$) has been used, the IFM solution appears to be, nevertheless, fairly accurate. Indeed, the accuracy attained for the lowest eigenvalue $\lambda_1$ (related to the so-called *fundamental natural frequency*) is at the level of 99.999%. The accuracy gradually drops down for the upper members of the spectrum, though remains at a relatively high level exceeding 99% for $\lambda_6$. The eigenfunctions were computed, in fact, with the same high accuracy level as the eigenvalues.

**Table 4.2:** Eigenvalues of the problem in eqns (4.32) and (4.33), computed by the influence function method (IFM) and finite difference method (FDM)

| Method used | Eigenvalue, $\lambda_m$ | | | | | |
|---|---|---|---|---|---|---|
| | $m=1$ | $m=2$ | $m=3$ | $m=4$ | $m=5$ | $m=6$ |
| IFM | 2.14158 | 6.28283 | 9.42164 | 12.5508 | 15.6508 | 18.6738 |
| FDM | 4.63605 | 7.52141 | 10.2228 | 12.6720 | 14.8237 | 16.6413 |
| exact | 2.14159 | 6.28319 | 9.42478 | 12.5664 | 15.7080 | 18.8496 |

The accuracy level potentially attainable by our version of the influence function method (IFM) has been controlled within a computational experiment, during which we compare its actual outcome against the results computed by the finite difference method (FDM).

When running the actual computation for the eigenvalue problem in eqns (4.36) and (4.33) by the FDM, we utilized the extended version

$$x_k = \frac{ak}{n}, \quad (k = \overline{-1, n+1})$$

of the uniform partition of the interval $[0, a]$. The two extra mesh-points $x_{-1}$ and $x_{n+1}$ are added to the partition. This is done for the sake of methodological convenience. Such a partition helps to obtain a consistent system of linear algebraic equations. In the development that follows, the partition step $a/n$ is denoted with $h$, and the approximate value of the deflection function $w(x)$ at $x = x_k$ is denoted with $w_k$.

One of the simplest finite difference schemes of the order of accuracy $O(h^2)$, which has been described earlier in Chapter 3 (see Section 3.2), reduces the boundary-value problem in eqns (4.36) and (4.33) to the well-posed eigenvalue problem of linear algebra. In doing so, eqn (4.36) is approximated at the interior $(k = \overline{1, n-1})$ mesh-points with the system

$$w_{k+2} - 4w_{k+1} + 6w_k - 4w_{k-1} + w_{k-2} - (\lambda h)^4 w_k = 0, \quad k = \overline{1, n-1}$$

It is clearly seen that the above system of linear algebraic equations is not well-posed. Indeed, the number of unknowns in it is four units greater than the number of equations. This inconsistency, however, can easily be rectified. Four additional equations are derived upon approximating the boundary conditions in eqn (4.33). This yields

$$w_0 = w_n = 0, \quad w_{-1} - 2w_0 + w_1 = 0, \quad w_{n-1} - 2w_n + w_{n+1} = 0$$

As soon as these relations are incorporated into the main system, we obtain a standard well-posed eigenvalue problem of linear algebra, which approximates the problem in eqns (4.32) and (4.33).

In Chapter 3, we explained that this primitive finite difference scheme has been chosen on purpose, as it is equivalent to the trapezoid rule of approximate integration in terms of the order of accuracy $O(h^2)$ provided. Hence, from the error estimation viewpoint, it follows that computed output of both the finite difference (FDM) and influence function (IFM) methods, used in this experiment, ought to be equivalently accurate. However, the data in Table 4.2 show a different result. The low accuracy level of the FDM solution is not acceptable at all with $n = 10$. Hence, the partition number required for more accurate results should be essentially increased for this method. Whereas the IFM, as we observe from Table 4.2, provides much higher accuracy. ▲

As the reader may recall from Chapter 3, we have discussed the comparison of the accuracy level practically attained by both the FDM and IFM procedures and explained why the latter is usually more accurate.

### 4.2.2 Various vibration problems

A number of illustrative examples is discussed in this section covering various peculiarities in vibration problems for elastic beams. We present some numerical results obtained by the influence function method in solving eigenvalue problems for single-span beams. Some of the problem settings include complicating factors (a variable flexural rigidity, axial forces and so on).

**Example 2.2:** Consider a beam of length $a$, having a uniform flexural rigidity $EI$. If, for example, the left edge $x = 0$ of the beam is simply supported while the right edge $x = a$ is clamped, then the problem reduces to the following standard eigenvalue formulation

$$\frac{d^4 w(x)}{dx^4} - \lambda^4 w(x) = 0, \quad x \in [0, a]$$

$$w(0) = \frac{d^2 w(0)}{dx^2} = 0, \quad w(a) = \frac{dw(a)}{dx} = 0 \tag{4.38}$$

To simplify notations in what follows, the above setting will be referred to as the S–C problem. Contrary to the previous case of a simply supported beam, the exact solution of the S–C problem is not available, but its approximate solution is well tabulated (see, for example, [13]) and will be used herein for testing purposes.

Tracing out the IFM procedure, we reduce the S–C problem again to the integral equation in (4.37), where the influence function $g(x, s)$ of a transverse unit concentrated force for the simply supported-clamped beam has earlier been obtained in Chapter 3 as

$$g(x, s) = \frac{1}{12a^3} \begin{cases} x(a-s)^2[s(a^2 - x^2) - 2a(x^2 - as)], & x \le s \\ s(a-x)^2[x(a^2 - s^2) - 2a(s^2 - ax)], & s \le x \end{cases}$$

The results of our experiment are shown in Table 4.3, where we exhibit approximate eigenvalues $\lambda_m$ to the integral equation in (4.37) for beams of a unit length, with four different types of the edge conditions imposed. The trapezoid rule is utilized with the partition parameter $n = 10$. The upper block in the table presents, in particular, the values of $\lambda_1$ through $\lambda_5$ for the simply supported–clamped beam (S–C problem).

In addition to the S–C problem, one can also find in this table data for: (i) C–Sd problem (one edge is clamped while the other is subject to the sliding conditions), (ii) S–Sd problem (one edge is simply supported while the other is sliding), and (iii) C–C problem (beam clamped at both edges). Each of these problems reduces to the integral equation (4.37) with a corresponding influence functions involved. The integral equation in (4.37) reduces, in turn, to an eigenvalue problem of linear algebra (by the trapezoid rule with $n = 10$) and was solved then numerically in the standard way.

Accuracy of the computed eigenvalues varies slightly from case to case, but remains at the relatively high level agreeing with the conclusions of the validation

**Table 4.3:** Approximate eigenvalues $\lambda_m$ of eqn (4.37), computed by the IFM

| Problem solved | Method used | Eigenvalue, $\lambda_m$ | | | | |
|---|---|---|---|---|---|---|
| | | $m=1$ | $m=2$ | $m=3$ | $m=4$ | $m=5$ |
| S–C | IFM | 2.92652 | 7.06841 | 10.2065 | 12.3182 | 16.4021 |
| | exact | 2.92660 | 7.06858 | 10.2102 | 12.3518 | 16.4934 |
| C–Sd | IFM | 2.35708 | 5.53224 | 8.82739 | 12.1274 | 15.3958 |
| | exact | 2.36502 | 5.49781 | 8.63937 | 11.7810 | 14.9226 |
| S–Sd | IFM | 1.56642 | 4.77013 | 8.06684 | 11.3506 | 14.6214 |
| | exact | 1.57080 | 4.71239 | 7.85398 | 10.9956 | 14.1372 |
| C–C | IFM | 4.72990 | 7.85172 | 10.9862 | 14.0972 | 17.1455 |
| | exact | 4.73004 | 7.85319 | 10.9956 | 14.1372 | 17.2788 |

problem discussed in Example 2.1. The lowest eigenvalues, for example, for all of the problems have been computed with the accuracy level that is well above 99.5%. It is worth noting again that one of the most primitive quadrature formulas has been used. This reveals the high computational potential of the IFM in the eigenvalue analysis.

All the influence functions, which have been utilized as the kernel of the integral equation in (4.37) in computing the data exhibited in Table 4.3, are available in Chapter 3. ▲

At this point in our discussion, we turn to the problem in eqns (4.27) and (4.28) that simulates natural vibrations for a beam of variable flexural rigidity. As the reader had learned earlier, the influence function formalism reduces that problem to the homogeneous integral equation shown in (4.31). Section 4.1 describes in detail a procedure for the numerical construction of a required influence function of a transverse point force for the beam. The procedure has been utilized in the next example.

**Example 2.3:** Compute natural frequencies of transverse vibrations for a simply supported beam of length $a$, with a rectangular cross-section $b \times h(x)$, whose height is a linear function of $x$. The material of which the beam is made is supposed to be isotropic and homogeneous. That is, $E = const$ and $m = const$ (see eqn (4.27)). The configuration of the beam in reference to the operative coordinate system is shown in Figure 4.3.

It is evident that the cross-sectional properties of the beam depicted in Figure 4.3 vary with $x$. As the reader is supposed to learn from the standard courses of either structural mechanics or a relevant discipline [21, 27, 63], the flexural rigidity of a rectangular cross-section represents a cubic function of the $x$ variable. This reads

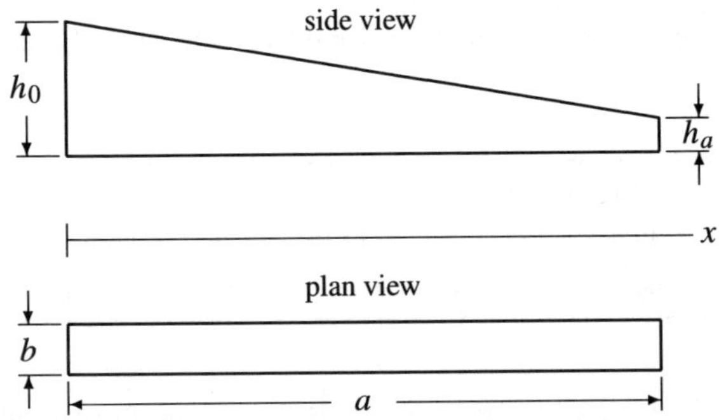


Figure 4.3: Configuration of a beam with a variable cross-section

in our case as

$$EI(x) = \frac{Eb}{12}\left(h_0 - \frac{h_0 - h_a}{a}x\right)^3$$

while the cross-sectional area is the following linear function of $x$

$$A(x) = b\left(h_0 - \frac{h_0 - h_a}{a}x\right)$$

While computing the results for this example, we assumed, for the sake of simplicity, that the beam has a unit length ($a = 1$) with unit flexural rigidity at the left edge, that is $EI(0) = 1$. The influence function of a transverse unit concentrated force for this beam has been obtained numerically by computing all the components of a fundamental set of solutions for the governing equation in (4.32) with the aid of a standard routine based on the Runge–Kutta method of the fourth order (see Chapter 1).

In Tables 4.4 and 4.5 the reader can find some results obtained for this beam. Two types of edge conditions have been considered. The exhibited data have been computed by implementing the standard trapezoid rule for eqn (4.31), with a limited number of uniform partitions ($n = 10$).

**Table 4.4:** Approximate eigenvalues computed for the simply-supported beam

| $h_0/h_a$ | 1 | 2 | 3 | 4 | 5 | 6 |
|---|---|---|---|---|---|---|
| $p_1$ | 9.8687 | 7.0226 | 5.6711 | 5.3015 | 4.9147 | 4.7118 |
| $p_2$ | 39.4805 | 27.9486 | 22.7583 | 21.6953 | 20.7092 | 20.5619 |
| $p_3$ | 88.8269 | 64.1278 | 54.8396 | 49.9238 | 46.9872 | 45.3128 |

**Table 4.5:** Approximate eigenvalues for the clamped–simply supported beam

| $h_0/h_a$ | 1 | 2 | 3 | 4 | 5 | 6 |
|---|---|---|---|---|---|---|
| $p_1$ | 15.4216 | 12.7739 | 10.8024 | 9.9846 | 9.9729 | 9.9678 |
| $p_2$ | 49.9682 | 37.4928 | 31.4519 | 28.3651 | 27.2130 | 27.1544 |
| $p_3$ | 104.2542 | 77.5147 | 65.9788 | 59.8972 | 55.3747 | 54.4665 |

Evidently, the case of $h_0/h_a = 1$ is related to a beam of a uniform flexural rigidity $EI \equiv 1$. The exact eigenvalues for a simply supported beam in this case are obtained as $p_k = (k\pi)^2$. Thus, the relative accuracy attained for the data of the first column in Table 4.4 ($p_1 = 9.8687$, $p_2 = 39.4805$, and $p_3 = 88.8269$) is greater than 99.9%. The rest of the results in this table also compare fairly well with data available in literature (see, for example, [13]).

Numerical results obtained by the IFM for another eigenvalue problem are presented in Table 4.5. We again consider the beam whose configuration is depicted in Figure 4.3. In this case, however, different boundary conditions are imposed. That is, a clamped–simply supported beam is considered. These results are also in a good agreement with corresponding data available in the relevant sources (see, for example, [13]). The same computational procedure as before has been utilized. ▲

Other types of natural vibration problems for a single-span beam could also be reduced to integral equations and tackled then by means of the influence function method. The example that follows is designed to illustrate this assertion. We consider a beam subject to axial forces as shown in Figure 4.4.

Note that tensile forces are shown in Figure 4.4, although compressive forces could also be considered, in which case the issue of buckling phenomenon comes probably to the reader's mind. This issue, however, is not going to be a subject in the present consideration. Some buckling problems are perfectly suitable for the influence function treatment. And we will focus on buckling problems in our next section.

While considering compressive forces, we will assume that their magnitude is well below the value of the so-called *Euler elastic buckling force*. Consequently, the question to discuss here is how axial forces acting on a beam, affect the spectrum of its natural frequencies. This classical statement is well known in mechanics and considered in standard texts. What we plan to do in this study is to answer this question within the scope of the influence function method.

**Example 2.4:** Consider a beam of uniform flexural rigidity that is subject to either tensile or compressive axial forces. To be specific, the tensile forces are shown in Figure 4.4 where a simply supported beam is depicted. However, this case is picked up conditionally and, in what follows, we also consider beams undergoing compressive forces and other types of the edge conditions imposed.

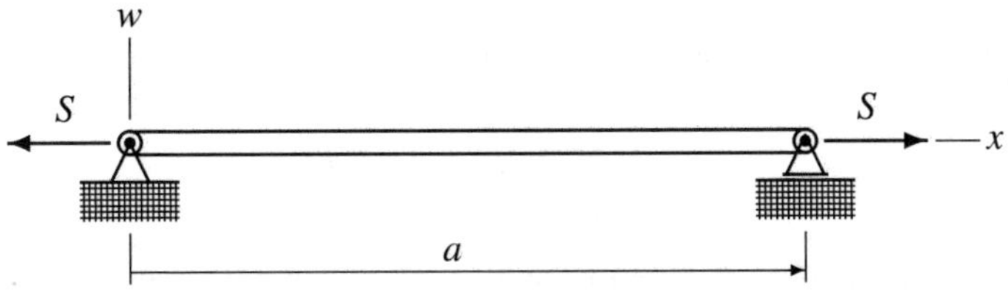


Figure 4.4: A beam subject to axial forces

As it can be learned from mechanics of structures [21, 27, 63], the following eigenvalue formulation

$$\frac{d^4w(x)}{dx^4} - \frac{S}{EI}\frac{d^2w(x)}{dx^2} = p^2\frac{mA}{EI}w(x), \qquad x \in (0, a) \tag{4.39}$$

$$w(0) = \frac{d^2w(0)}{dx^2} = 0, \quad w(a) = \frac{d^2w(a)}{dx^2} = 0 \tag{4.40}$$

models the natural vibrations of the simply supported beam, with $S$ being the magnitude of tensile axial forces. If compressive forces are applied instead, then the sign of the second term in eqn (4.39) has to be changed to a plus.

The exact solution (eigenvalues and corresponding eigenfunctions) of this classical problem setting can be easily obtained by inspection. Indeed, each of the following functions

$$w_l(x) = C\sin\frac{l\pi x}{a}, \qquad l = 1, 2, 3, \ldots \tag{4.41}$$

where $C$ is an arbitrary constant, satisfies all of the boundary conditions in eqn (4.40). Upon substituting these functions into eqn (4.39), one obtains the following algebraic equation

$$\left(\frac{l\pi}{a}\right)^4 + \frac{S}{EI}\left(\frac{l\pi}{a}\right)^2 = p_l^2\frac{mA}{EI}$$

in $p_l$. Solving this equation for $p_l$, one obtains

$$p_l = \left(\frac{l\pi}{a}\right)^2\sqrt{R\left(1 + \left(\frac{qa}{l\pi}\right)^2\right)} \tag{4.42}$$

where $R$ and $q$ are introduced as $R = EI/mA$ and $q = \sqrt{S/EI}$.

Thus, eqn (4.41) presents the eigenfunctions (natural modes of the beam vibrations), while the eigenvalues $p_l$ (angular frequencies of the vibrations) for a tensile force $S$ are computed by eqn (4.42).

For compressive forces $S$, the sign of the second term in eqn (4.39) ought to be changed to a plus. Thus, the sign in the radicand in eqn (4.42) has to be consequently changed to a minus. That is

$$p_l = \left(\frac{l\pi}{a}\right)^2 \sqrt{R\left(1 - \left(\frac{qa}{l\pi}\right)^2\right)} \tag{4.43}$$

At this point in our presentation, an important comment ought to be made. It is concerned with the magnitude of the compressive force acting on the beam. From equation (4.43), it clearly follows that

$$1 - \left(\frac{qa}{l\pi}\right)^2 = 1 - \frac{S}{EI}\left(\frac{a}{l\pi}\right)^2 > 0$$

This relation implies that the magnitude of the compressive forces is bounded from above. That is,

$$S < EI\left(\frac{l\pi}{a}\right)^2$$

Physical interpretation of this inequality is obvious. It means that if the value $S$ of the compressive forces exceeds the magnitude of the *Euler elastic buckling force* (the right-hand side of the above inequality actually represents this force for the beam under consideration), then the statement of the corresponding eigenvalue problem becomes physically meaningless.

Before proceeding any further with the development of the influence function method for the eigenvalue problem posed by eqns (4.39) and (4.40), we present the Green's function

$$g(x, s) = \frac{1}{aq^3 \sinh qa} \begin{cases} qx(s-a)\sinh qa - a\sinh qx \sinh q(s-a), & x \le s \\ qs(x-a)\sinh qa - a\sinh qs \sinh q(x-a), & x \ge s \end{cases} \tag{4.44}$$

for the following boundary-value problem

$$\frac{d^4w(x)}{dx^4} - q^2\frac{d^2w(x)}{dx^2} = 0, \quad x \in (0, a)$$

$$w(0) = \frac{d^2w(0)}{dx^2} = 0, \quad w(a) = \frac{d^2w(a)}{dx^2} = 0$$

The Green's function in eqn (4.44) represents the influence function of a transverse unit point force for a simply supported beam undergoing tensile axial forces $S$.

In compliance with Theorem 2.4 in Chapter 2, the problem in eqns (4.39) and (4.40) reduces to the following homogeneous integral equation

$$w(x) = -\lambda^4 \int_0^a g(x, s)w(s)\, ds \tag{4.45}$$

in $w(x)$, where

$$\lambda^4 = \frac{p^2 mA}{EI} = \frac{p^2}{R}$$

This notation enables us to match results of the current development with those of Example 2.1 considered earlier in this section.

For compressive forces $S$, the influence function method again yields the integral equation in (4.45). However, its kernel $g(x, s)$ represents, in this case, the Green's function

$$g(x, s) = \frac{1}{aq^3 \sin qa} \begin{cases} qx(a-s)\sin qa + a \sin qx \sin q(s-a), & x \leq s \\ qs(a-x)\sin qa + a \sin qs \sin q(x-a), & s \leq x \end{cases} \tag{4.46}$$

of the boundary-value problem written as

$$\frac{d^4 w(x)}{dx^4} + q^2 \frac{d^2 w(x)}{dx^2} = 0, \quad x \in (0, a)$$

$$w(0) = \frac{d^2 w(0)}{dx^2} = 0, \quad w(a) = \frac{d^2 w(a)}{dx^2} = 0$$

The Green's function presented in eqn (4.46) is referred to as the influence function of a transverse unit concentrated force for a simply supported beam undergoing compressive axial forces $S$.

To construct the influence functions presented in eqns (4.44) and (4.46), fundamental sets of solutions for different governing equations have been used. Namely, the first of the influence functions was routinely derived with the fundamental set of solutions written as

$$1, \quad x, \quad \sinh qx, \quad \cosh qx$$

whereas for the influence function shown in eqn (4.46), the fundamental set of solutions was obtained as

$$1, \quad x, \quad \sin qx, \quad \cos qx$$

The integral equation in (4.45) has been, in both cases, solved numerically. We used the quadrature formulae method. The trapezoid rule with a uniform partition of the interval $[0, a]$ has been implemented. Note again that we are purposely using, in this text, such a primitive numerical routine (where, in addition, a limited number of mesh-points is used). And the purpose is to bring to the reader's attention the high potential of the influence function approach. Indeed, if one of the most primitive numerical schemes is that effective, then the approach has many unused resources.

The eigenvalue setting for the integral equation in (4.45) (to which the original boundary-value problem reduces) has been approximately replaced with a corresponding eigenvalue problem of linear algebra. The trapezoid rule-based quadrature formulae method was used, with $n = 10$. The data in Tables 4.6 through 4.9 are obtained for a beam of a unit length with the parameter $R = 1$, with two different types of the edge conditions imposed.

Note that the accuracy level of the approximate eigenvalues presented in Tables 4.6 and 4.7 is well above 99.5%. This brings another confirmation of a high potential that the IFM attains in the numerical eigenvalue analysis. ▲

**Table 4.6:** Approximate eigenvalues for the S–S beam (tensile forces)

| Method used | Parameter $q^2$ | Eigenvalue, $\lambda_l$ $l=1$ | $l=2$ | $l=3$ | $l=4$ | $l=5$ |
|---|---|---|---|---|---|---|
| IFM | 1 | 2.2183 | 6.3224 | 9.4458 | 12.5535 | 15.6659 |
| | 10 | 2.7421 | 6.6476 | 9.6759 | 12.7440 | 15.8050 |
| | 100 | 5.7382 | 8.6125 | 11.3723 | 14.1783 | 17.0188 |
| exact | 1 | 2.2183 | 6.3226 | 9.4512 | 12.5862 | 15.7239 |
| | 10 | 2.7422 | 6.6480 | 9.6795 | 12.7608 | 15.8648 |
| | 100 | 5.7384 | 8.6142 | 11.3802 | 14.2060 | 17.1026 |

**Table 4.7:** Approximate eigenvalues for the S–S beam (compressive forces)

| Method used | Parameter $q^2$ | Eigenvalue, $\lambda_l$ $l=1$ | $l=2$ | $l=3$ | $l=4$ | $l=5$ |
|---|---|---|---|---|---|---|
| IFM | 1 | 2.0588 | 6.2427 | 9.3950 | 12.5310 | 15.6352 |
| | 3 | 2.8695 | 6.1599 | 9.3411 | 12.4911 | 15.6037 |
| | 9 | 1.7116 | 5.8896 | 9.1742 | 12.3769 | 15.5076 |
| exact | 1 | 2.0588 | 6.2430 | 9.3981 | 12.5465 | 15.6920 |
| | 3 | 2.8695 | 6.1603 | 9.3442 | 12.5063 | 15.6600 |
| | 9 | 1.7116 | 5.8896 | 9.1764 | 12.3834 | 15.5627 |

**Example 2.5:** In addition to the eigenvalue problems already considered, we have conducted another computational experiment based on the influence function method. A simply supported–sliding beam was examined subject to either tensile axial forces (as shown in Figure 4.5) or compressive forces. Exact eigenvalues in these cases are also available in the existing literature [13].

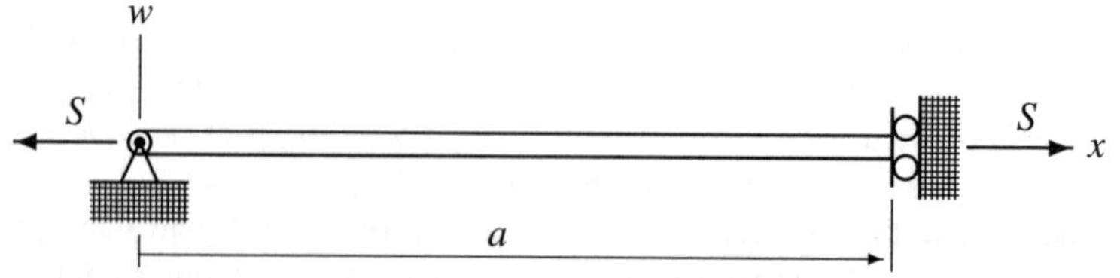


Figure 4.5: A simply supported–sliding beam

The boundary eigenvalue problems, which are associated with the natural vibrations of a simply supported–sliding beam subject to axial forces, are also converted to the homogeneous integral equation in (4.45). The kernel function $g(x, s)$ in that equation represents the corresponding influence function of a transverse unit point force. It depends on the type of axial forces applied. The required influence functions are shown below. For the tensile forces (the case depicted in Figure 4.5), we have

$$g(x, s) = -\frac{1}{q^3 \sinh qa} \begin{cases} qx \sinh qa + \sinh qx \cosh q(s-a), & x \le s \\ qs \sinh qa + \sinh qs \cosh q(x-a), & s \le x \end{cases} \tag{4.47}$$

while in the case of compressive forces, we obtain

$$g(x, s) = \frac{1}{q^3 \sin qa} \begin{cases} qx \cos qa - \sin qx \cos q(s-a), & x \le s \\ qs \cos qa - \sin qs \cos q(x-a), & s \le x \end{cases} \tag{4.48}$$

When solving for eigenvalues, the homogeneous integral equation in (4.45) has been reduced to an eigenvalue problem of linear algebra with the aid of the quadrature formulae method (trapezoid rule, where the actual computations have been conducted with $n = 10$). Some results of this computation for the beam subject to tensile axial forces are exhibited in Table 4.8, while Table 4.9 contains results obtained for the beam that is subject to compressive axial forces.

**Table 4.8:** Approximate eigenvalues for the S–Sd beam (tensile forces)

| Method used | Parameter $q^2$ | Eigenvalue, $\lambda_l$ | | | | |
|---|---|---|---|---|---|---|
| | | $l=1$ | $l=2$ | $l=3$ | $l=4$ | $l=5$ |
| IFM | 1 | 1.7066 | 4.8116 | 8.0882 | 11.3580 | 14.6361 |
| | 10 | 2.3486 | 5.1849 | 8.3012 | 11.3281 | 14.7684 |
| | 100 | 2.9767 | 7.2213 | 10.0653 | 12.9815 | 15.9876 |
| exact | 1 | 1.7103 | 4.7646 | 7.8856 | 11.0182 | 14.1374 |
| | 10 | 2.3551 | 5.1714 | 8.1546 | 10.9994 | 14.3119 |
| | 100 | 2.9876 | 7.2176 | 9.9934 | 12.7838 | 15.6463 |

Note that the accuracy of approximate eigenvalues exhibited in Tables 4.8 and 4.9 drops slightly down compared to those shown in Tables 4.6 and 4.7. Nevertheless, it still remains at a level that exceeds 99.5% for the lowest eigenvalues $\lambda_1$ and 97% for the highest eigenvalues exposed. ▲

The direct influence function method-based computational procedure can also be applied to other problems of natural vibrations for beams with more complicated statements. This assertion is supported below by the case of a beam of

**Table 4.9:** Approximate eigenvalues for the S–Sd beam (compressive forces)

| Method used | Parameter $q^2$ | Eigenvalue, $\lambda_l$ $l=1$ | $l=2$ | $l=3$ | $l=4$ | $l=5$ |
|---|---|---|---|---|---|---|
| IFM | 1 | 1.3756 | 4.7334 | 8.0467 | 11.3334 | 14.6068 |
| | 2 | 1.0334 | 4.7089 | 8.0201 | 11.3067 | 14.5023 |
| | 2.3 | 0.7995 | 4.6259 | 8.0128 | 11.3119 | 14.5126 |
| exact | 1 | 1.3794 | 4.6584 | 7.8220 | 10.9500 | 14.1018 |
| | 2 | 1.0363 | 4.6025 | 7.7895 | 10.9043 | 14.0663 |
| | 2.3 | 0.8017 | 4.5853 | 7.7797 | 10.9956 | 14.0963 |

variable flexural rigidity $EI(x)$, which is subject to a continuously distributed axial force $S(x)$. A governing boundary eigenvalue problem can, in this case, be written in the form

$$\frac{d^2}{dx^2}\left(EI(x)\frac{d^2w(x)}{dx^2}\right) - \frac{d}{dx}\left(S(x)\frac{dw(x)}{dx}\right) - p^2m(x)A(x)w(x) = 0 \quad (4.49)$$

$$B_{0,k}[w(0)] = 0, \quad B_{a,k}[w(a)] = 0, \quad k = 1, 2 \quad (4.50)$$

where the boundary conditions in eqn (4.50) are not specified (any physically feasible setting is well-posed from a mathematics standpoint).

Assuming that $g(x, s)$ is the Green's function for the homogeneous boundary-value problem of eqn (4.50) stated for the differential equation

$$\frac{d^2}{dx^2}\left(EI(x)\frac{d^2w(x)}{dx^2}\right) - \frac{d}{dx}\left(S(x)\frac{dw(x)}{dx}\right) = 0 \quad (4.51)$$

and following Theorem 2.4 of Chapter 2, we customarily obtain the following homogeneous Fredholm integral equation of the second kind

$$w(x) = -p^2 \int_0^a g(x, s)m(s)A(s)w(s)\, ds$$

for the solution $w(x)$ of the problem in eqns (4.49) and (4.50). This equation contains a parameter ($p^2$) and poses therefore an eigenvalue problem whose solution (the eigenvalues and eigenfunctions) can be computed with the aid of standard numerical procedures. Note that the quadrature formulae method, which is being repeatedly used in this study, is not the only possible option for such problems. It represents just one of the available options [60].

The Green's function $g(x, s)$ of the boundary-value problem in eqns (4.50) and (4.51) represents, in fact, the influence function of a transverse unit concentrated force for a beam under consideration. One can readily construct $g(x, s)$

numerically by using the procedure introduced earlier in Section 4.1 for equations with variable coefficients.

From the above, it follows that the influence function method is readily adaptive to eigenvalue problems posed by differential equations with variable coefficients. Such an adaptability ought to make this method attractive to users of numerical methods in engineering. This assertion is well grounded due to a potentially high accuracy level attainable within our approach.

In the next section, we turn the reader's attention to another class of problems for which the influence function method appears to be fairly effective. An integral equation version of the classical (Euler formulation) buckling problems in the beam theory will be discussed in some detail.

## 4.3 Euler Buckling Problems

The discussion in the preceding two sections of this chapter touched upon some beam problems for which the influence function method is found to be more efficient compared to other more traditional approaches (like the finite difference method, for instance). In this section, we will further extend the range of possible applications of the IFM. Another standard class of problems from Kirchhoff beam theory is explored, a one for which influence functions of a unit transverse point force could also be productive.

Consider, as an example, a clamped–simply supported elastic beam of a variable flexural rigidity $EI(x)$. The beam is subject to the axial compressive force $N$ as shown in Figure 4.6. Other physically feasible types of the edge conditions can also be assumed within the scope of our approach. The depicted particular case is chosen for the sake of certainty.

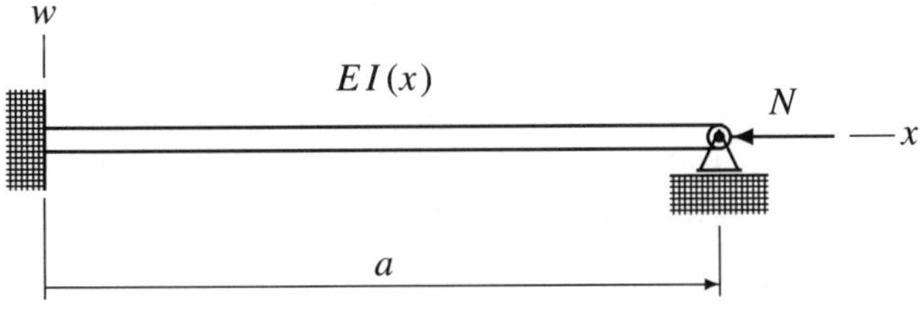


Figure 4.6: A clamped–simply supported beam

Static equilibrium of the depicted beam can be modeled by the following boundary-value problem

$$\frac{d^2}{dx^2}\left(EI(x)\frac{d^2w(x)}{dx^2}\right)+N\frac{d^2w(x)}{dx^2}=0, \quad x\in[0,a] \tag{4.52}$$

$$w(0)=\frac{dw(0)}{dx}=0, \quad w(a)=\frac{d^2w(a)}{dx^2}=0 \tag{4.53}$$

If different supports are in place at the end-points, then different boundary conditions ought to be imposed at $x = 0$ and $x = a$ replacing accordingly those in eqn (4.53). As the reader has learned from our earlier discussion in this text, boundary conditions in a statement are taken care of by the influence function of a transverse point force, which can routinely be constructed for any physically feasible edge conditions.

In order for the statement in eqns (4.50) and (4.52) to represent a true mathematical model of the phenomenon under consideration, a certain limitation ought to be put on the value of the parameter $N$ in eqn (4.52). Indeed, it is essential that the magnitude $N$ of the axial force does not exceed a certain critical value. Let us denote it with $N_{cr}$. The latter is usually referred to, in mechanics, as the *critical value* of the compressive force or the *Euler elastic buckling force*, which yields a sudden jump of the neutral stable equilibrium state $w(x) \equiv 0$ to a bent unstable one.

We denote the bent unstable equilibrium state caused by the critical value of the compressive force with $w_{cr}(x)$. It is usually referred to as the *buckling failure shape*, which is defined up to a constant multiple.

The search for the value of $N_{cr}$ is, subsequently, crucial in reality. In order to find $N_{cr}$, one reduces the problem in eqns (4.50) and (4.52) to a corresponding eigenvalue problem of linear algebra. This is usually accomplished by a partition of the interval $[0, a]$. The lowest component of the eigenvalue spectrum of that problem is directly related to $N_{cr}$. The entries of the eigenvector corresponding to the lowest eigenvalue represents a set of values of the buckling failure shape of the beam at the partition points.

The eigenvalue problem of linear algebra, which emerges as an approximate analogue of an original differential eigenvalue setting, is usually derived, in this study, through the conversion of the original problem to an integral equation. The kernel of the latter equation represents the corresponding influence function of a transverse unit concentrated force. We will follow this strategy for the setting in eqns (4.50) and (4.52).

Before going any further with the setting in eqns (4.50) and (4.52), we consider a sample case which allows an exact solution. That is a simply supported beam of a uniform flexural rigidity ($EI(x) \equiv EI_0 = const$). In this case eqn (4.52) simplifies to the equation with constant coefficients

$$\frac{d^4w(x)}{dx^4} + \frac{N}{EI_0}\frac{d^2w(x)}{dx^2} = 0, \quad x \in [0, a] \tag{4.54}$$

and the boundary conditions read as

$$w(0) = \frac{d^2w(0)}{dx^2} = 0, \quad w(a) = \frac{d^2w(a)}{dx^2} = 0 \tag{4.55}$$

From observation, it follows that the set of functions

$$w_{cr}(x) = C \sin\frac{\pi x}{a}$$

where $C$ is an arbitrary constant, represent the buckling failure shape and

$$N_{cr} = \frac{\pi^2}{a^2} EI_0$$

is the Euler elastic buckling force for the beam.

Hence, in the case of a simply supported beam of a uniform flexural rigidity, we face the eigenvalue problem in eqns (4.54) and (4.55) which is analytically solvable. This setting will be used later as a validation example for a numerical algorithm that is developed for the formulation in eqns (4.50) and (4.52).

If the beam's flexural rigidity is, however, variable and/or more complex edge conditions are implied, then an analytic solution of the eigenvalue problem in (4.50) and (4.52) is, generally speaking, impossible and a numerical approach is the only option in obtaining an approximate value of $N_{cr}$ and the corresponding buckling failure shape.

There exist a number of approximate methods that are developed for the solution of eigenvalue problems of the type in eqns (4.50) and (4.52). It is worth noting, however, that the influence function method has not been listed yet as an option in this area. Filling out this unfortunate gap, we are going to show how various versions of this method can be naturally implemented for buckling problems. A computational experiment will be conducted in this section to reveal the potential of this method.

### 4.3.1 Influence function algorithm

Two different influence function-based algorithms are offered herein for solving the eigenvalue problem posed by eqns (4.50) and (4.52). These algorithms are equivalently accurate, although they are based on slightly different ideas and utilize different influence functions in solving the same problem.

Presenting the first of these algorithms, let $g(x, s)$ represent the Green's function for a boundary-value problem posed by the equation

$$\frac{d^2}{dx^2}\left(EI(x)\frac{d^2w(x)}{dx^2}\right) = 0, \quad x \in [0, a]$$

subject to a corresponding set of boundary conditions. In other words, by this we again assume that $g(x, s)$ represents the influence function of a unit transverse concentrated force for the beam which is not subject to a compressive force $N$.

Using the influence function $g(x, s)$ just mentioned and treating formally the second term of eqn (4.52) as its right-hand side, one expresses the solution of the problem in eqns (4.50) and (4.52), in compliance with Theorem 2.4 in Chapter 2, by the following integral

$$w(x) = \int_0^a g(x, s)\left[N\frac{d^2w(s)}{ds^2}\right] ds$$

The constant parameter $N$ can be taken out of the integral sign, leading us to the integral representation

$$w(x) = N \int_0^a g(x, s)\frac{d^2w(s)}{ds^2}\, ds \tag{4.56}$$

of the deflection function $w(x)$ in terms of itself. This relation represents the so-called *integro-differential equation* in $w(x)$. Due to the presence of the second order derivative of $w(s)$ in the integrand, it cannot be reduced to an eigenvalue problem of linear algebra by directly applying some quadrature formulae. This deficiency can, however, be fixed upon reducing eqn (4.56) to a regular homogeneous integral equation of the second kind. This implies that the computational procedure described earlier can finally be utilized.

To reduce the relation in (4.56) to a regular integral equation, to its integral term we apply integration by parts twice successively. For the first time, the integrand in (4.56) is taken apart in the following manner

$$g(x, s) = u(s), \qquad \frac{d^2w(s)}{ds^2}\, ds = dv(s)$$

which consequently yields

$$du(s) = \frac{\partial g(x, s)}{\partial s}\, ds, \quad v(s) = \frac{dw(s)}{ds}$$

reducing the integral in eqn (4.56), in compliance with the integration by parts formula, to

$$\int_0^a g(x, s)\frac{d^2w(s)}{ds^2}\, ds = g(x, s)\frac{dw(s)}{dx}\bigg|_0^a - \int_0^a \frac{\partial g(x, s)}{\partial s}\frac{dw(s)}{ds}\, ds$$

From the defining properties of the Green's function $g(x, s)$, for either clamped–simply supported, or clamped-clamped, or simply supported (S–S) beam, for example, the boundary conditions in eqn (4.50) suggest

$$g(x, 0) = g(x, a) = 0$$

These relations imply that the beam does not deflect if a transverse concentrated force is applied at either of its edge points $s = 0$ or $s = a$. Thus, the first integration by parts yields for the integral term of eqn (4.56)

$$\int_0^a g(x, s)\frac{d^2w(s)}{ds^2}\, ds = -\int_0^a \frac{\partial g(x, s)}{\partial s}\frac{dw(s)}{ds}\, ds$$

Performing the integration by parts for the second time, we partition the integrand of the above right-hand side integral as

$$\frac{\partial g(x, s)}{\partial s} = u(s), \qquad \frac{dw(s)}{ds}\, ds = dv(s)$$

which yields

$$du(s) = \frac{\partial^2 g(x, s)}{\partial s^2}\, ds, \qquad v(s) = w(s)$$

resulting in

$$\int_0^a \frac{\partial g(x, s)}{\partial s} \frac{dw(s)}{ds}\, ds = \frac{\partial g(x, s)}{\partial s} w(s)\bigg|_0^a - \int_0^a \frac{\partial^2 g(x, s)}{\partial s^2} w(s)\, ds$$

The boundary conditions in eqn (4.50) imply that the non-integral term in the above relation vanishes. Thus, the integro-differential equation in (4.56) is written finally as

$$w(x) = N \int_0^a \frac{\partial^2 g(x, s)}{\partial s^2} w(s)\, ds \tag{4.57}$$

which is a homogeneous Fredholm integral equation of the second kind in $w(x)$. This opens a way for a standard treatment and makes it possible, in particular, to reduce eqn (4.57) to a corresponding eigenvalue problem of linear algebra by a direct application of the quadrature formulae method.

The situation is, in this case, even less demanding compared to that for the natural vibrations problem discussed in Section 4.2. Indeed, we are not required to compute the eigenvalue spectrum of the coefficient matrix, because the critical value $N_{cr}$ of the compressive force, which we are looking for, is associated with only the lowest eigenvalue of that matrix. That is why obtaining the critical force should not be problematic at all.

Examples that we consider below equip the reader with some data on the computational potential of the influence function method in solving buckling problems for elastic beams.

**Example 3.1:** This is a validation example dealing with a problem that has already been examined in the opening segment of this section. We consider the buckling problem (Euler formulation) for a simply supported (S–S) beam of a uniform flexural rigidity $EI_0$.

Remember that this setting is analytically solvable and the exact value $N_{cr}$ of the Euler elastic buckling force and the buckling failure shape $w_{cr}(x)$ were earlier presented, for this problem, in this section.

The influence function of a transverse concentrated unit force for the simply supported beam of length $a$

$$g(x, s) = \frac{1}{6a} \begin{cases} x(a-s)(x^2 + s^2 - 2as), & x \le s \\ s(a-x)(s^2 + x^2 - 2ax), & s \le x \end{cases}$$

is, in this case, the kernel of the equation in (4.56). Taking the second order partial derivative of $g(x, s)$ with respect to $s$, one obtains the kernel for (4.57) as

$$\frac{\partial^2 g(x, s)}{\partial s^2} = \frac{1}{a} \begin{cases} x(a-s), & x \le s \\ s(a-x), & s \le x \end{cases}$$

We build the computational procedure for (4.57) by uniformly breaking the interval $[0, a]$ with the set of grid-points $s_j$, $(j = \overline{0, n})$ as shown in Figure 4.7.

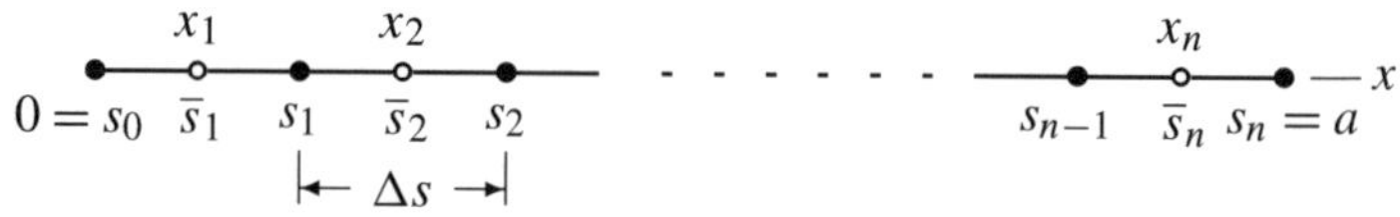


Figure 4.7: Setup for the partition of the interval $[0, a]$

Upon using this partition, we break down the integral in eqn (4.57) into $n$ elementary integrals and rewrite it as

$$w(x) = N \sum_{j=0}^{n-1} \int_{s_j}^{s_{j+1}} \frac{\partial^2 g(x, s)}{\partial s^2} w(s)\, ds$$

Applying then the second mean value theorem of a definite integral [60] to each of the elementary integrals, we rewrite the above relation as

$$\frac{w(x)}{N} = \sum_{j=0}^{n-1} w(\overline{s}_{j+1}) \int_{s_j}^{s_{j+1}} \frac{\partial^2 g(x, s)}{\partial s^2}\, ds$$

where $\overline{s}_{j+1}$ is an arbitrary point in the $j$-th elementary subinterval $[s_j, s_{j+1}]$.

To be certain, we place $\overline{s}_{j+1}$ at the midpoints of $[s_j, s_{j+1}]$. Let then the observation point $x$ go through the same set of points (we denote them with $x_i$, $(i = \overline{1, n})$ and plot both $\overline{s}_j$ and $x_i$ with the empty dots in Figure 4.7). This yields the following homogeneous system of linear algebraic equations

$$\frac{w_i}{N} = \sum_{j=0}^{n-1} w_{j+1} \int_{s_j}^{s_{j+1}} \frac{\partial^2 g(x_i, s)}{\partial s^2}\, ds, \quad (i = \overline{1, n}) \tag{4.58}$$

in the approximate values $w_i$ of the deflection function $w(x)$ defined at the grid-points $x_i$, $(i = \overline{1, n})$.

The elementary integrals that represent entries of the coefficient matrix of the system in (4.58), are approximated as

$$A_{i,j} = \int_{s_j}^{s_{j+1}} \frac{\partial^2 g(x_i, s)}{\partial s^2}\, ds \approx (G_{i,j+1} + G_{i,j}) \frac{\Delta s}{2} \tag{4.59}$$

where, for the notational convenience, we denote

$$G_{i,j} = \frac{\partial^2 g(x_i, s_j)}{\partial s^2}$$

Evidently, for a beam of uniform flexural rigidity, with any standard type of edge conditions imposed, the integrals defining $A_{i,j}$ in eqn (4.59) do not require a numerical treatment, because analytic expressions for corresponding influence functions are available. For a beam of variable rigidity, for which the influence function $g(x, s)$ values are to be numerically obtained, only approximate methods can be applied to compute the coefficient matrix for the system in (4.58). Eqn (4.59) suggests one of the possible ways of doing so.

The homogeneous system of linear algebraic equations in eqn (4.58) can be written in the standard (for eigenvalue problems) form

$$(A - \mu I)W = 0 \tag{4.60}$$

where $A$ represents a square $n \times n$ matrix whose entries $A_{i,j}$ can be obtained, in general, by eqn (4.59), $I$ is an identity matrix of the same $n \times n$ dimension, and $W$ represents a vector whose components are approximate values $w_i$ of the deflection function $w(x)$, which were earlier introduced, while the parameter $\mu$ represents the reciprocal value of $N$.

Thus, following our algorithm, the original homogeneous integral equation in (4.57) is approximately reduced to the eigenvalue problem of linear algebra in eqn (4.60), whose lowest eigenvalue $\mu_1$ represents the reciprocal of the critical force $N_{cr}$ for the beam subject to an axial compressive force. Numerical solution of such a problem is definitely a routine procedure.

A computational experiment has been conducted in order to test the described algorithm (to estimate the accuracy level attainable and the rate of its convergence). Approximate values of the critical force $N_{cr}$ have been computed for the simply supported beam of a unit length and unit flexural rigidity. Three different values of the partition parameter $n = 4$, 6, and 10 have been chosen. The results of this experiment are shown in Table 4.10.

**Table 4.10:** Relative error of the critical force $N_{cr}$ for the S–S beam

| | Partition parameter, $n$ | | |
|---|---|---|---|
| | 4 | 6 | 10 |
| Approx. values of $N_{cr}$ | 9.8696037 | 9.8696043 | 9.8696044 |
| Relative error, % | $0.7 \times 10^{-4}$ | $0.1 \times 10^{-5}$ | $0.1 \times 10^{-8}$ |

The data in Table 4.10 bring two evident observations. That is:

(i) the influence function-based algorithm described in this section provides an exceptionally high accuracy (even for a very limited value of the partition parameter $n = 4$ the relative error is at the level of a fraction of a percent);

(ii) the convergence rate of the algorithm is fairly high (the error drastically drops down with a modest increase of $n$). ▲

### 4.3.2 An alternative algorithm

Continuing our discussion on the setting of the type depicted in Figure 4.6, we present another influence function-based algorithm for computing the critical value

$N_{cr}$ of the compressive axial force acting on a beam and the buckling failure shape. In doing so, we add formally the term $-w(x)$ to both sides of the governing differential equation in (4.52). This yields

$$\frac{d^2}{dx^2}\left(EI(x)\frac{d^2w(x)}{dx^2}\right)+N\frac{d^2w(x)}{dx^2}-w(x)=-w(x), \quad x\in[0,a] \tag{4.61}$$

Let $g(x, s; N)$ represent the Green's function of the boundary-value problem

$$\frac{d^2}{dx^2}\left(EI(x)\frac{d^2w(x)}{dx^2}\right)+N\frac{d^2w(x)}{dx^2}-w(x)=0, \quad x\in[0,a] \tag{4.62}$$

$$B_{0,k}[w(0)]=0, \quad B_{a,k}[w(a)]=0, \quad k=1,2 \tag{4.63}$$

It should be observed that the Green's function $g(x, s; N)$ is a nonlinear function of $N$. This feature of $g(x, s; N)$ will be taken into account when a numerical procedure is specified.

Tracing out the Green's function approach, we reduce the boundary-value problem in eqns (4.61) and (4.63) to a regular homogeneous integral equation

$$w(x)=\int_0^a g(x,s;N)w(s)\,ds \tag{4.64}$$

with respect to $w(x)$.

This is another integral formulation of the eigenvalue problem in eqns (4.52) and (4.63). However, contrary to the integral equations in (4.45) or (4.57), the equation in (4.64) poses a non-standard eigenvalue problem, since its kernel $g(x, s; N)$ represents a nonlinear function of $N$. The nonlinearity raises an obvious hardship in computing components of the eigenvalue spectrum for this problem. But since we are interested in only the lowest eigenvalue $N_{cr}$, the hardship can be overcome. Indeed, based on the partition depicted in Figure 4.7 and using the approach described earlier in this section, one can approximate the relation in (4.64) with the following homogeneous system of linear algebraic equations

$$w_i=\sum_{j=0}^{n-1} w_{j+1}\int_{s_j}^{s_{j+1}} g(x_i,s;N)\,ds, \quad (i=\overline{1,n})$$

in the approximate values $w_i$ of the deflection function $w(x)$, defined on the set of grid-points $x_i$ that is depicted with empty dots in Figure 4.7. The above system can be written in a matrix form as

$$(A(N)-I)W=0 \tag{4.65}$$

where $W$ represents a vector whose components are $w_i$ and the right-hand side is the zero-vector. It is evident that entries of the coefficient matrix in this system depend on the parameter $N$ in a nonlinear fashion, making (4.65) a non-standard eigenvalue problem of linear algebra. This implies that the relation

$$\det(A(N) - I) = 0 \tag{4.66}$$

does not represent, in this case, an algebraic equation in $N$ (as it would be in standard eigenvalue problems of linear algebra), but is rather a transcendental equation in $N$. Thus, the eigenvalues of the matrix $A(N) - I$ represent roots of the transcendental equation in (4.66) with the smallest of them being an approximate value of $N_{cr}$.

Since we are dealing with a non-standard eigenvalue problem, standard linear algebra routines are not directly applicable, and the value of $N_{cr}$ can be found to a required accuracy level by means of successive approximations, with a direct evaluation of the determinant in (4.66) for fixed values of $N$, with a subsequent minimization of the discrepancy. One can always make an educated guess as to the initial approximation for $N_{cr}$.

**Example 3.2:** By implementing the algorithm described above, determine approximate values of the critical force $N_{cr}$ for beams of a uniform flexural rigidity $EI_0$, with various edge conditions imposed.

The key element in our algorithm is the Green's function $g(x, s; N)$ for equation (4.62) which, in this case, simplifies to

$$\frac{d^4w(x)}{dx^4} + \frac{N}{EI_0}\frac{d^2w(x)}{dx^2} - \frac{1}{EI_0}w(x) = 0, \quad x \in [0, a] \tag{4.67}$$

Since this is the fourth order linear equation with constant coefficients, which allows an analytic solution, the Green's function can be constructed analytically. Indeed, by a routine algebra, a fundamental set of solutions for the above equation can be obtained as

$$\sinh px, \quad \cosh px, \quad \sin qx, \quad \text{and} \quad \cos qx \tag{4.68}$$

where

$$p = \frac{\sqrt{2}}{2}\sqrt{Q - P}, \quad q = \frac{\sqrt{2}}{2}\sqrt{Q + P}$$

with $P = N/EI_0$ and $Q = \sqrt{(N/EI_0)^2 + 4/EI_0}$.

Based on the set in eqn (4.68) and following our standard procedures, a Green's function can be constructed for any well-posed boundary-value problem for eqn (4.67). As an example, we expose below the case of a simply supported beam. The branch of the Green's function, which is valid for $0 \le x \le s \le a$, is found, in this case, as

$$g(x, s; N) = \frac{1}{pq(p^2 + q^2)}\left(\frac{\sinh p(a - s)}{\sinh pa}\sinh px - \frac{\sin q(a - s)}{\sin qa}\sin qx\right)$$

Due to the self-adjointness of the boundary-value problem, the other branch of $g(x, s; N)$, valid for $0 \le s \le x \le a$, can be obtained from that above by interchanging $x$ with $s$.

Approximate values of $N_{cr}$ are exhibited in Table 4.11, as computed for the simply supported (S–S), simply supported–clamped (S–C), and clamped (C–C) beam. The partition parameter is fixed at $n = 10$. We assume a unit length and a unit flexural rigidity of beams.

**Table 4.11:** Approximate values of $N_{cr}$ for beams of uniform flexural rigidity

| Critical force, $N_{cr}$ | Type of a beam | | |
|---|---|---|---|
| | S–S | S–C | C–C |
| Approximate value | 9.8696044 | 20.1997290 | 39.4784176 |
| Exact value | 9.8696044 | 20.1997 | 39.4784176 |
| Relative error, % | $0.1 \times 10^{-8}$ | – | $0.5 \times 10^{-7}$ |

Note that, since either exact or sufficiently accurate value of $N_{cr}$ is known for each of the three problems considered, the choice of the initial approximation for $N_{cr}$ was not an issue.

The accuracy of approximate values of $N_{cr}$ in Table 4.11 varies slightly from case to case, remaining nevertheless at a very high level for the relatively coarse partition ($n = 10$) which we used. Notice that for a simply supported–clamped (S–C) beam the exact value of the critical force is not available and the value of 20.1997 represents a justified approximate value which is available in literature (see, for example, [13]). Subsequently, the relative error for the S–C beam is not shown. ▲

Consider another buckling problem to show the potential of the influence function method-based algorithm. A beam of variable flexural rigidity $EI(x)$, resting on a simple elastic foundation (with the elastic constant $k_0$), is subject to an axial compressive force $N$ as shown in Figure 4.8. Note that the algorithm is equally applicable to any other single-span beam and the cantilever one is chosen just to be specific. Similarly to the problem posed by eqns (4.50) and (4.51), we are looking for a critical value $N_{cr}$ of the compressive force that causes the loss of stability for the beam.

The static equilibrium of the beam depicted in Figure 4.8 can be simulated with the following boundary-value problem

$$\frac{d^2}{dx^2}\left(EI(x)\frac{d^2w(x)}{dx^2}\right) + N\frac{d^2w(x)}{dx^2} + k_0w(x) = 0, \quad x \in [0, a] \tag{4.69}$$

$$w(0) = \frac{dw(0)}{dx} = 0, \quad \frac{d^2w(a)}{dx^2} = \frac{d^3w(a)}{dx^3} = 0 \tag{4.70}$$

If the value of the elastic constant $k_0$ is fixed, then the formulation in eqns (4.69) and (4.70) represents an eigenvalue problem with respect to $N$ and the lowest

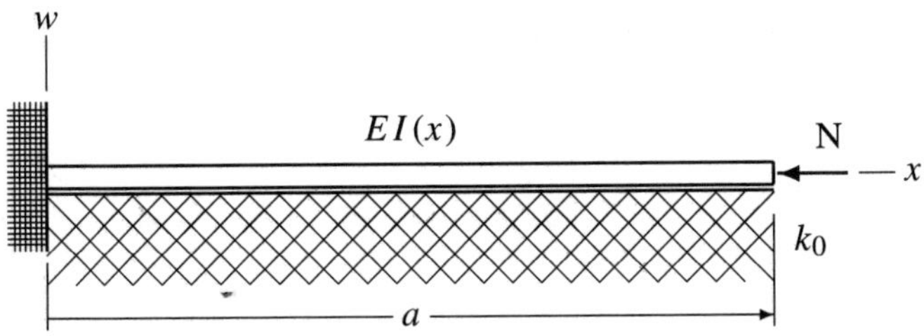


Figure 4.8: Buckling of a beam on elastic foundation

eigenvalue $N_{cr}$ is, in such a case, to find. We will reduce this problem to a homogeneous integral equation, the kernel of which depends on $N$.

In doing so, let $g(x, s; N)$ represent the Green's function of the boundary-value problem in eqn (4.70) stated for the homogeneous equation

$$\frac{d^2}{dx^2}\left(EI(x)\frac{d^2w(x)}{dx^2}\right) + N\frac{d^2w(x)}{dx^2} = 0, \quad x \in [0, a] \tag{4.71}$$

It is worth noting that there does not exist a single elementary function $EI(x)$ for which the above equation allows an analytic solution. Whereas in the case of a uniform flexural rigidity $EI_0$, Green's functions $g(x, s; N)$ to eqn (4.71) can analytically be obtained for any feasible set of boundary conditions. The case of a simply supported beam, for example, has already been handled earlier in Section 4.2 (see eqn (4.44)). Green's functions for other boundary conditions could also be obtained in a compact form. For a beam whose left-hand end is clamped

$$w(0) = \frac{dw(0)}{dx} = 0$$

while the right-hand end is sliding

$$\frac{dw(a)}{dx} = \frac{d^3w(a)}{dx^3} = 0$$

the Green's functions is found as

$$g(x, s; N) = \frac{1}{\Delta}\begin{cases} [a\cos qa - \cos q(a-s)](1-\cos qx) \\ \qquad + a\sin qa(qx - \sin qx), & x \le s \\ [a\cos qa - \cos q(a-x)](1-\cos qs) \\ \qquad + a\sin qa(qs - \sin qs), & s \le x \end{cases}$$

where $\Delta = q^3 \sin qa$ and $q = \sqrt{N/EI_0}$.

For a cantilever beam whose left-hand edge is clamped, the Green's function is found in the form

$$g(x, s; N) = \frac{1}{q^3}\begin{cases} qx\cos qs - \sin qs - \sin q(x-s), & x \le s \\ qs\cos qx - \sin qx - \sin q(s-x), & s \le x \end{cases}$$

In compliance with Theorem 2.4 of Chapter 2, solution of the problem in eqns (4.69) and (4.70) is expressed in terms of $g(x, s; N)$ as

$$w(x) = k_0 \int_0^a g(x, s; N)w(s)\,ds \tag{4.72}$$

Thus, the eigenvalue problem in eqns (4.69) and (4.70) reduces to a linear homogeneous integral equation that represents a non-standard eigenvalue formulation with $N$ being a parameter. To explore the accuracy level attainable when the problem in eqn (4.72) is solved numerically, we consider a validation example that follows.

**Example 3.3:** Compute an approximate value of $N_{cr}$ for a simply supported beam of a uniform flexural rigidity $EI_0$ resting on a elastic foundation whose elastic constant is $k_0$.

A boundary eigenvalue problem, which is associated with the setting in this example, can be written as

$$EI_0\frac{d^4w(x)}{dx^4} + N\frac{d^2w(x)}{dx^2} + k_0w(x) = 0, \quad x \in [0, a] \tag{4.73}$$

$$w(0) = \frac{d^2w(0)}{dx^2} = 0, \quad w(a) = \frac{d^2w(a)}{dx^2} = 0 \tag{4.74}$$

It can be shown that the function

$$w(x) = C\sin\frac{\pi x}{a} \tag{4.75}$$

where $C$ is an arbitrary constant, could represent a nontrivial solution to the problem in eqns (4.73) and (4.74) or, in other words, it could be the buckling failure shape for the beam. Indeed, the function in (4.75) satisfies the boundary conditions in (4.74), while its substitution in (4.73) yields

$$C\left[EI_0\left(\frac{\pi}{a}\right)^4 - N\left(\frac{\pi}{a}\right)^2 + k_0\right]\sin\frac{\pi x}{a} = 0$$

The above statement is identically true on the interval $[0, a]$ if the factor in brackets is zero. That is

$$EI_0\left(\frac{\pi}{a}\right)^4 - N\left(\frac{\pi}{a}\right)^2 + k_0 = 0$$

Upon solving this equation for $N$, one finds the critical value of the compressive force for the beam under consideration as

$$N_{cr} = \frac{EI_0\pi^4 + a^4k_0}{(a\pi)^2} \tag{4.76}$$

which implies that eqn (4.75) brings the buckling failure shape for the beam if the parameter $N$ takes on the value just found.

Thus, the buckling problem for a simply supported beam (eqns (4.73) and (4.74)) can readily be used as a validation example to test the influence function-based algorithm. The influence function $g(x, s; N)$ that serves in this case as a kernel of eqn (4.72), has been presented earlier (see eqn (4.44), where $q$ ought to be replaced with $\sqrt{N/EI_0}$).

**Table 4.12:** Approximate values of the critical force $N_{cr}$ for the S–S beam of a uniform flexural rigidity, resting on the elastic ($k_0$) foundation

| | Elastic constant, $k_0$ | | | |
|---|---|---|---|---|
| | 50 | 100 | 150 | 200 |
| Approximate values, $N_{cr}$ | 14.9356 | 20.0016 | 25.0675 | 30.1333 |
| Exact values, eqn (4.76) | 14.9357 | 20.0017 | 25.0678 | 30.1338 |
| Relative error, % | $0.3 \times 10^{-3}$ | $0.1 \times 10^{-2}$ | $0.3 \times 10^{-2}$ | $0.6 \times 10^{-2}$ |

Some data on the solution of the validation problem can be found in Table 4.12. They were computed for the setting in eqns (4.73) and (4.74) for the beam of unit length with $EI_0 = 1$. The lowest eigenvalue $N_{cr}$ of eqn (4.72) has been computed by the algorithm described earlier in this section, with the partition parameter $n = 10$. From these data it follows, in particular, that the accuracy of computing the values of $N_{cr}$ depends on the elastic constant $k_0$ slightly dropping as $k_0$ increases, but remaining at a high level of a fraction of a percent. ▲

So far in this chapter (Sections 4.2 and 4.3), we have been concerned with some nontraditional implementations of the influence function method for beam problems. The following section deals with another class of beam problems which might also benefit from the influence function treatment. That is, the bending of multi-span elastic beams, where we use the development of Chapter 2, where the notion of matrix of Green's type was introduced for a piecewise homogeneous media. Implementing finite weighted graphs, we will construct influence matrices of a point force for multi-span beams.

## 4.4 Bending of Multi-span Beams

The developments in this section have a bearing upon the theory of multi-point posed boundary-value problems stated for a specific type of systems of ordinary linear differential equations. This theory has been presented in Chapter 1, where we have appropriately extended the Green's function formalism to piecewise homogeneous media. Matrices of Green's type were introduced as a natural extension of the Green's function notion. Here we will work out this topic further towards its application to the static equilibrium of Kirchhoff beams comprised of more than one span, where each of which might have a different flexural rigidity.

### 4.4.1 Influence function as a matrix of Green's type

In drawing a relation between the influence function of a point force of a multi-span bean and the matrix of Green's type of the corresponding multi-point posed boundary-contact value problem, we consider a compound cantilever beam overhanging an intermediate simple support. The beam is comprised of two spans having uniform flexural rigidities $EI_1$ and $EI_2$ as shown in Figure 4.9.

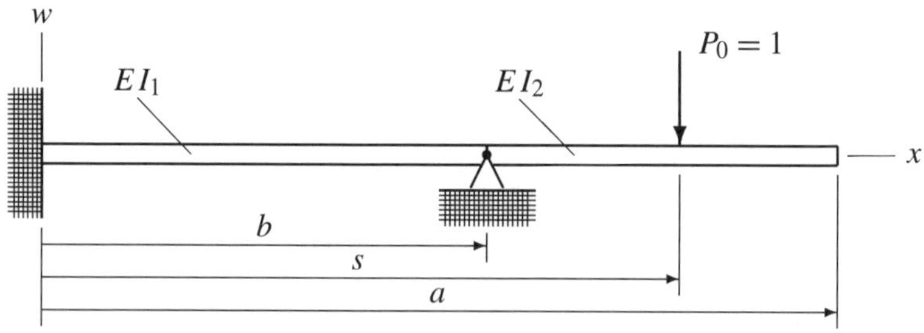


Figure 4.9: A compound beam overhanging simple support

To determine the influence function of a transverse unit force (this could be applied to either span) for such a beam, let us formulate the following three-point posed boundary-contact value problem

$$\frac{d^4w_1(x)}{dx^4} = -\frac{q_1(x)}{EI_1} = -f_1(x), \quad x \in (0, b) \tag{4.77}$$

$$\frac{d^4w_2(x)}{dx^4} = -\frac{q_2(x)}{EI_2} = -f_2(x), \quad x \in (b, a) \tag{4.78}$$

$$w_1(0) = \frac{dw_1(0)}{dx} = 0, \quad \frac{d^2w_2(a)}{dx^2} = \frac{d^3w_2(a)}{dx^3} = 0 \tag{4.79}$$

$$w_1(b) = w_2(b) = 0, \quad \frac{dw_1(b)}{dx} = \frac{dw_2(b)}{dx} \tag{4.80}$$

$$EI_1\frac{d^2w_1(b)}{dx^2} = EI_2\frac{d^2w_2(b)}{dx^2} \tag{4.81}$$

where $q_1(x)$ and $q_2(x)$ in eqns (4.77) and (4.78) represent arbitrary transverse continuously distributed loads applied to the left-hand and the right-hand span, respectively. The matrix of Green's type to the above problem represents the influence function of a unit point force to the beam under consideration.

The variation of parameters method-based procedure will be used to obtain the matrix of Green's type. From our preceding discussions on this method, the reader may recall that $q_1(x)$ and $q_2(x)$ are not required to be specified. However, at a certain stage of the procedure, they aid us in determining the influence matrix that we are assigned to find. The functions $f_1(x)$ and $f_2(x)$, in turn, are introduced here for the sake of notational convenience in the development that follows.

For an easier digestion of the forthcoming material, the discussion that we have had in Sections 2.4 and 2.5 of Chapter 2 is essential, with a specific emphasis on Example 5.1. The only difference of the current statement compared to the problems analyzed in Sections 2.4 and 2.5 is that we are posing a problem for equations of higher order. This will definitely result in a more cumbersome and time consuming computation.

We let $G(x, s)$ represent the matrix of Green's type for the homogeneous boundary-contact value problem corresponding to that in eqns (4.77)–(4.81). It is evident that $G(x, s)$ is the influence function of a point force that we are looking for. That is, the entries $g_{11}(x, s)$ and $g_{12}(x, s)$ in $G(x, s)$ represent the deflection in the left-hand span ($0 \le x \le b$) of the beam, caused by a unit force applied within that same span ($s \in (0, b)$) and the right-hand span ($s \in (b, a)$), respectively. The entries $g_{21}(x, s)$ and $g_{22}(x, s)$, in turn, show how the right-hand span ($b \le x \le a$) responds to a unit force applied within the left-hand and right-hand span, respectively.

To solve the boundary-contact value problem in eqns (4.77)–(4.81), we recall the standard technique of Lagrange's method of variation of parameters, repeatedly used in Chapter 1 and in the preceding sections of the current chapter. In doing so, one represents the general solutions of eqns (4.77) and (4.78) as

$$w_i(x) = A_i(x) + B_i(x)x + C_i(x)x^2 + D_i(x)x^3, \quad i = 1, 2 \tag{4.82}$$

The system of linear equations in $A_i'(x)$, $B_i'(x)$, $C_i'(x)$, and $D_i'(x)$ $(i = 1, 2)$, which results from the standard procedure of Lagrange's method, is found to be, in this case, in the form

$$\begin{pmatrix} 1 & x & x^2 & x^3 \\ 0 & 1 & 2x & 3x^2 \\ 0 & 0 & 2 & 6x \\ 0 & 0 & 0 & 6 \end{pmatrix} \times \begin{pmatrix} A_i'(x) \\ B_i'(x) \\ C_i'(x) \\ D_i'(x) \end{pmatrix} = \begin{pmatrix} 0 \\ 0 \\ 0 \\ -f_i(x) \end{pmatrix}$$

The upper triangular form of the coefficient matrix makes the solution of the system a simple backward substitution, which yields

$$A_i'(x) = \frac{x^3}{6} f_i(x), \quad B_i'(x) = -\frac{x^2}{2} f_i(x)$$

$$C_i'(x) = \frac{x}{2} f_i(x), \quad D_i'(x) = -\frac{1}{6} f_i(x)$$

Hence, the coefficients $A_1(x), \ldots, D_1(x)$ for $w_1(x)$ in eqn (4.82) can be obtained by integrating the above derivatives over the interval $[0, x]$. This yields

$$A_1(x) = \int_0^x \frac{s^3}{6} f_1(s)\, ds + H_1, \quad B_1(x) = -\int_0^x \frac{s^2}{2} f_1(s)\, ds + K_1$$

$$C_1(x) = \int_0^x \frac{s}{2} f_1(s)\, ds + L_1, \quad D_1(x) = -\int_0^x \frac{1}{6} f_1(s)\, ds + M_1$$

Similarly to the above, the coefficients $A_2(x), \ldots, D_2(x)$ for $w_2(x)$ can be found as integrals over the interval $[b, x]$, that is

$$A_2(x) = \int_b^x \frac{s^3}{6} f_2(s)\, ds + H_2, \quad B_2(x) = -\int_b^x \frac{s^2}{2} f_2(s)\, ds + K_2$$

$$C_2(x) = \int_b^x \frac{s}{2} f_2(s)\, ds + L_2, \quad D_2(x) = -\int_b^x \frac{1}{6} f_2(s)\, ds + M_2$$

Upon substituting the coefficient just found into eqn (4.82) and grouping all the integral terms together, one obtains the following expressions

$$w_1(x) = \int_0^x \frac{(s-x)^3}{6} f_1(s)\, ds + H_1 + K_1 x + L_1 x^2 + M_1 x^3 \tag{4.83}$$

$$w_2(x) = \int_b^x \frac{(s-x)^3}{6} f_2(s)\, ds + H_2 + K_2 x + L_2 x^2 + M_2 x^3 \tag{4.84}$$

for the general solutions ($w_1(x)$ as $x \in [0, b]$ and $w_2(x)$ as $x \in [b, a]$) of eqns (4.77) and (4.78). The parameters $H_1, \ldots, M_2$ are arbitrary constants. To compute the latter, we take advantage of the boundary and contact conditions as of eqns (4.79)–(4.81). In doing so, the first condition $w_1(0) = 0$ of a clamped edge in eqn (4.79) yields $H_1 = 0$. While the second condition $w'(0) = 0$ of a clamped edge in eqn (4.79) analogously results in $K_1 = 0$.

Satisfying then the first condition of a free edge at $x = a$ in eqn (4.79), which is associated with vanishing of the bending moment, one obtains

$$2L_2 + 6M_2 a = -\int_b^a (s-a) f_2(s)\, ds \tag{4.85}$$

while the second free edge condition imposed at $x = a$, which is associated with vanishing of the shear force, yields

$$M_2 = \int_b^a \frac{1}{6} f_2(s)\, ds$$

Substituting then the value of $M_2$ just found in eqn (4.85), one obtains

$$L_2 = -\int_b^a \frac{s}{2} f_2(s)\, ds$$

The first contact condition $w_1(b) = 0$ from eqn (4.80) yields

$$L_1 b^2 + M_1 b^3 = -\int_0^b \frac{(s-b)^3}{6} f_1(s)\, ds \tag{4.86}$$

while the second condition $w_2(b) = 0$ provides

$$H_2 + K_2 b + L_2 b^2 + M_2 b^3 = 0$$

Taking into account the values of $L_2$ and $M_2$ found earlier, we rewrite the above relation in the form

$$H_2 + K_2 b = -\int_b^a \frac{b^2(b-3s)}{6} f_2(s)\, ds \tag{4.87}$$

Satisfying the third contact condition $w_1'(b) = w_2'(b)$ in eqn (4.80), which spells out the continuity of the slope of the deflection function at $x = b$, we obtain

$$2L_1 b + 3M_1 b^2 - \int_0^b \frac{(s-b)^2}{2} f_1(s)\, ds = K_2 + \int_b^a \frac{b(b-2s)}{2} f_2(s)\, ds \tag{4.88}$$

The contact condition in eqn (4.81), which articulates the continuity of the bending moment at $x = b$, yields

$$EI_1\left[2L_1 + 6M_1 b + \int_0^b (s-b) f_1(s)\, ds\right] = EI_2 \int_b^a (b-s) f_2(s)\, ds \tag{4.89}$$

It is evident that the relations in eqns (4.86)–(4.89) form a well-posed system of linear algebraic equations in the four unknowns $L_1$, $M_1$, $H_2$, and $K_2$. To solve this system, we realize that equations (4.86) and (4.89) form a two-by-two system in $L_1$ and $M_1$. Hence, these equations are independent of those in (4.87) and (4.88) and their solution can be found as

$$L_1 = \int_0^b \frac{1}{4b^2}[s(b-s)(s-2b)] f_1(s)\, ds - \int_b^a \frac{\lambda}{4}(b-s) f_2(s)\, ds$$

and

$$M_1 = \int_0^b \frac{1}{12b^3}[(s-b)(s^2 - 2bs - 2b^2)]\, ds + \int_b^a \frac{\lambda}{4b}(b-s) f_2(s)\, ds$$

where $\lambda$ represents the ratio of the flexural rigidities $EI_2$ and $EI_1$, that is $\lambda = EI_2/EI_1$.

Since the values of the parameters $H_1$, $K_1$, $L_1$, and $M_1$ are already available, one can write an explicit expression for the deflection function $w_1(x)$ for the left-hand span of the beam. This is caused by a combination of the two continuously distributed transverse loads $q_1(x)$ and $q_2(x)$ applied to the left-hand and right-hand span respectively. Substituting these parameters into eqn (4.83) and performing some elementary transformations, one obtains

$$w_1(x) = \int_0^b \frac{x^2}{12b^3}(s-b)[s(3b-x)(2b-s) - 2b^2x]f_1(s)\,ds + \int_b^a \frac{\lambda x^2}{4b}(b-s)(x-b)f_2(s)\,ds + \int_0^x \frac{(s-x)^3}{6}f_1(s)\,ds \quad (4.90)$$

While obtaining the deflection function $w_2(x)$ for the right-hand span, notice that, since the values of $L_1$ and $M_1$ are already found, eqn (4.88) provides the value of $K_2$

$$K_2 = \int_0^b \frac{1}{4b}s^2(b-s)f_1(s)\,ds + \int_b^a \frac{b}{4}(b-s)(\lambda-2)f_2(s)\,ds$$

Based on this, eqn (4.87) can be solved to provide the value of $H_2$. This yields

$$H_2 = -\int_0^b \frac{1}{4}s^2(b-s)f_1(s)\,ds + \int_b^a \frac{b^2}{12}[3\lambda(s-b) - 2(3s-2b)]f_2(s)\,ds$$

Upon substituting values of the parameters $H_2$, $K_2$, $L_2$, and $M_2$ into eqn (4.84) and performing trivial algebra, one obtains an explicit expression for the deflection $w_2(x)$ for the right-hand span of the beam. This is caused by the combination of the two continuously distributed loads $q_1(x)$ and $q_2(x)$, and found as

$$w_2(x) = \int_0^b \frac{1}{4b}s^2(b-s)(b-x)f_1(s)\,ds + \int_b^x \frac{(s-x)^3}{6}f_2(s)\,ds + \int_b^a \frac{1}{12}(x-b)[2(x-b)^2 - 3(s-b)(b(\lambda-2)+2x)]f_2(s)\,ds \quad (4.91)$$

We combine now the first and the last integrals in eqn (4.90) and recall the relations between the functions $f_1(x)$ and $f_2(x)$, on one hand, and the loading functions $q_1(x)$ and $q_2(x)$, on the other hand (see eqns (4.77) and (4.78)). This allows us to formally rewrite (4.90) in the form

$$w_1(x) = \int_0^b g_{11}(x,s)q_1(s)\,ds + \int_b^a g_{12}(x,s)q_2(s)\,ds$$

where

$$g_{11}(x,s) = \frac{1}{12b^3EI_1}\begin{cases} x^2(s-b)[s(3b-x)(2b-s) - 2b^2x], & \text{for } x \le s \\ s^2(x-b)[x(3b-s)(2b-x) - 2b^2s], & \text{for } s \le x \end{cases} \quad (4.92)$$

with both variables $x$ and $s$ ranging between 0 and $b$. Whereas, for $g_{12}(x, s)$, with $0 \le x \le b$ and $b \le s \le a$, we have

$$g_{12}(x, s) = \frac{\lambda x^2}{4bEI_2}(b - s)(x - b) \tag{4.93}$$

To obtain the entries $g_{21}(x, s)$ and $g_{22}(x, s)$ of the matrix of Green's type that we are working on, we turn to the representation for the deflection function $w_2(x)$ in eqn (4.91). Combining the second and the last integrals in eqn (4.91), we rewrite the latter as

$$w_2(x) = \int_0^b g_{21}(x, s)q_1(s)\,ds + \int_b^a g_{22}(x, s)q_2(s)\,ds$$

where the kernel function $g_{21}(x, s)$ is defined by the single piece expression

$$g_{21}(x, s) = \frac{s^2}{4bEI_1}(b - s)(x - b) \tag{4.94}$$

with the variables ranging as $b \le x \le a$ and $0 \le s \le b$. For the kernel function $g_{22}(x, s)$, with both variables $x$ and $s$ ranging between $b$ and $a$, we obtain the expression in two pieces

$$g_{22}(x, s) = \frac{1}{12EI_2}\begin{cases}(x - b)[2(x - b)^2 + 3(b - s)(2x - b(2 - \lambda))], & x \le s\\ (s - b)[2(s - b)^2 + 3(b - x)(2s - b(2 - \lambda))], & s \le x\end{cases} \tag{4.95}$$

From the definition introduced in Section 2.5 in Chapter 2, it follows that the functions $g_{ij}(x, s)$ presented in eqns (4.92)–(4.95) represent the entries of the matrix of Green's type, $G(x, s)$, for the homogeneous boundary-contact value problem corresponding to that in eqns (4.77)–(4.81). That is, $G(x, s)$ represents the influence matrix (in the sense described earlier in this section) of a transverse concentrated unit force for the compound beam shown in Figure 4.9.

It is worth noting that there is a specific match between the variables and the subscripts in $g_{ij}(x, s)$. One ought to keep in mind that the first subscript $i$ stays for the observation point's span number, whereas the second subscript $j$ – for the force application point's span number. This implies that the $x$ variable in $g_{ij}(x, s)$ is located in the $i$-th span, while the $s$ variable belongs to the $j$-th span.

Components of the stress-strain state generated by a reasonable (physical and geometrical linearity ought to always be taken care of) combination of conventional loads applied to this beam, can readily be computed by using the influence function method, given that the corresponding influence matrix is available. We will be more specific on that when considering particular examples below.

### 4.4.2 Beams undergoing transverse loads

**Example 4.1:** Suppose a double-span cantilever beam overhanging a simple support is loaded with a concentrated bending moment of magnitude $M$ and a transverse force of magnitude $P$ as shown in Figure 4.10. By using the entries of the influence matrix obtained in eqns (4.92) through (4.95), compute the components of the stress-strain state for this beam.

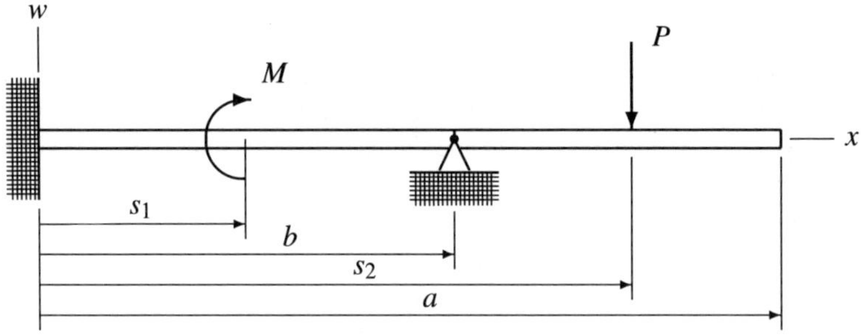


Figure 4.10: Double-span beam subject to a combination of loads

The entries of the influence matrix to this beam have just been constructed. And it is evident that the deflection $w_{1,p}(x)$ in the left-hand span of the beam, caused by the force $P$, can be computed in terms of $g_{12}(x, s_2)$ as

$$w_{1,p}(x) = -Pg_{12}(x, s_2) = -P\frac{\lambda x^2}{4b}(b - s_2)(x - b)$$

while the deflection $w_{1,m}(x)$ in the left-hand span, caused by the bending moment $M$, equals

$$w_{1,m}(x) = -M\frac{\partial g_{11}(x, s_1)}{\partial s}$$

Due to the piecewise definition of $g_{11}(x, s)$ (see eqn (4.90)), the deflection $w_{1,m}(x)$ is to be computed by different formulae for $x \leq s_1$ (to the left of $s_1$) and for $x \geq s_1$ (to the right of that point). This yields

$$w_{1,m}(x) = -\frac{M}{4b^2EI_1}\begin{cases} x^2[s_1(3b - x)(2b - s_1) - 2b^3], & x \leq s_1 \\ s_1[x^2(3b - x)(2b - s_1) - 2b^3(2x - s_1)], & s_1 \leq x \end{cases}$$

and summing up $w_{1,p}(x)$ and $w_{1,m}(x)$, one obtains the deflection in the left-hand span that is caused by both $P$ and $M$.

The deflection $w_{2,p}(x)$ in the right-hand span, caused by $P$, is expressed in terms of the entry $g_{22}(x, s_2)$ as

$$w_{2,p}(x) = \frac{-P}{12EI_2}\begin{cases} (x - b)[2(x - b)^2 + 3(b - s_2)(2x - b(2 - \lambda))], & x \leq s_2 \\ (s_2 - b)[2(s_2 - b)^2 + 3(b - x)(2s_2 - b(2 - \lambda))], & s_2 \leq x \end{cases}$$

while the deflection function $w_{2,m}(x)$ in the right-hand span, caused by $M$, is computed as

$$w_{2,m}(x) = -M\frac{\partial g_{21}(x, s_2)}{\partial s} = -\frac{Ms}{4b}(x - b)(2b - s_2)$$

and summing up $w_{2,p}(x)$ and $w_{2,m}(x)$, one obtains the deflection in the right-hand span that is caused by both loads $P$ and $M$.

Explicit expressions for the bending moment $M(x)$ and the shear force $Q(x)$ can readily be found for any cross-section of the beam in terms of the deflection function $w(x)$. Hence, obtaining $M(x)$ and $Q(x)$ is a matter of taking appropriate derivatives of the expressions for the deflection function that have just been obtained. ▲

**Example 4.2:** We are going to obtain components of the stress-strain state for the double-span simply supported compound ($EI_1$ and $EI_2$) beam subject to a combination of loads as depicted in Figure 4.11.

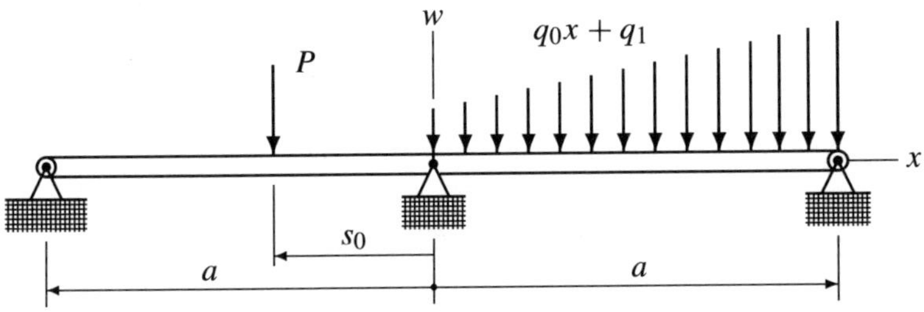


Figure 4.11: A compound double-span beam

Tracing out the construction procedure used earlier in this section, we obtain the entries $g_{ij}(x, s)$ of the influence matrix $G(x, s)$ of a transverse concentrated unit force for this beam. For $g_{11}^{+}(x, s)$, representing the branch of $g_{11}(x, s)$ which is valid for $-a \le x \le s \le 0$, we have

$$g_{11}^{+}(x, s) = \frac{s(x+a)}{p}\{2a^2[(x+a)^2 + (s^2 - a^2)] + \lambda s[2a^2 s - x(s+3a)(x+2a)]\}$$

where the parameters $p$ and $\lambda$ are defined as

$$p = 12a^3(EI_1 + EI_2) \quad \text{and} \quad \lambda = EI_2/EI_1$$

The branch $g_{11}^{-}(x, s)$ of $g_{11}(x, s)$, with $-a \le s \le x \le 0$, is found as

$$g_{11}^{-}(x, s) = \frac{x(s+a)}{p}\{2a^2[(s+a)^2 + (x^2 - a^2)] + \lambda x[2a^2 x - s(x+3a)(s+2a)]\}$$

The entry $g_{12}(x, s)$ of $G(x, s)$, with $-a \le x \le 0$ and $0 \le s \le a$, is obtained as

$$g_{12}(x, s) = \frac{xs}{p}(s-a)(x+a)(x+2a)(s-2a)$$

while for the entry $g_{21}(x, s)$, with $-a \le s \le 0$ and $0 \le x \le a$, we obtain

$$g_{21}(x, s) = \frac{xs}{p}(x-a)(s+a)(s+2a)(x-2a)$$

Finally, the branch $g_{22}^{+}(x, s)$ of $g_{22}(x, s)$, which is valid for $0 \le x \le s \le a$, is presented in the form

$$g_{22}^{+}(x, s) = \frac{x(s-a)}{\lambda p}\{2\lambda a^2[(s-a)^2 + (x^2 - a^2)] \\ + x[2a^2x - s(2a-s)(3a-x)]\}$$

while the branch $g_{22}^{-}(x, s)$, with $0 \le s \le x \le a$, is found as

$$g_{22}^{-}(x, s) = \frac{s(x-a)}{\lambda p}\{2\lambda a^2[(x-a)^2 + (s^2 - a^2)] \\ + s[2a^2s - x(2a-x)(3a-s)]\}$$

When computing the resultant deflection of the beam that is caused by both the distributed and the concentrated load in the statement, one ought to account for a piecewise form of the entries $g_{11}(x, s)$ and $g_{22}(x, s)$ of the influence matrix. For the resultant deflection $w_1(x)$ in the left-hand span, to the left of $s_0$, we eventually obtain

$$w_1(x) = Pg_{11}^{+}(x, s_0) + \int_0^a g_{12}(x, s)(q_0 s + q_1)\, ds$$

while to the right of $s_0$, $w_1(x)$ is obtained as

$$w_1(x) = Pg_{11}^{-}(x, s_0) + \int_0^a g_{12}(x, s)(q_0 s + q_1)\, ds$$

The resultant deflection $w_2(x)$ caused by the concentrated force and the distributed load in the right-hand span can be computed as

$$w_2(x) = Pg_{21}(x, s_0) + \int_0^x g_{22}^{-}(x, s)(q_0 s + q_1)\, ds \\ + \int_x^a g_{22}^{+}(x, s)(q_0 s + q_1)\, ds$$

Upon accomplishing the above integration and performing routine algebra, one obtains explicit analytic expressions for $w_1(x)$ and $w_2(x)$. These can be utilized to compute expressions for the bending moment and the shear force at any cross-section of the beam. ▲

### 4.4.3 Other influence functions

In both cases considered so far in this section, we have dealt with beams having a simple intermediate support (see Figures 4.10 and 4.11). Multi-span beams with intermediate elastic supports can also be treated by means of the influence function method. Influence functions for beams having more than two spans can also be treated. To address these points, we consider the following illustrative examples.

**Example 4.3:** Compute the influence matrix (of Green's type) of a transverse unit concentrated force for the compound ($EI_1$ and $EI_2$) cantilever beam overhanging an intermediate elastic support, with the elastic spring constant $k^*$, as shown in Figure 4.12.

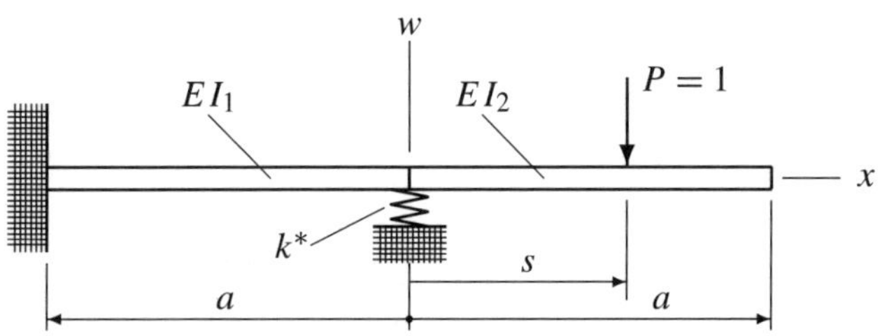


Figure 4.12: A beam overhanging an elastic support

The influence matrix that we are looking for represents the matrix of Green's type to the homogeneous problem corresponding to the following three-point posed boundary-contact value problem

$$\frac{d^4 w_1(x)}{dx^4} = -\frac{q_1(x)}{EI_1} = -f_1(x), \quad x \in (-a, 0) \tag{4.96}$$

$$\frac{d^4 w_2(x)}{dx^4} = -\frac{q_2(x)}{EI_2} = -f_2(x), \quad x \in (0, a) \tag{4.97}$$

$$w_1(-a) = \frac{dw_1(-a)}{dx} = 0, \quad \frac{d^2 w_2(a)}{dx^2} = \frac{d^3 w_2(a)}{dx^3} = 0 \tag{4.98}$$

$$w_1(0) = w_2(0), \quad \frac{dw_1(0)}{dx} = \frac{dw_2(0)}{dx} \tag{4.99}$$

$$EI_1 \frac{d^2 w_1(0)}{dx^2} = EI_2 \frac{d^2 w_2(0)}{dx^2} \tag{4.100}$$

$$EI_1 \frac{d^3 w_1(0)}{dx^3} - kw_1(0) = EI_2 \frac{d^3 w_2(0)}{dx^3} + kw_2(0), \quad k = 2k^* \tag{4.101}$$

for the deflection functions $w_1(x)$ and $w_2(x)$ to be determined on the intervals $[-a, 0]$ and $[0, a]$, respectively.

Using the technique described in detail earlier in this section, we present $w_1(x)$ and $w_2(x)$ in the form

$$w_1(x) = \int_{-a}^{x} \frac{(s-x)^3}{6} f_1(s)\, ds + H_1 + K_1 x + L_1 x^2 + M_1 x^3$$

and

$$w_2(x) = \int_{0}^{x} \frac{(s-x)^3}{6} f_2(s)\, ds + H_2 + K_2 x + L_2 x^2 + M_2 x^3$$

Values of the constants $H_i$, $K_i$, $L_i$ and $M_i$, $(i = 1, 2)$ can be determined by taking advantage of the set of boundary and contact conditions imposed by eqns (4.98)–(4.101). This yields a well-posed system of eight linear algebraic equations in $H_i$, $K_i$, $L_i$ and $M_i$. When these are obtained and substituted into the above expressions for $w_1(x)$ and $w_2(x)$, the latter read as

$$\begin{aligned} w_1(x) = &\int_{-a}^{x} \frac{(s-x)^3}{6} f_1(s)\, ds \\ &+ \int_{-a}^{0} \frac{(a+x)^2}{6p} \{ks[x(s^2-a^2) - 2a(s^2+ax)] \\ &+ 3EI_1[(x+a) - 3(s+a)]\} f_1(s)\, ds \\ &+ \lambda \int_{0}^{a} \frac{(a+x)^2}{2p} \{EI_1[(a+x) - 3(a+s)] - ka^2 xs\} f_2(s)\, ds \end{aligned} \quad (4.102)$$

and

$$\begin{aligned} w_2(x) = &\int_{-a}^{0} \frac{(a+s)^2}{2p} \{EI_1[(a+s) - 3(x+a)] - ka^2 sx\} f_1(s)\, ds \\ &+ \int_{0}^{x} \frac{(s-x)^3}{6} f_2(s)\, ds + \int_{0}^{a} \frac{1}{6p} \{px^2(x-3s) - 3\lambda ka^4 xs \\ &- 3EI_2 a[a(2a+3s) + 3x(a+2s)]\} f_2(s)\, ds \end{aligned} \quad (4.103)$$

where the parameters $p$ and $\lambda$ are defined as $p = (2a^3k + 3EI_1)$ and $\lambda = EI_2/EI_1$.

Observing eqns (4.102) and (4.103), which present the solution to the boundary-contact value problem posed by eqns (4.96)–(4.101), and recalling the definition of the matrix of Green's type introduced in Chapter 1, we conclude that

$$\begin{aligned} g_{11}^{+}(x, s) = &\frac{(a+x)^2}{6pEI_1} \{ks[x(s^2-a^2) - 2a(s^2+ax)] \\ &+ 3EI_1[(x+a) - 3(s+a)]\}, \end{aligned}$$

represents the branch of the entry $g_{11}(x, s)$ of the matrix of Green's type to the homogeneous problem corresponding to that in eqns (4.96)–(4.101), which is valid for $-a \le x \le s \le 0$, while for the other branch valid for $-a \le s \le x \le 0$ we have

$$g_{11}^{-}(x, s) = \frac{(a+s)^2}{6pEI_1}\{kx[s(x^2 - a^2) - 2a(x^2 + as)] + 3EI_1[(s+a) - 3(x+a)]\}$$

The entry $g_{12}(x, s)$ defined for $x \in [-a, 0]$ and $s \in [0, a]$ is found as

$$g_{12}(x, s) = \frac{(a+x)^2}{2pEI_1}\{EI_1[(a+x) - 3(a+s)] - ka^2xs\}$$

and for $g_{21}(x, s)$ where $s \in [-a, 0]$ and $x \in [0, a]$, we have

$$g_{21}(x, s) = \frac{(a+s)^2}{2pEI_1}\{EI_1[(a+s) - 3(x+a)] - ka^2sx\}$$

Finally, for $g_{22}^{+}(x, s)$, with both variables $x$ and $s$ belonging to the interval $[0, a]$ and $x \le s$, we obtain

$$g_{22}^{+}(x, s) = \frac{1}{6pEI_2}\{px^2(x - 3s) - 3\lambda ka^4xs - 3EI_2a[a(2a+3s) + 3x(a+2s)]\}$$

while for $g_{22}^{-}(x, s)$ with $x \ge s$, we have

$$g_{22}^{-}(x, s) = \frac{1}{6pEI_2}\{ps^2(s - 3x) - 3\lambda ka^4xs - 3EI_2a[a(2a+3x) + 3s(a+2x)]\}$$

By using the influence matrix of a transverse concentrated unit force, whose entries have just been obtained, one can compute the response of the beam depicted in Figure 4.12 to any reasonable combination of loads applied. ▲

**Example 4.4:** We will determine the response of a compound triple-span beam shown in Figure 4.13 to a transverse concentrated force of magnitude $P$ applied at a point $s_0$ in the left-hand span.

The influence matrix for such a beam can be found by tracing out the procedure of the method of variation of parameters described earlier in this section. In doing so we pose the following four-point posed boundary-value problem

$$\frac{d^4w_1(x)}{dx^4} = -\frac{q_1(x)}{EI_1}, \quad x \in (0, a) \tag{4.104}$$

$$\frac{d^4w_2(x)}{dx^4} = -\frac{q_2(x)}{EI_2}, \quad x \in (a, 2a) \tag{4.105}$$

$$\frac{d^4w_3(x)}{dx^4} = -\frac{q_3(x)}{EI_3}, \quad x \in (2a, 3a) \tag{4.106}$$

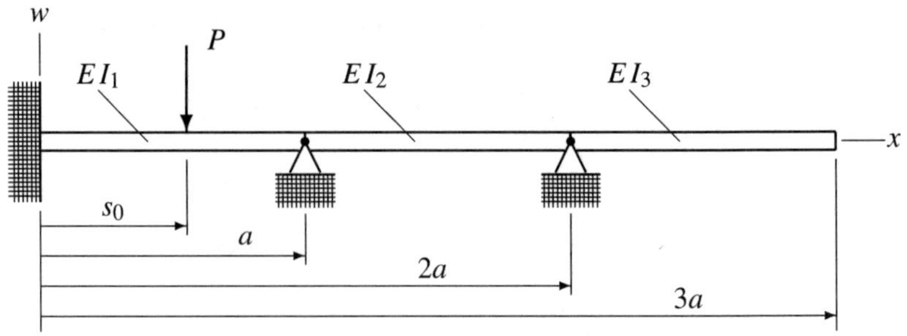


Figure 4.13: A compound beam overhanging two supports

$$w_1(0) = \frac{dw_1(0)}{dx} = 0, \qquad \frac{d^2w_3(3a)}{dx^2} = \frac{d^3w_3(3a)}{dx^3} = 0 \tag{4.107}$$

$$w_1(a) = w_2(a) = 0, \qquad \frac{dw_1(a)}{dx} = \frac{dw_2(a)}{dx} \tag{4.108}$$

$$EI_1\frac{d^2w_1(a)}{dx^2} = EI_2\frac{d^2w_2(a)}{dx^2}, \qquad EI_2\frac{d^2w_2(2a)}{dx^2} = EI_3\frac{d^2w_3(2a)}{dx^2} \tag{4.109}$$

$$w_2(2a) = w_3(2a) = 0, \qquad \frac{dw_2(2a)}{dx} = \frac{dw_3(2a)}{dx} \tag{4.110}$$

in the deflection function whose values are denoted as: $w_1(x)$ for $x \in (0, a)$, $w_2(x)$ for $x \in (a, 2a)$, and $w_3(x)$ for $x \in (2a, 3a)$.

Proceeding with the variation of parameters algorithm, we obtain the deflection function for the $(0, a)$ span of the beam as

$$w_1(x) = \int_0^a g_{11}(x, s)q_1(s)\,ds + \int_a^{2a} g_{12}(x, s)q_2(s)\,ds + \int_{2a}^{3a} g_{13}(x, s)q_3(s)\,ds \tag{4.111}$$

where $g_{11}(x, s)$, $g_{12}(x, s)$, and $g_{13}(x, s)$ represent entries of the first row in a $3 \times 3$ matrix of Green's type $G(x, s)$ to the homogeneous problem corresponding to that which appeared in eqns (4.104)–(4.110). In other words, these entries represent the response of the beam to the unit concentrated force shown in Figure 4.13. We omit details of the algorithm and present just the final result. For the branch $g_{11}^{+}(x, s)$ of $g_{11}(x, s)$, which is valid for $0 \leq x \leq s \leq a$, we obtain

$$g_{11}^{+}(x, s) = \frac{x^2(a-s)}{6pa^3}\{2[s(2a-s)(x-3a) + 2a^2x] + 3\lambda_1(a-s)[x(a+s) + s(x-3a)]\}$$

while the branch $g_{11}^{-}(x, s)$ of $g_{11}(x, s)$, which is valid for $0 \leq s \leq x \leq a$, is found as

$$g_{11}^{-}(x, s) = \frac{s^2(a-x)}{6pa^3}\{2[x(2a-x)(s-3a)+2a^2s] + 3\lambda_1(a-x)[s(a+x)+x(s-3a)]\}$$

where $p = 4EI_1 + 3EI_2$ and $\lambda_1 = EI_2/EI_1$.

Since the arguments $x \in [a, 2a]$ and $s \in [0, a]$ in $g_{21}(x, s)$ have different domains, this entry is defined in one piece as

$$g_{21}(x, s) = \frac{1}{2pa^3}s^2(s-a)(3a-x)(a-x)(2a-x)$$

while for $g_{31}(x, s)$, with $x \in [2a, 3a]$ and $s \in [0, a]$, we similarly have

$$g_{31}(x, s) = \frac{1}{2pa}s^2(a-s)(2a-x)$$

It is evident that scalar multiples $Pg_{11}(x, s_0)$, $Pg_{21}(x, s_0)$, and $Pg_{31}(x, s_0)$ represent the deflection functions $w_1(x)$, $w_2(x)$, and $w_3(x)$, in the left-hand, intermediate, and right-hand span of the beam, respectively, caused by the transverse force of magnitude $P$ concentrated at an arbitrary point $s_0$ in the left-hand span. By appropriately differentiating the deflection function, one can readily obtain explicit expressions for the bending moments and the shear forces generated in the beam by $P$. ▲

Thus, the problem posed in Example 4.4 is formally solved. Indeed, the response of this beam to the force $P$ applied at an arbitrary point in the left-hand span is already found. If, however, an external load would also be applied to other span (or spans) of the beam under consideration, then the rest of the entries of the influence matrix ought to be available. Therefore, we present all of them.

The entry $g_{12}(x, s)$, with $x \in [0, a]$ and $s \in [a, 2a]$, is obtained in the form

$$g_{12}(x, s) = \frac{1}{2pa^3}x^2(a-s)(3a-s)(x-a)(2a-s)$$

while the branch $g_{22}^{+}(x, s)$ of $g_{22}(x, s)$, with both variables $x$ and $s$ belonging to $[a, 2a]$, as $x \leq s$, is expressed as

$$g_{22}^{+}(x, s) = \frac{1}{6\lambda_1 pa^3}(2a-s)(a-x)\{2(x-a)[s(s-4a)(x-4a) + a^2(x-10a)] - 3\lambda_1 a^2[(s-3a)(s-a)+(x-a)^2]\}$$

The other branch $g_{22}^{-}(x, s)$, with $s \leq x$, is found as

$$g_{22}^{-}(x, s) = \frac{1}{6\lambda_1 pa^3}(2a-x)(a-s)\{2(s-a)[x(x-4a)(s-4a) + a^2(s-10a)] - 3\lambda_1 a^2[(x-3a)(x-a)+(s-a)^2]\}$$

and the entry $g_{32}(x, s)$, with $x \in [2a, 3a]$ and $s \in [a, 2a]$, reads as

$$g_{32}(x, s) = \frac{1}{2\lambda_1 pa}(s - a)(x - 2a)(s - 2a)[2(a - s) - \lambda_1 s]$$

For the entry $g_{13}(x, s)$, with $x \in [0, a]$ and $s \in [2a, 3a]$, we obtain

$$g_{13}(x, s) = \frac{1}{2pa}x^2(a - x)(2a - s)$$

The entry $g_{23}(x, s)$, with $x \in [a, 2a]$ and $s \in [2a, 3a]$, is expressed as

$$g_{23}(x, s) = \frac{1}{2\lambda_1 pa}(x - 2a)(x - a)(s - 2a)[2(a - x) - \lambda_1 x]$$

and for the branch $g_{33}^{+}(x, s)$ of $g_{33}(x, s)$, with $2a \le x \le s \le 3a$, we obtain

$$g_{33}^{+}(x, s) = (x - 2a)\left\{\frac{1 + \lambda_1}{\lambda_1 p}a(2a - s) + \frac{(x - 2a)}{6EI_3}[(x + a) + 3(a - s)]\right\}$$

while the other branch $g_{33}^{-}(x, s)$ of $g_{33}(x, s)$, which is defined for $2a \le s \le x \le 3a$, is found as

$$g_{33}^{-}(x, s) = (s - 2a)\left\{\frac{1 + \lambda_1}{\lambda_1 p}a(2a - x) + \frac{(s - 2a)}{6EI_3}[(s + a) + 3(a - x)]\right\}$$

The influence matrix just presented allows one to analytically obtain components of the stress-strain state caused by any reasonable combination of transverse and bending loads applied to the beam depicted in Figure 4.13. Indeed, when the deflection function is explicitly available, one can routinely compute bending moments and shear forces in any cross-section of the beam by correspondingly differentiating the integral representations of the deflection function. If the loading functions $q_1(x)$, $q_2(x)$, and $q_3(x)$ have simple form (polynomial, exponential, trigonometric, or their elementary combinations), then the integration (see, for example, eqn (4.111)) can be conducted analytically, otherwise it can be accomplished approximately by applying appropriate quadrature formulae.

The examples analyzed in this section are helpful in comprehending the material but they cannot clear up all possible peculiarities of the influence function method as applied to multi-span compound beams. The reader is therefore encouraged to go through the End Chapter Exercises to gain an experience in obtaining influence matrices for compound multi-span beams as well as in utilizing influence matrices for computing components of stress-strain states in beams undergoing various combinations of loads.

## 4.5 End Chapter Exercises

**4.1** For a cantilever beam of length $a$, whose left-hand edge $x = 0$ is clamped, with $EI(x) \equiv mx + b$ representing a variable flexural rigidity, determine the response to different loads shown below:

**a)** a uniform transverse load $q_0$ applied to the entire beam length;

**b)** two concentrated forces $P_1$ and $P_2$, spaced at $x = a_1$ and $x = a_2$, respectively, with $a_2 > a_1$;

**c)** two concentrated bending moments $M_1$ and $M_2$, spaced at $x = a_1$ and $x = a_2$, respectively, with $a_2 > a_1$.

**4.2** Construct the influence function for a cantilever beam of length $a$, whose right-hand edge $x = a$ is clamped, with $EI(x) \equiv mx$ representing the flexural rigidity.

**4.3** For the beam in Exercise 4.2, determine its response to a combination of the transverse concentrated force $P_0$ spaced at $x = a_1$, and the concentrated bending moment $M_0$ spaced at $x = a_2$, with $a_1 < a_2$.

**4.4** Construct the influence function for a cantilever beam of length $a$, whose edge $x = 0$ is clamped, with $EI(x) \equiv me^{bx}$ representing the flexural rigidity.

**4.5** For the beam in Exercise 4.4, determine its response to a combination of the concentrated force $P_0$ spaced at $x = a/4$ and the uniform transverse load $q_0$ distributed over the interval $[a/2, a]$.

**4.6** For a double-span cantilever beam overhanging a simple support (see Figure 4.14), utilize the influence matrix, obtained in Section 4.4 (with $EI_1 = EI_2 = EI$), and compute the deflection, bending moment, and shear force caused by the loads shown below:

**a)** continuously distributed as $q(x) = \dfrac{q_0}{a}(x - a)$, for $0 < x < 2a$;

**b)** piecewise constant as: $q(x) = q_1 = const$, for $0 < x < a$ and $q(x) = q_2 = const$, for $a < x < 2a$;

**c)** continuously distributed as $q(x) = \dfrac{q_0}{a^2}x(2a - x)$, for $0 < x < 2a$.

**d)** constant as: $q(x) = q_0 = const$, for $\alpha < x < \beta$ where $a < \alpha,\ \beta < 2a$;

**e)** constant as: $q(x) = q_0 = const$, for $\alpha < x < \beta$ where $0 < \alpha,\ \beta < a$;

**f)** continuously distributed as $q(x) = \dfrac{q_0}{a^3}x^2(2a - x)$, for $0 < x < 2a$.

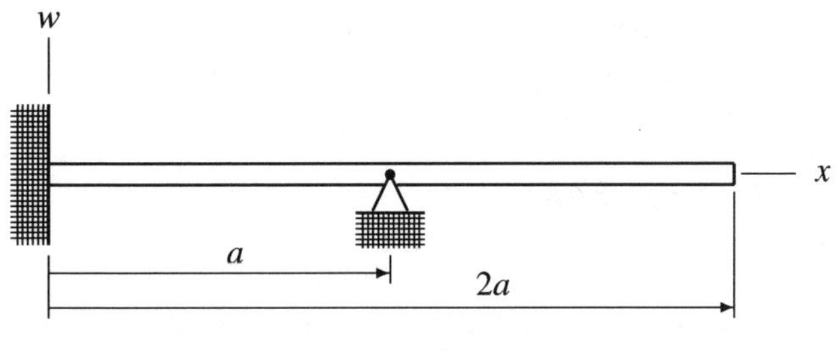


Figure 4.14: Double-span cantilever beam

**4.7** Utilize the entries of the influence matrix presented in Example 4.2 to determine the basic components (deflection, bending moment, and shear force) of the stress-strain state for the double-span beam depicted in Figure 4.11. Let the beam's spans be made of different elastic materials ($EI_1$ and $EI_2$ represent the flexural rigidities of the left-hand and the right-hand span, respectively), and let the load be applied as shown:

**a)** continuously distributed as $q(x) = q_0x^2 + q_1x + q_2$, for $-a < x < a$;

**b)** continuously distributed as $q(x) = q_0e^{-mx}$, for $-a < x < a$;

**c)** continuously distributed as $q(x) = (q_0x + q_1)e^{-mx}$, for $-a < x < 0$;

**d)** continuously distributed as $q(x) = q_0xe^{-mx}$, for $0 < x < a$;

**e)** piecewise constant as: $q(x) = q_1 = const$, for $-a < x < 0$ and $q(x) = q_2 = const$, for $0 < x < a$;

**4.8** Use the procedure described in Section 4.4 to construct the influence matrix of a transverse unit point concentrated force for the double-span clamped compound ($EI_1$ and $EI_2$) beam having an intermediate simple support right at the midpoint as depicted in Figure 4.15.

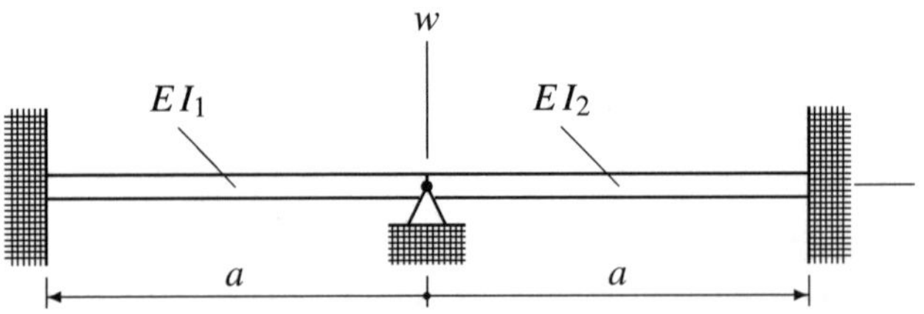


Figure 4.15: A clamped beam having an intermediate support

**4.9** Utilize the influence matrix constructed in Exercise 4.8 and determine the basic components (deflection, bending moment and shear force) of the stress-strain state for the double-span compound beam having an intermediate support and loaded as shown below:

**a)** $q(x) = q_0x + q_1$, for $0 < x < a$;

**b)** $q(x) = q_0x + q_1$, for $-a < x < a$;

**4.10** Construct the influence matrix of a transverse unit point concentrated force for the double-span clamped compound beam having an intermediate simple support at the midpoint as depicted in Figure 4.16. The beam's spans are made of different ($EI_1$ and $EI_2$) isotropic homogeneous elastic materials.

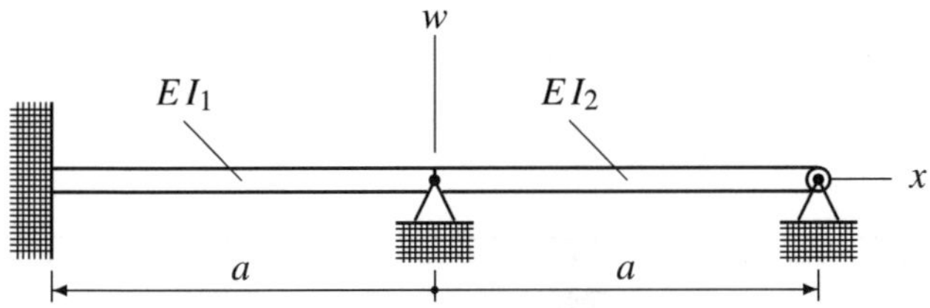


Figure 4.16: A beam with an intermediate support

**4.11** Utilize the influence matrix constructed in Exercise 4.10 and determine the basic components (deflection, bending moment and shear force) of the stress-strain state for the clamped–simply supported compound beam having an intermediate support and loaded as shown below:

**a)** piecewise constant as: $q(x) = q_1 = const$, for $-a < x < 0$ and $q(x) = q_2 = const$, for $0 < x < a$;

**b)** piecewise defined as: $q(x) = q_0 \cos(\pi x/2a)$, for $-a < x < 0$ and $q(x) = \dfrac{q_0}{a}(a - x)$, for $0 < x < a$;

In Exercises 4.12 and 4.13, the reader will gain an experience of going through a quite cumbersome algebra associated with the extension of our procedure to the construction of influence functions to triple-span beams.

**4.12** Construct the influence matrix of a transverse unit concentrated force for the triple-span cantilever compound beam overhanging a simple support as depicted in Figure 4.17.

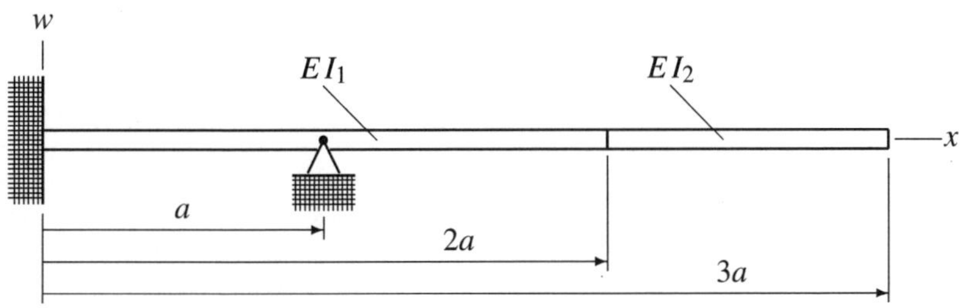


Figure 4.17: A compound cantilever beam overhanging one support

**4.13** Construct the influence matrix of a transverse unit concentrated force for the triple-span beam of a uniform flexural rigidity $EI$, having two simple supports as depicted in Figure 4.18.

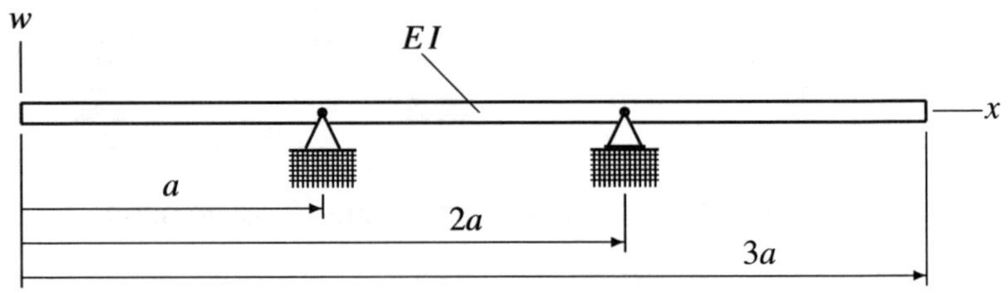


Figure 4.18: A triple-span beam having two supports

# Chapter 5

# Bending of Plates and Shells

Most of the Kirchhoff beam problems, which we have been involved with in Chapters 3 and 4, are governed with boundary-value problems for fourth order linear ordinary differential equations. It is assumed that the reader has already learned that influence functions of a transverse point concentrated force for a beam are associated with some Green's functions of differential equations that govern the bending phenomenon for beams.

Our involvement in this chapter is with thin plates and shells made in most cases of elastic isotropic homogeneous materials. Bending of a thin plate or shell is described with a boundary-value problem for either a single partial differential equation (as it occurs in the case of Poisson–Kirchhoff plates) or for a system of partial differential equations (Reissner plates or shells, for example). Consequently, the influence function of a transverse point force for an element of a thin-walled structure represents either a Green's function (in the case of a single governing differential equation) or a Green's matrix, if the problem is governed by a system of equations.

One of the objectives in this chapter is to develop and validate a methodology for the construction of influence functions of a transverse point force for Poisson–Kirchhoff plates. Mathematically those influence functions represent Green's functions to some boundary-value problems stated for the biharmonic equation over regions of standard shape. We will also extend the methodology to mathematical formulations of the Reissner plate theory, within which the modeling is further advanced compared to the classical Poisson–Kirchhoff theory. Bending problems for shells of revolution are visited later.

In a brief section that opens this chapter, we give a special definition of the Green's matrix to a boundary-value problem for a system of elliptic partial differential equations. Our definition is not traditional from the standpoint of mathematics. We define the Green's matrix in a way to create an appropriate basis for the practical construction of influence functions of a point force for a broad variety of plate and shell problems.

Section 5.2 is devoted to the construction of influence functions for Poisson–Kirchhoff plates of a standard shape with various boundary conditions imposed. In Section 5.3, we present some results on influence functions of a point force for Poisson–Kirchhoff plates resting on elastic foundation. Section 5.4 extends our interest to influence functions for Reissner plates, while in Section 5.5 we take a close look at the construction of influence functions for thin shells of revolution.

## 5.1 Influence Matrices for Plates and Shells

Problems of static equilibrium will be considered in this chapter for thin plates and shells. The presentation is limited to physically and geometrically linear problem settings. Since input and output functions depend on two spacial variables, governing differential equations are partial, except for the case with axially symmetric statement of problems for shells of revolution where high order systems of ordinary differential equations are considered.

As of this point in our text the reader has learned that the influence matrix of a transverse point concentrated force for a thin plate or shell represents the Green's matrix of a boundary-value problem that governs the bending phenomenon for the object. To maximally accommodate the presentation in this chapter to the construction of influence functions, we introduce the notion of Green's matrix in a specific manner. This will make our presentation more pragmatic and allows us to avoid cumbersome and lengthy developments that are required when this topic is covered in mathematics. Instead, the focus is made herein on the construction procedure itself.

Let $\Omega$ represent a simply connected region (bounded or unbounded) in the two-dimensional Euclidean space and let $\Gamma$ denote a piecewise smooth contour of $\Omega$. Consider the boundary-value problem

$$L[U(P)] = -F(P), \quad P \in \Omega \tag{5.1}$$

$$M[U(P)] = 0, \quad P \in \Gamma \tag{5.2}$$

where $L$ and $M$ represent matrix-operators of the governing system of partial differential equations and of the boundary conditions, respectively. $U(P)$ and $F(P)$ are the solution vector-function (the output) and a vector-function of the right-hand side (the loading function or input), respectively. If the matrix-operators $L$ and $M$ degenerate to a scalar operator form, then the above setting reduces to a boundary-value problem for a single partial differential equation for a single scalar function $U(P)$.

We assume that the boundary-value problem posed by eqns (5.1) and (5.2) has a unique solution. This implies that the corresponding homogeneous problem (where $F(P) \equiv 0$) has only the trivial $U(P) \equiv 0$ solution. The above assumption is absolutely crucial for both mathematics and mechanics standpoints, because it: (i) ensures the existence and uniqueness of the Green's matrix of the homogeneous problem corresponding to that in eqns (5.1) and (5.2); and (ii) makes the setting in (5.1) and (5.2) feasible from a physics point of view.

**Definition:** When, for any allowable right-hand side vector $F(P)$, the solution vector $U(P)$ to the problem in (5.1) and (5.2) is expressed in the form

$$U(P) = \iint_{\Omega} G(P, Q)F(Q)\, d\Omega(Q) \tag{5.3}$$

then the kernel-matrix $G(P, Q)$ of the above integral is said to be the *Green's matrix* of the homogeneous problem

$$L[U(P)] = 0, \quad P \in \Omega$$
$$M[U(P)] = 0, \quad P \in \Gamma$$

corresponding to that of eqns (5.1) and (5.2). In other words, the kernel-matrix $G(P, Q)$ of (5.3) represents the influence matrix of a point force for the plate or shell whose bending is simulated by the boundary-value problem in (5.1) and (5.2). This spells that $G(P, Q)$ represents the plate's or shell's response at a point $P$ due to a unit transverse point force applied at a point $Q$.

Notice that the term "allowable", with regard to the right-hand side function $F(P)$, implies that the integral of $F(P)$ over $\Omega$ is bounded. That is

$$\left| \iint_{\Omega} F(P)\, d\Omega(P) \right| < \infty$$

This condition reflects an obvious physical limitation on the statement in eqns (5.1) and (5.2). Namely, the amount of energy, delivered to the plate under consideration by the loading function $F(P)$, ought to be a finite quantity.

Properties of the influence matrix $G(P, Q)$ are individually defined for every particular problem setting in (5.1) and (5.2). The operators $L$ and $M$ actually determine the properties. Whereas, the structure of $G(P, Q)$ is standard for any elliptic system and is viewed as

$$G(P, Q) = S(P, Q) + R(P, Q) \tag{5.4}$$

where the additive component $S(P, Q)$ represents the *fundamental solution matrix* of the governing system and is referred to as the *singular component* of $G(P, Q)$. The other additive term $R(P, Q)$ in (5.4) is the *regular component* of $G(P, Q)$ and, as a function of $P$, it represents such a solution of the homogeneous system

$$L[R(P, Q)] = 0$$

that enables $G(P, Q)$ to satisfy the boundary conditions imposed with eqn (5.2), for every fixed location of the source point $Q \in \Omega$.

Two different models for the bending of thin plates are considered in this text. In Sections 5.2 and 5.3, we construct influence functions of a transverse point force for Poisson–Kirchhoff plates of various shape and edge conditions, while influence matrices for Reissner plates are constructed in Section 5.4.

## 5.2 Poisson–Kirchhoff Plates

In this section, we begin our development for influence functions for a variety of problems for thin plates. The simplest plate model is considered herein. That is, thin plates of a uniform thickness are treated within the scope of the classical

Poisson–Kirchhoff theory [21, 27, 58, 65] which results in a boundary-value problem for the biharmonic equation

$$D\Delta\Delta w(P) = -q(P) \tag{5.5}$$

where $\Delta$ represents the Laplace operator written in terms of the coordinates of $P$, while $w(P)$ and $q(P)$ represent the lateral deflection of the plate's middle plane and the transverse distributed load applied to the plate, respectively. The coefficient $D$, which is called the plate's *flexural rigidity*, is defined through physical and geometrical parameters of the plate as

$$D = \frac{Eh^3}{12(1-\sigma^2)}$$

with $E$ and $\sigma$ representing the elasticity modulus and the Poisson ratio, respectively, of the material of which the plate is made. The parameter $h$ represents the thickness of the plate.

Notice that a single partial differential equation governs the problem in this case. We will thus be dealing herein with Green's functions rather than with Green's matrices. Physical interpretation of the Green's function $G(P, Q)$ of the homogeneous equation corresponding to (5.5) is the lateral deflection of the middle plane of the plate at $P$ due to a transverse unit point force acting at $Q$.

Since the fundamental solution of the homogeneous biharmonic equation

$$\frac{1}{8\pi}|P-Q|^2 \ln \frac{1}{|P-Q|} \tag{5.6}$$

(which represents the singular component $S(P, Q)$ of the Green's function $G(P, Q)$ in eqn (5.4)) is well known [65], the construction procedure for the Green's function could exclusively be focused on the regular component $R(P, Q)$.

Note, however, that the procedure advocated in this text for obtaining Green's functions and matrices does not target their regular component alone. Instead, it allows us to obtain both singular and regular components at the same time. It is also important to note that the term "singular" is conditionally applied to the term in (5.6). The point is that, as it can easily be proved by the L'Hospital rule, the limit of the function in (5.6), as the force application point $Q$ approaches the observation point $P$, equals zero. The term spells a singularity of the second order derivatives of $S(P, Q)$ with respect to the coordinates of the observation point $P$. These derivatives are directly proportional to bending moments which are unbounded when $S \to P$.

Only a limited number of influence functions for Poisson–Kirchhoff plates are available in the existing literature. The most comprehensive list of those can be found in [45, 65]. The most productive traditional methods for obtaining influence functions for plate problems are: (i) expansion in trigonometric series (in the plate and shell theory [28, 65], it is often referred to as either *Navier* (double series) or *Levy* (single series) method), (ii) the reflection or the image method, and (iii) the complex variables method based on the presentation of a biharmonic function by means of two harmonic functions.

### 5.2.1 Rectangular-shaped plates

A number of particular Poisson–Kirchhoff plate problems will be examined in this subsection as examples illustrating the procedure that appears to be productive in the construction of influence functions of a point force. Standard rectangular-shaped plates (infinite strip, semi-infinite strip and rectangle) are considered with a variety of edge conditions imposed.

**Example 2.1:** We begin with a classical problem by considering a rectangular plate whose middle plane occupies the region $\Omega = \{(x, y) : 0 < x < a, 0 < y < b\}$ and all the edges are simply-supported.

Note that the double Fourier series form

$$G(x, y; s, t) = -\frac{4}{ab}\sum_{m=1}^{\infty}\sum_{n=1}^{\infty}\frac{\sin\mu x \sin\mu s \sin\nu y \sin\nu t}{(\mu^2+\nu^2)^2}$$

(where: $\mu = m\pi/b$ and $\nu = n\pi/b$) of the influence function for the simply-supported rectangular plate is available in every text dealing with mathematical models of plate problems (see, for example, [38, 65]). We will show how the above expression can be derived in light of the definition given earlier in Section 5.1.

The setting in this example translates the boundary-value problem of eqns (5.1) and (5.2) to

$$\frac{\partial^4 w(x, y)}{\partial x^4} + 2\frac{\partial^4 w(x, y)}{\partial x^2 \partial y^2} + \frac{\partial^4 w(x, y)}{\partial y^4} = -f(x, y) \tag{5.7}$$

$$w = \left.\frac{\partial^2 w}{\partial y^2}\right|_{y=0,b} = 0, \quad w = \left.\frac{\partial^2 w}{\partial x^2}\right|_{x=0,a} = 0 \tag{5.8}$$

where the right-hand side function in eqn (5.7) is defined in terms of the loading function $q(x, y)$ of eqn (5.5) as

$$f(x, y) = \frac{q(x, y)}{D}$$

A specific form of the boundary conditions makes it possible to apply the Navier approach [58, 65] to the problem in (5.7) and (5.8). In compliance with this approach the solution is expressed in a form of the double Fourier sine-series

$$w(x, y) = \sum_{m=1}^{\infty}\sum_{n=1}^{\infty} w_{mn} \sin\mu x \sin\nu y, \quad \mu = \frac{m\pi}{a}, \quad \nu = \frac{n\pi}{b} \tag{5.9}$$

It is evident that the above representation for $w(x, y)$ satisfies all the boundary conditions as imposed in (5.8).

We also expend the right-hand side function $f(x, y)$ of eqn (5.7) in the identical double sine-series form

$$f(x, y) = \sum_{m=1}^{\infty} \sum_{n=1}^{\infty} f_{mn} \sin \mu x \sin \nu y \tag{5.10}$$

Substituting then expansions from (5.9) and (5.10) in (5.7) and combining the like terms in the left-hand side yields the equation

$$\sum_{m=1}^{\infty} \sum_{n=1}^{\infty} (\mu^4 + 2\mu^2\nu^2 + \nu^4) w_{mn} \sin \mu x \sin \nu y = -\sum_{m=1}^{\infty} \sum_{n=1}^{\infty} f_{mn} \sin \mu x \sin \nu y$$

from which, equating the corresponding coefficients of the above series and performing a trivial algebra, one obtains

$$w_{mn} = -\frac{f_{mn}}{(\mu^2 + \nu^2)^2}$$

Substitution of this expression in (5.9) yields

$$w(x, y) = -\sum_{m=1}^{\infty} \sum_{n=1}^{\infty} \frac{f_{mn}}{(\mu^2 + \nu^2)^2} \sin \mu x \sin \nu y \tag{5.11}$$

Recall the Euler–Fourier formula from eqn (1.26) in Chapter 1 and adjust it to the Fourier double-series environment. This yields for the Fourier coefficients $f_{mn}$ of the right-hand side term in eqn (5.7)

$$f_{mn} = \frac{4}{ab} \int_0^a \int_0^b f(s, t) \sin \mu s \sin \nu t \, ds \, dt$$

Substitute this expression for $f_{mn}$ in (5.11) and assume that the summation and integration in it can be interchanged. This delivers the ultimate solution to the boundary-value problem of eqns (5.7) and (5.8) as

$$w(x, y) = -\frac{4}{ab} \int_0^a \int_0^b \sum_{m=1}^{\infty} \sum_{n=1}^{\infty} \frac{\sin \mu x \sin \mu s \sin \nu y \sin \nu t}{(\mu^2 + \nu^2)^2} f(s, t) \, ds \, dt$$

The change of order of the summation and integration operations that we just made is well justified in theory [60], but we will not go to specifics of the justification itself because it stays beyond the scope of this text.

The above expression for $w(x, y)$ can be rewritten in terms the loading function $q(x, y)$ as

$$w(x, y) = -\frac{4}{abD}\int_0^a\int_0^b \sum_{m=1}^{\infty}\sum_{n=1}^{\infty} \frac{\sin \mu x \sin \mu s \sin \nu y \sin \nu t}{(\mu^2+\nu^2)^2} q(s, t)\, ds\, dt \tag{5.12}$$

Thus, in light of the definition introduced earlier in this Chapter (see eqn (5.3)), we conclude that the kernel-function

$$G(x, y; s, t) = -\frac{4}{abD}\sum_{m=1}^{\infty}\sum_{n=1}^{\infty} \frac{\sin \mu x \sin \mu s \sin \nu y \sin \nu t}{(\mu^2+\nu^2)^2} \tag{5.13}$$

in (5.12) is indeed the influence function of a transverse point force for the simply-supported rectangular plate that was presented earlier. The points $(x, y)$ and $(s, t)$ represent the observation and the force application point, respectively.

Note that the series in eqn (5.13) converges at a relatively high rate. This implies that the plate's deflection at a point $(x, y)$, caused by a transverse point force acting at $(s, t)$, can be accurately computed by an appropriate truncation of the series. This is true with no regard to a mutual location of the observation and the force application points.

As to the stress components caused in the plate by a point force, the situation is not that propitious. According to the Poisson–Kirchhoff theory [65], the bending and twisting moments caused in the plate by a transverse point force have to infinitely increase as the observation point approaches the force application point. This is supported by the series in eqn (5.13). Indeed, stress components are defined in terms of derivatives of the deflection function, but differentiation of a functional series is going to worsen its convergence. We can readily articulate this assertion.

From the plate theory (see, for example, [65]), it follows that the bending moments $M_x(x, y)$ and $M_y(x, y)$ in the plate are expressed in terms of the deflection function $w(x, y)$ as

$$M_x(x, y) = -D\left(\frac{\partial^2 w(x, y)}{\partial x^2} + \sigma\frac{\partial^2 w(x, y)}{\partial y^2}\right)$$

and

$$M_y(x, y) = -D\left(\frac{\partial^2 w(x, y)}{\partial y^2} + \sigma\frac{\partial^2 w(x, y)}{\partial x^2}\right)$$

Hence, second order partial derivatives of the influence function $G(x, y; s, t)$ with respect to the coordinates of the observation point $(x, y)$ are required to get the bending moments at $(x, y)$ caused by a transverse point force at $(s, t)$. Upon differentiating the series in (5.13) in the term-by-term manner, one arrives, for the bending moment $M_x(x, y; s, t)$, for example, at the series

$$M_x(x, y; s, t) = \frac{4}{ab} \sum_{m=1}^{\infty} \sum_{n=1}^{\infty} (\mu^2 + \sigma \nu^2) \frac{\sin \mu x \sin \mu s \sin \nu y \sin \nu t}{(\mu^2 + \nu^2)^2}$$

whose slow convergence can be easily disclosed. In doing so, we add and subtract the term of $\nu^2$ to the factor of $(\mu^2 + \sigma \nu^2)$ transforming the above series as

$$\begin{aligned} M_x(x, y; s, t) &= \frac{4}{ab} \sum_{m=1}^{\infty} \sum_{n=1}^{\infty} [(\mu^2 + \nu^2) + (\sigma - 1)\nu^2] \frac{\sin \mu x \sin \mu s \sin \nu y \sin \nu t}{(\mu^2 + \nu^2)^2} \\ &= \frac{4}{ab} \left\{ \sum_{m=1}^{\infty} \sum_{n=1}^{\infty} (\mu^2 + \nu^2) \frac{\sin \mu x \sin \mu s \sin \nu y \sin \nu t}{(\mu^2 + \nu^2)^2} \right. \\ &\quad \left. + (\sigma - 1) \sum_{m=1}^{\infty} \sum_{n=1}^{\infty} \nu^2 \frac{\sin \mu x \sin \mu s \sin \nu y \sin \nu t}{(\mu^2 + \nu^2)^2} \right\} \\ &= \frac{4}{ab} \left\{ \sum_{m=1}^{\infty} \sum_{n=1}^{\infty} \frac{\sin \mu x \sin \mu s \sin \nu y \sin \nu t}{\mu^2 + \nu^2} \right. \\ &\quad \left. + (\sigma - 1) \sum_{m=1}^{\infty} \sum_{n=1}^{\infty} \nu^2 \frac{\sin \mu x \sin \mu s \sin \nu y \sin \nu t}{(\mu^2 + \nu^2)^2} \right\} \end{aligned}$$

Hence, the series representing $M_x(x, y; s, t)$ has been broken onto two, where the second series is rapidly convergent, whereas the first one

$$\sum_{m=1}^{\infty} \sum_{n=1}^{\infty} \frac{\sin \mu x \sin \mu s \sin \nu y \sin \nu t}{\mu^2 + \nu^2}$$

can be recognized as the classical double-series representation to the Green's function for the Dirichlet problem for the Laplace equation on the rectangle $\Omega = \{(x, y) : 0 < x < a, 0 < y < b\}$. It is well-known [60] that the above series converges at a slow rate if the observation point $(x, y)$ is close to the force application point $(s, t)$ and even diverges when these points coincide. So, special attention is required when computing stress components in the simply-supported rectangular plate subject to a transverse point force.

In contrast to the case of a transverse point force, the convergence of the series representing $G(x, y; s, t)$ is not an issue when components of the stress-strain state are computed with eqn (5.12) for a plate subject to a distributed load $q(x, y)$. We will support this statement by considering a few examples for a simply-supported rectangular plate.

It is evident that, since the series in (5.13) is convergent, the integration in (5.12) can be carried out on the term-by-term basis. As the first example, let the whole plate be loaded with a uniform transverse load of magnitude $Q_0$. The deflection function in eqn (5.13) reads, in this case, as

$$w(x, y) = -\frac{4Q_0}{abD}\int_0^a\int_0^b \sum_{m=1}^{\infty}\sum_{n=1}^{\infty} \frac{\sin\mu x \sin\mu s \sin\nu y \sin\nu t}{(\mu^2+\nu^2)^2}\, ds\, dt$$

and a trivial integration yields

$$w(x, y) = -\frac{4Q_0}{abD}\sum_{m=1}^{\infty}\sum_{n=1}^{\infty} \frac{\sin\mu x \sin\nu y(1-\cos\mu a)(1-\cos\nu b)}{\mu\nu(\mu^2+\nu^2)^2}$$

Evidently, convergence of this series is much higher compared to the series in eqn (5.13) and the perspective of an accurate computation of stress-related components, with its aid, is viewed quite attractive. To find the bending moment $M_x$, we differentiate the above expression in compliance with the recently presented relation as

$$M_x(x, y) = -D\left(\frac{\partial^2 w(x, y)}{\partial x^2} + \sigma\frac{\partial^2 w(x, y)}{\partial y^2}\right)$$

$$= \frac{4Q_0}{abD}\sum_{m=1}^{\infty}\sum_{n=1}^{\infty} \frac{(\mu^2+\sigma\nu^2)\sin\mu x \sin\nu y(1-\cos\mu a)(1-\cos\nu b)}{\mu\nu(\mu^2+\nu^2)^2}$$

while for the bending moment $M_y$ we have

$$M_y(x, y) = -D\left(\frac{\partial^2 w(x, y)}{\partial y^2} + \sigma\frac{\partial^2 w(x, y)}{\partial x^2}\right)$$

$$= \frac{4Q_0}{abD}\sum_{m=1}^{\infty}\sum_{n=1}^{\infty} \frac{(\nu^2+\sigma\mu^2)\sin\mu x \sin\nu y(1-\cos\mu a)(1-\cos\nu b)}{\mu\nu(\mu^2+\nu^2)^2}$$

It is clearly seen that the above series converges at the same rate as the series representing the influence function itself. This even makes it possible to accurately compute shear forces in the plate. Indeed, as it follows from the plate theory [65], the shear forces $Q_x$ and $Q_y$ are expressed in terms of the deflection function as

$$Q_x(x, y) = -D\frac{\partial}{\partial x}\left(\frac{\partial^2 w(x, y)}{\partial x^2} + \frac{\partial^2 w(x, y)}{\partial y^2}\right)$$

and

$$Q_y(x, y) = -D\frac{\partial}{\partial y}\left(\frac{\partial^2 w(x, y)}{\partial x^2} + \frac{\partial^2 w(x, y)}{\partial y^2}\right)$$

Hence, in the plate uniformly loaded with the transverse load $Q_0$, the force $Q_x$, for example, is found as

$$Q_x(x, y) = \frac{4Q_0}{abD}\sum_{m=1}^{\infty}\sum_{n=1}^{\infty} \frac{\cos\mu x \sin\nu y(1-\cos\mu a)(1-\cos\nu b)}{\nu(\mu^2+\nu^2)}$$

For another illustration of a practicality of the solution in eqn (5.12), consider the simply-supported rectangular plate with a transverse uniform load $Q_0$ applied to the rectangular ($a_1 \leq x \leq a_2,\ b_1 \leq y \leq b_2$) region on the plate. The relation in eqn (5.13) transforms, in this case, to

$$w(x, y) = -\frac{4Q_0}{abD}\int_{a_1}^{a_2}\int_{b_1}^{b_2}\sum_{m=1}^{\infty}\sum_{n=1}^{\infty}\frac{\sin\mu x \sin\mu s \sin\nu y \sin\nu t}{(\mu^2+\nu^2)^2}\,ds\,dt$$

$$= -\frac{4Q_0}{abD}\sum_{m=1}^{\infty}\sum_{n=1}^{\infty}\frac{\sin\mu x \sin\nu y(\cos\mu a_1 - \cos\mu a_2)(\cos\nu b_1 - \cos\nu b_2)}{\mu\nu(\mu^2+\nu^2)^2}$$

whose high convergence creates a basis for accurate computation of the stress-related components. In the End Chapter Exercises, the reader is encouraged to obtain explicit expressions for the bending moments and shear force for this load.

We complete the discussion on the simply-supported rectangular plate with considering the case of the load $q(x, y) = Q_0 x(a - x)$. This transforms the relation in (5.12) to

$$w(x, y) = -\frac{4Q_0}{abD}\int_0^a\int_0^b\sum_{m=1}^{\infty}\sum_{n=1}^{\infty}\frac{\sin\mu x \sin\mu s \sin\nu y \sin\nu t}{(\mu^2+\nu^2)^2}s(a-s)\,ds\,dt$$

where the integration yields a rapidly convergent series

$$w(x, y) = -\frac{4Q_0}{abD}\sum_{m=1}^{\infty}\sum_{n=1}^{\infty}\frac{\sin\mu x \sin\nu y(1-\cos\nu b)[2(1-\cos\mu a) - \mu a \sin\mu a]}{\mu^3\nu(\mu^2+\nu^2)^2}$$

allowing a term-by-term differentiation for accurate computation of the stress-strain state.

Many other distributed loads allow analytic integration in (5.12). Some of such cases are included in the End Chapter Exercises, where the reader is required to go through the integration routine. If it appears that a loading function $q(x, y)$ is too intricate for analytic integration in (5.12), then any standard numerical integration procedure can be employed. ▲

It is evident that the double-series Navier-type method (used in Example 2.1) is effective in obtaining the influence function for a simply-supported rectangular plate. It is also clear that this method cannot be applied to plates with different edge conditions imposed. That is why, in a broad set of examples that are considered later in this text, an approach based on the single-series Levy-type method [58, 65] is developed. It appears to be productive when other than simple support boundary conditions are imposed.

**Example 2.2:** Consider a semi-infinite strip-shaped plate whose middle plane occupies the region $\Omega = \{(x, y) : 0 < x < \infty, 0 < y < b\}$, with the edges $y = 0$ and $y = b$ being simply-supported while the edge $x = 0$ is clamped. This yields a boundary-value problem for the governing differential equation in (5.7) with edge conditions written as

$$w = \left.\frac{\partial^2 w}{\partial y^2}\right|_{y=0,b} = 0, \quad w = \left.\frac{\partial w}{\partial x}\right|_{x=0} = 0 \tag{5.14}$$

In addition to the boundary conditions in (5.14), the components of the stress-strain state of the plate are required to be bounded as $x$ approaches infinity. This comment is important to ensure a unique solvability of the problem.

To obtain the solution to the boundary-value problem in eqns (5.7) and (5.14), we express $w(x, y)$ in a form of the Fourier sine-series

$$w(x, y) = \sum_{n=1}^{\infty} w_n(x) \sin \nu y, \quad \nu = \frac{n\pi}{b} \tag{5.15}$$

and also express the loading function $f(x, y)$ of eqn (5.7) in the identical sine-series form

$$f(x, y) = \sum_{n=1}^{\infty} f_n(x) \sin \nu y \tag{5.16}$$

The representation for $w(x, y)$ in (5.15) satisfies the first two boundary conditions in eqn (5.14). Similarly to the development in the previous example, we substitute the expansions from (5.15) and (5.16) in (5.7) and equate then the corresponding coefficients of the two Fourier sine-series that arise on the right and on the left of the equal sign. This brings the following set ($n = 1, 2, 3, \ldots$) of boundary-value problems

$$\frac{d^4 w_n(x)}{dx^4} - 2\nu^2 \frac{d^2 w_n(x)}{dx^2} + \nu^4 w_n(x) = -f_n(x), \quad x \in (0, \infty) \tag{5.17}$$

$$w_n(0) = \frac{dw_n(0)}{dx} = 0, \quad |w_n(\infty)| < \infty, \quad \left|\frac{dw_n(\infty)}{dx}\right| < \infty \tag{5.18}$$

in the coefficients $w_n(x)$ of the series from eqn (5.15).

The Green's function $g_n(x, s)$ of the homogeneous ($f_n(x) \equiv 0$) problem corresponding to that of eqns (5.17) and (5.18) has earlier been derived in Chapter 2 (see Example 3.7). We rewrite it here using our current notations as

$$g_n(x, s) = \frac{1}{4\nu^3}\{[1 + \nu(x + s) + 2\nu^2 xs]e^{-\nu(x+s)} - (1 + \nu|x - s|)e^{-\nu|x-s|}\} \tag{5.19}$$

Notice that this expression for $g_n(x, s)$ is written, in contrast to that of eqn (2.114) in Chapter 2, in a compact single-piece form, which is valid for any mutual location of the variables $x$ and $s$. This becomes possible due to the symmetry of $g_n(x, s)$ with the introduction of the absolute value function.

Theorem 2.4 of Chapter 2 suggests that the solution of the boundary-value problem posed by eqns (5.17) and (5.18) can be written in terms of $g_n(x, s)$ from eqn (5.19) as the improper integral

$$w_n(x) = \int_0^\infty g_n(x, s) f_n(s)\, ds$$

This integral converges because, as we earlier assumed, $f(x, y)$ is integrable over $\Omega$, which implies that $f_n(x)$ ought to be integrable over $(0, \infty)$.

For the series expansion of eqn (5.16), the Fourier coefficients $f_n(s)$ of the right-hand side term $f(x, y)$ in eqn (5.7) can be written by means of the Euler–Fourier formula as

$$f_n(s) = \frac{2}{b} \int_0^b f(s, t) \sin \nu t\, dt$$

providing for $w_n(x)$ the integral representation

$$w_n(x) = \frac{2}{b} \int_0^b \int_0^\infty g_n(x, s) \sin \nu t\ f(s, t)\, ds\, dt$$

Upon substituting this in eqn (5.15) and interchanging the summation and the integration, the solution of the boundary-value problem posed by eqns (5.7) and (5.14) is ultimately found as

$$w(x, y) = \int_0^b \int_0^\infty \left( \frac{2}{b} \sum_{n=1}^\infty g_n(x, s) \sin \nu y \sin \nu t \right) f(s, t)\, ds\, dt$$

In light of the definition introduced earlier in Section 5.1, the kernel-function

$$G(x, y; s, t) = \frac{2}{bD} \sum_{n=1}^\infty g_n(x, s) \sin \nu y \sin \nu t \tag{5.20}$$

in the above integral represents the influence function of a transverse point force for the plate under consideration, with $(x, y)$ and $(s, t)$ being the observation and the force application point, respectively.

So, the series representation in (5.20) is the influence function that we are looking for. That form would be satisfactory if the series in it is convenient in computer implementations. This is not, however, the case for the series in (5.20) because its convergence rate is too low (it is of the order of $1/n$). This notably diminishes the practical value of the representation and hardens its direct numerical use especially if both the variables $x$ and $s$ are close to the edge $x = 0$ of the plate.

A two-step approach is proposed and used in eliminating the deficiency of the representation in (5.20). First, a low convergent component of the series is spotted and isolated and, second, that component is summed up analytically. This results in a part analytic-part series form of the influence function, where the series component is rapidly convergent.

To accomplish this plan, let us substitute $g_n(x, s)$ from eqn (5.19) in the series of eqn (5.20) and break down the first exponential term in $g_n(x, s)$ onto two. This reduces $G(x, y; s, t)$ to

$$G(x, y; s, t) = \frac{1}{bD}\sum_{n=1}^{\infty}\left(\frac{xs}{\nu}e^{-\nu(x+s)} + \widetilde{g}_n(x, s)\right)\sin \nu y \sin \nu t \tag{5.21}$$

where

$$\widetilde{g}_n(x, s) = \frac{1+\nu(x+s)}{2\nu^3}e^{-\nu(x+s)} - \frac{1-\nu|x-s|}{2\nu^3}e^{-\nu|x-s|} \tag{5.22}$$

Of the two series in (5.21) the second (the one with the coefficient $\widetilde{g}_n(x, s)$) converges at the rate of $1/n^2$, whereas the convergence rate of the first series in (5.21) is of the order of $1/n$. So, the first step of the approach in increasing the practicality of the representation in eqn (5.20) is a success. Indeed, its low convergent component is already isolated. To complete the job, we ought to sum up the first series in (5.21). In doing so, take out the factor of $xs$ and transform (5.21) by applying the standard trigonometric identity

$$\sin\alpha \sin\beta = \tfrac{1}{2}[\cos(\alpha-\beta) - \cos(\alpha+\beta)]$$

This yields

$$\begin{aligned}
&\sum_{n=1}^{\infty}\frac{e^{-\nu(x+s)}}{\nu}\sin \nu y \sin \nu t \\
&= \frac{1}{2}\left(\sum_{n=1}^{\infty}\frac{e^{-\nu(x+s)}}{\nu}\cos \nu(y-t) - \sum_{n=1}^{\infty}\frac{e^{-\nu(x+s)}}{\nu}\cos \nu(y+t)\right) \\
&= \frac{1}{2p}\left(\sum_{n=1}^{\infty}\frac{(e^{-p(x+s)})^n}{n}\cos np(y-t) - \sum_{n=1}^{\infty}\frac{(e^{-p(x+s)})^n}{n}\cos np(y+t)\right)
\end{aligned} \tag{5.23}$$

where $p = \pi/b$.

The two series in (5.23) are completely summable. This can be shown with the aid of the summation formula

$$\sum_{n=1}^{\infty}\frac{r^n}{n}\cos n\varphi = -\ln\sqrt{1-2r\cos\varphi + r^2} \tag{5.24}$$

that we have derived in Chapter 1 (see eqn (1.42)). Indeed, as it follows from the above relation, the parameters $r$ and $\varphi$ are defined for the series in (5.23) as

$$r = e^{-p(x+s)} \quad \text{and} \quad \varphi = p(y \pm t)$$

and it is evident that they meet the limitations

$$r^2 < 1 \quad \text{and} \quad 0 \le \varphi < 2\pi$$

required for the summation formula in (5.24).

Thus, the series in (5.23) sums up as

$$= \frac{1}{2p}\left(\sum_{n=1}^{\infty}\frac{(e^{-p(x+s)})^n}{n}\cos np(y-t) - \sum_{n=1}^{\infty}\frac{(e^{-p(x+s)})^n}{n}\cos np(y+t)\right)$$

$$= \frac{1}{2p}\left(\ln\sqrt{1-2e^{-p(x+s)}\cos p(y+t)+e^{-2p(x+s)}}\right.$$

$$\left. - \ln\sqrt{1-2e^{-p(x+s)}\cos p(y-t)+e^{-2p(x+s)}}\right)$$

$$= \frac{1}{2p}\ln\sqrt{\frac{1-2e^{-p(x+s)}\cos p(y+t)+e^{-2p(x+s)}}{1-2e^{-p(x+s)}\cos p(y-t)+e^{-2p(x+s)}}}$$

Multiplying the numerator and the denominator of the radicand by $e^{2p(x+s)}$, we reduce the above expression to

$$\frac{1}{2p}\ln\sqrt{\frac{1-2e^{p(x+s)}\cos p(y+t)+e^{2p(x+s)}}{1-2e^{p(x+s)}\cos p(y-t)+e^{2p(x+s)}}}$$

This function can be written in a more compact form by introducing the complex variables $z = x + iy$ and $\zeta = s + it$ for the observation and the force application point, respectively. Indeed, it reads as the real-valued function of $z$ and $\zeta$

$$\frac{1}{2p}\ln\frac{|1-e^{p(z+\zeta)}|}{|1-e^{p(z+\overline{\zeta})}|} \tag{5.25}$$

where the bar on $\zeta$ denotes the conjugate of $\zeta$, that is $\overline{\zeta} = s - it$.

Thus, the function in (5.25) represents the sum of the first series in (5.21), so that the entire expression for the influence function of a point force for the plate in Example 2.2 is ultimately found as

$$G(x, y; s, t) = \frac{xs}{2\pi D}\ln\frac{|1-e^{p(z+\zeta)}|}{|1-e^{p(z+\overline{\zeta})}|} + \frac{1}{bD}\sum_{n=1}^{\infty}\tilde{g}_n(x, s)\sin \nu y \sin \nu t \tag{5.26}$$

It is evident that this form has greater practical merit compared to the form in eqn (5.20), because the series in (5.26) converges at the faster rate of $1/n^2$ (see the expression for $\tilde{g}_n(x, s)$ in eqn (5.22)). This makes it possible to accurately compute values of $G(x, y; s, t)$ in (5.26) by appropriately truncating its series. ▲

Later in this section, we will examine series of the type in eqn (5.26) and address the convergence issue in more detail. An analysis will be provided of differential properties that are of great importance in obtaining stress-related components of the stress-strain state of the plate undergoing transverse loads.

**Example 2.3:** The Levy method-based technique that we developed above can successfully be used in the construction of influence functions for the semi-infinite strip-shaped plate with other types of edge conditions imposed. If, for example, the edges $y=0$ and $y=b$ are simply-supported as in the previous case, while the edge $x=0$ is free of tension, then the boundary conditions at $x=0$ in the corresponding boundary-value problem for the biharmonic equation that models the bending of the plate are written as

$$\left.\left(\frac{\partial^2 w}{\partial x^2}+\sigma\frac{\partial^2 w}{\partial y^2}\right)\right|_{x=0}=0$$

$$\left.\frac{\partial}{\partial x}\left(\frac{\partial^2 w}{\partial x^2}+(2-\sigma)\frac{\partial^2 w}{\partial y^2}\right)\right|_{x=0}=0$$

where $\sigma$ represents the Poisson ratio of the material of which the plate is made. These conditions assign to zero the bending moment $M_x$ and the shear force $Q_x$ on the plate's edge $x=0$ (see [58, 65]).

The influence function of a transverse point concentrated force for such a plate can also be written in a form of the series expansion from eqn (5.20). The coefficient $g_n(x, s)$ of that expansion represents, in this case, the Green's function for the homogeneous boundary-value problem

$$\frac{d^4 w_n(x)}{dx^4}-2\nu^2\frac{d^2 w_n(x)}{dx^2}+\nu^4 w_n(x)=0, \quad x\in(0,\infty)$$

$$\frac{d^2 w(x)}{dx^2}-\sigma\nu^2 w_n(0)=\frac{d^3 w_n(0)}{dx^3}-(2-\sigma)\nu^2\frac{dw_n(0)}{dx}=0$$

$$|w_n(\infty)|<\infty, \quad \left|\frac{dw_n(\infty)}{dx}\right|<\infty$$

where the parameter $\nu=n\pi/b$.

Leaving the details of the derivation procedure as an exercise for the reader, we write down the final expression for the coefficient of the series in (5.20) as

$$g_n(x,s)=-\frac{1+\nu|x-s|}{4\nu^3}e^{-\nu|x-s|}$$
$$-\frac{(4+(1+\sigma)^2)+\nu(1-\sigma)^2(x+s+2\nu xs)}{4\nu^3(1-\sigma)(3+\sigma)}e^{-\nu(x+s)} \tag{5.27}$$

Clearly, the improvement of the convergence of the series in (5.20) can, in this case, be achieved in a manner similar to that used in Example 2.2. The reader is recommended to explore this issue in the End Chapter Exercises. ▲

The influence function of a concentrated unit force for the semi-infinite strip-shaped plate with all the edges being simply-supported is again given by the expansion from eqn (5.20), with $g_n(x, s)$ being, for $x \le s$, defined as

$$g_n(x,s)=\frac{1+\nu(x+s)}{4\nu^3}e^{-\nu(x+s)}-\frac{1+\nu|x-s|}{4\nu^3}e^{-\nu|x-s|}$$

This problem to be included in the End Chapter Exercises.

At this point in our presentation, we revisit the rectangular plate where the Levy method-based approach allows one to obtain influence functions for other than simple support edge conditions imposed.

**Example 2.4:** Let, for example, a plate occupy the region $\Omega = \{(x, y) : 0 < x < a, 0 < y < b\}$. Let the edge $x = 0$ be simply-supported and the edge $x = a$ clamped, while both the edges $y = 0$ and $y = b$ are simply-supported. That is, boundary conditions for the biharmonic equation simulating the bending of the plate are imposed, in this case, as

$$w = \left.\frac{\partial^2 w}{\partial y^2}\right|_{y=0,b} = 0, \quad w = \left.\frac{\partial^2 w}{\partial x^2}\right|_{x=0} = 0, \quad w = \left.\frac{\partial w}{\partial x}\right|_{x=a} = 0$$

For such a plate, the influence function of a transverse concentrated point force also reduces to the series of eqn (5.20), whose coefficient $g_n(x, s)$ represents the Green's function of the following boundary-value problem

$$\frac{d^4 w_n(x)}{dx^4} - 2\nu^2 \frac{d^2 w_n(x)}{dx^2} + \nu^4 w_n(x) = 0, \quad x \in (0, a) \tag{5.28}$$

$$w_n(0) = \frac{d^2 w_n(0)}{dx^2} = 0, \quad w_n(a) = \frac{dw_n(a)}{dx} = 0 \tag{5.29}$$

where $\nu = n\pi/b$.

Tracing out a routine but quite cumbersome procedure, expression of the Green's function to the above boundary-value problem is found, for $x \leq s$, as

$$\begin{aligned} g_n(x, s) = \frac{1}{2\nu^3 \Delta^*} \{ & \nu x \cosh \nu x [2\nu(s-a) \cosh \nu s - \sinh \nu(s-2a) \\ & - \sinh \nu s] - \sinh \nu x [\nu s \cosh \nu(s-2a) - \sinh \nu(s-2a) \\ & + \nu(s-2a) \cosh \nu s + (2\nu^2 a(s-a) - 1) \sinh \nu s] \} \end{aligned} \tag{5.30}$$

where $\Delta^* = \sinh 2\nu a - 2\nu a$.

Note that expression for $g_n(x, s)$ valid for $x \geq s$, can be obtained from that above by interchanging of $x$ with $s$. ▲

We turn now to the issue of convergence for the series in eqn (5.20). To be certain, the semi-infinite strip-shaped plate is considered, with all edges simply-supported. The coefficient $g_n(x, s)$ in (5.20) is, in this case, expressed as

$$g_n(x, s) = \frac{1 + \nu(x+s)}{4\nu^3} e^{-\nu(x+s)} - \frac{1 + \nu|x-s|}{4\nu^3} e^{-\nu|x-s|}, \quad \nu = \frac{n\pi}{b} \tag{5.31}$$

Notice that for any location of the observation $(x, y)$ and the force application $(s, t)$ point, the expansion in eqn (5.20) represents a uniformly convergent series. It rapidly converges so that its first order partial derivatives can be taken on the term-by-term basis. That is, the series obtained from that of eqn (5.20) by the term-by-term partial differentiation with respect to either $x$ or $y$ also converges uniformly. Hence, the computational implementations based on either the series in eqn (5.20) itself or on its first order partial derivatives can be accurately carried out by an appropriate truncation of the series.

Let us show that computing of higher order derivatives of the expression in eqn (5.20) is not simple. Recall that the influence function $G(x, y; s, t)$ of a plate is viewed as the deflection $w(x, y)$ at the observation point $(x, y)$ due to a transverse concentrated unit force applied at the force application point $(s, t)$. From the Poisson–Kirchhoff theory, it follows that the bending moments $M_x(x, y)$ and $M_y(x, y)$ in the plate

$$M_x(x, y) = -D\left(\frac{\partial^2 w(x, y)}{\partial x^2} + \sigma\frac{\partial^2 w(x, y)}{\partial y^2}\right)$$

and

$$M_y(x, y) = -D\left(\frac{\partial^2 w(x, y)}{\partial y^2} + \sigma\frac{\partial^2 w(x, y)}{\partial x^2}\right)$$

are expressed in terms of the second order partial derivatives of the deflection function $w(x, y)$. Hence, the second order partial differentiation of the influence function $G(x, y; s, t)$ is required to get the bending moments caused by a transverse point concentrated force.

The term-by-term second order partial differentiation of the series in eqn (5.20) yields, however, a non-uniformly convergent series that diverges logarithmically when the observation point approaches the force application point. This agrees with our expectation. Indeed, for a Poisson–Kirchhoff plate loaded with a transverse point concentrated force, the bending moments possess logarithmic singularity at a point of the force application.

To address the non-uniform convergence issue in more detail, let us determine the bending moments $M_x$ and $M_y$ in the semi-infinite strip-shaped plate whose edges are simply-supported and which is subject to a transverse unit force $P_0 = 1$ concentrated at a point $(s, t)$. Upon interpreting the influence function $G(x, y; s, t)$ of the plate as its deflection at the point $(x, y)$ due to $P_0$, we obtain the following representations

$$M_x(x, y; s, t) = -D\left(\frac{\partial^2 G(x, y; s, t)}{\partial x^2} + \sigma\frac{\partial^2 G(x, y; s, t)}{\partial y^2}\right)$$

and

$$M_y(x, y; s, t) = -D\left(\frac{\partial^2 G(x, y; s, t)}{\partial y^2} + \sigma\frac{\partial^2 G(x, y; s, t)}{\partial x^2}\right)$$

for the bending moments caused by the unit point force.

After one substitutes the expression for $G(x, y; s, t)$ from eqn (5.20), with the coefficients $g_n(x, s)$ presented in eqn (5.31), into the above representations, the latter convert to

$$M_x(x, y; s, t) = -\frac{1}{2b}\sum_{n=1}^{\infty}\left\{\left[\frac{1+\sigma}{\nu} - (1-\sigma)|x-s|\right]e^{-\nu|x-s|}\right.$$
$$\left. + \left[(1-\sigma)(x+s) - \frac{1+\sigma}{\nu}\right]e^{-\nu(x+s)}\right\}\sin\nu y \sin\nu t$$

and

$$M_y(x, y; s, t) = -\frac{1}{2b}\sum_{n=1}^{\infty}\left\{\left[\frac{1+\sigma}{\nu} + (1-\sigma)|x-s|\right]e^{-\nu|x-s|}\right.$$
$$\left. - \left[(1-\sigma)(x+s) + \frac{1+\sigma}{\nu}\right]e^{-\nu(x+s)}\right\}\sin\nu y \sin\nu t$$

These series representations for the bending moments can readily be summed up with the aid of the standard summation formulae derived in Chapter 1 (see eqns (1.28) and (1.32)). This finally yields

$$M_x(x, y; s, t) = -\frac{1}{4b}\left\{\frac{b(1+\sigma)}{\pi}\ln\frac{E(z-\bar{\zeta})E(z+\bar{\zeta})}{E(z-\zeta)E(z+\zeta)}\right.$$
$$+ (1-\sigma)\left[|x-s|\left(\frac{R(z-\bar{\zeta})}{E^2(z-\bar{\zeta})} - \frac{R(z-\zeta)}{E^2(z-\zeta)}\right)\right.$$
$$\left.\left. - (x+s)\left(\frac{R(-(z+\bar{\zeta}))}{E^2(-(z+\bar{\zeta}))} - \frac{R(-(z+\zeta))}{E^2(-(z+\zeta))}\right)\right]\right\} \tag{5.32}$$

and

$$M_y(x, y; s, t) = -\frac{1}{4b}\left\{\frac{b(1+\sigma)}{\pi}\ln\frac{E(z-\bar{\zeta})E(z+\bar{\zeta})}{E(z-\zeta)E(z+\zeta)}\right.$$
$$- (1-\sigma)\left[|x-s|\left(\frac{R(z-\bar{\zeta})}{E^2(z-\bar{\zeta})} - \frac{R(z-\zeta)}{E^2(z-\zeta)}\right)\right.$$
$$\left.\left. - (x+s)\left(\frac{R(-(z+\bar{\zeta}))}{E^2(-(z+\bar{\zeta}))} - \frac{R(-(z+\zeta))}{E^2(-(z+\zeta))}\right)\right]\right\} \tag{5.33}$$

with $z$ and $\zeta$ denoting the observation and the force application point, respectively. The real-valued functions $E(w)$ and $R(w)$ of a complex variable $w$ are defined as

$$E(w) = \left|1 - \exp\left(\frac{\pi w}{b}\right)\right|$$

and

$$R(w) = \mathrm{Re}\left(1 - \exp\left(\frac{\pi w}{b}\right)\right)$$

Thus, in the semi-infinite strip-shaped simply-supported plate, the bending moments $M_x$ and $M_y$, caused by a point concentrated force, are obtained in a closed easy computable form.

It is evident that the representations for the bending moments in (5.32) and (5.33) possess logarithmic singularity if the observation point $z$ coincides with the force application point $\zeta$. The singularity is caused by the logarithmic terms $\ln(E(z-\zeta))$. No other singularity is available. Indeed, the non-logarithmic terms containing $E(z-\zeta)$ in the denominators in (5.32) and (5.33) represent a removable singularity, because of the factor $|x-s|$.

### 5.2.2 Circular-shaped plates

A number of examples examined above should raise the reader's confidence in our approach to the construction of influence functions of a point force for rectangular shaped plates. The intention, in this subsection, is to show that the technique that we succeeded with so far can also be productive in considering circular-shaped plates with a variety of edge conditions imposed.

**Example 2.5:** We first turn the reader's attention to a classical problem. That is, a clamped circular plate of radius $a$ and of a uniform thickness. Assume that the plate is made of an isotropic homogeneous material and occupies the region $\Omega = \{(r,\varphi) : 0 < r < a, 0 \leq \varphi < 2\pi\}$. For this plate, the influence function

$$G(z,\zeta) = \frac{1}{8\pi D}\left[\frac{1}{2a^2}(a^2-|z|^2)(a^2-|\zeta|^2) - |z-\zeta|^2 \ln \frac{|a^2 - z\bar{\zeta}|}{a|z-\zeta|}\right] \tag{5.34}$$

of a transverse point concentrated force is available in the existing literature. It can be found, for instance, in [38, 45, 65].

$G(z,\zeta)$ in (5.34) represents the deflection of the plate that occurs at the observation point $z = r(\cos\varphi + i\sin\varphi)$ due to a transverse concentrated unit force applied at a point $\zeta = \rho(\cos\psi + i\sin\psi)$.

Let us show that the technique, described earlier in this section, provides an alternative way for deriving the representation for $G(z,\zeta)$ shown in eqn (5.34). In doing so, we consider the following boundary-value problem

$$\left(\frac{1}{r}\frac{\partial}{\partial r}\left(r\frac{\partial}{\partial r}\right) + \frac{1}{r^2}\frac{\partial^2}{\partial\varphi^2}\right)^2 w(r,\varphi) = -f(r,\varphi), \quad (r,\varphi) \in \Omega \tag{5.35}$$

$$w(a,\varphi) = 0, \qquad \frac{\partial w(a,\varphi)}{\partial r} = 0 \tag{5.36}$$

written for the biharmonic equation in polar coordinates $r$ and $\varphi$. This problem simulates the bending of a clamped circular plate caused by a transverse distributed load

$$q(r,\varphi) = Df(r,\varphi)$$

applied to the plate under consideration, where $D$ represents the plate's flexural rigidity.

Since the formulation in eqns (5.35) and (5.36) is $2\pi$-periodic with respect to the $\varphi$ variable, we assume for the deflection function $w(r,\varphi)$ the following trigonometric Fourier expansion

$$w(r,\varphi)=\frac{1}{2}w_0(r)+\sum_{n=1}^{\infty}(w_n^c(r)\cos n\varphi+w_n^s(r)\sin n\varphi) \tag{5.37}$$

Let also the right-hand side term in eqn (5.35) be expressed by the general Fourier series

$$f(r,\varphi)=\frac{1}{2}f_0(r)+\sum_{n=1}^{\infty}(f_n^c(r)\cos n\varphi+f_n^s(r)\sin n\varphi) \tag{5.38}$$

Upon substituting the above representations in eqn (5.35), one obtains, for the coefficients $w_n(r)$ of the expansion in eqn (5.37), the following set ($n = 0, 1, 2, \ldots$) of ordinary differential equations

$$\left(\frac{d^4}{dr^4}+\frac{2}{r}\frac{d^3}{dr^3}-\frac{1+2n^2}{r^2}\frac{d^2}{dr^2}+\frac{1+2n^2}{r^3}\frac{d}{dr}+\frac{n^2(n^2-4)}{r^4}\right)w_n(r)=-f_n(r)$$

where, for notational convenience, we omit the superscripts on $w_n(r)$ and $f_n(r)$, because the cosine and the sine modes in (5.37) and (5.38) will be treated similarly until a certain stage in the development.

It appears that if Green's function is constructed for the above differential equation, then it would not be in a symmetric form. But the property of symmetry can be restored by implementing the integrating factor of $r$. Indeed, the homogeneous boundary-value problem

$$\left(r\frac{d^4}{dr^4}+2\frac{d^3}{dr^3}-\frac{1+2n^2}{r}\frac{d^2}{dr^2}+\frac{1+2n^2}{r^2}\frac{d}{dr}+\frac{n^2(n^2-4)}{r^3}\right)w_n(r)=0 \tag{5.39}$$

$$|w_n(0)|<\infty,\quad \left|\frac{d^2w_n(0)}{dr^2}\right|<\infty,\quad w_n(a)=\frac{dw_n(a)}{dr}=0 \tag{5.40}$$

is in a form for which its Green's function ought to be symmetric.

Note that the first two relations in eqn (5.40) are imposed to ensure that the solution $w(r,\varphi)$ of the original boundary-value problem posed by eqns (5.35) and (5.36) is bounded at the origin, while the third and the fourth relations in eqn (5.40) directly follow from the clamped edge conditions of eqn (5.36) and from the Fourier expansion shown in eqn (5.37).

An important comment ought to be offered before we get down to the construction of the Green's function to the boundary-value problem in eqns (5.39) and (5.40). The point is that it is impossible to obtain a single fundamental set of solutions of the governing equation in (5.39), the set that is valid for the entire range of the parameter $n = 0, 1, 2, \ldots$. Indeed, three individual cases ($n = 0$, $n = 1$, and $n \geq 2$) of eqn (5.39) must be distinguished and treated separately, because their fundamental sets of solutions are different.

Consider first the equation in (5.39) in the case of $n = 0$

$$\left( r\frac{d^4}{dr^4} + 2\frac{d^3}{dr^3} - \frac{1}{r}\frac{d^2}{dr^2} + \frac{1}{r^2}\frac{d}{dr} \right) w_0(r) = 0 \tag{5.41}$$

This equation represents a well-known type of Cauchy–Euler [15, 23, 60, 68]. Its solution can be written as $w_0(r) = r^k$, where values of $k$ are to be determined by substituting this form of $w_0(r)$ into (5.41). This yields the auxiliary equation

$$k(k-1)(k-2)(k-3) + 2k(k-1)(k-2) - k(k-1) + k = 0 \tag{5.42}$$

in $k$, which can be rewritten as

$$k(k-1)(k-2)(k-3) + 2k(k-1)(k-2) - k(k-2) = 0$$

or

$$k(k-2)[(k-1)(k-3) + 2(k-1) - 1] = 0$$

reducing finally eqn (5.42) to the compact directly solvable form

$$k^2(k-2)^2 = 0$$

which implies that $k = 0$ and $k = 2$ each represents a double-root for the auxiliary equation in (5.42). Hence, taking into account the multiplicity of roots of the auxiliary equation (see Chapter 1), a fundamental set of solutions of eqn (5.41) can be formed with the functions

$$1, \quad \ln r, \quad r^2, \quad \text{and} \quad r^2 \ln r$$

with which the Green's function $g_0(r, \rho)$ can be derived to the boundary-value problem in eqns (5.41) and (5.40). Either the procedure based on the defining properties of Green's functions (see Section 2.1) or the method of variation of parameters (see Section 2.3) can be used. We omit details of the derivation procedure and deliver just the final expression for $g_0(r, \rho)$ as

$$g_0(r, \rho) = \frac{1}{8}\left[ \frac{1}{a^2}(a^2 - \rho^2)(a^2 + r^2) + 2(r^2 + \rho^2) \ln \frac{\rho}{a} \right], \quad r \le \rho \tag{5.43}$$

In the second of the three individual cases that we are going through (that is the case of $n = 1$), the equation in (5.39) reads as

$$\left( r\frac{d^4}{dr^4} + 2\frac{d^3}{dr^3} - \frac{3}{r}\frac{d^2}{dr^2} + \frac{3}{r^2}\frac{d}{dr} + \frac{3}{r^3} \right) w_1(r) = 0 \tag{5.44}$$

also representing the Cauchy–Euler type equation whose auxiliary equation is different of that in (5.42) and appears as

$$k(k-1)(k-2)(k-3) + 2k(k-1)(k-2) - 3k(k-1) + 3k - 3 = 0$$

By a trivial algebra, which resembles that applied to eqn (5.42), four real roots of the above equation are found as

$$k = 1 \text{ (double-root)}, \quad k = -1 \quad \text{and} \quad k = 3$$

This allows the fundamental set of solutions of eqn (5.44) to be represented with the functions

$$r^{-1}, \quad r, \quad r^3, \quad \text{and} \quad r \ln r$$

bringing the expression for the Green's function $g_1(r, \rho)$, which is valid for $r \leq \rho$, for this case in the form

$$g_1(r, \rho) = \frac{r(\rho^2 - a^2)}{16a^4\rho}[r^2(a^2 - \rho^2) + 2a^2\rho^2] - \frac{1}{4}r\rho \ln \frac{\rho}{a} \tag{5.45}$$

In the third of the three individual cases (for $n \geq 2$ in eqn (5.39)), we arrive at the auxiliary equation as

$$\begin{aligned} &k(k-1)(k-2)(k-3) + 2k(k-1)(k-2) \\ &\quad - (1+2n^2)k(k-1) + (1+2n^2)k + n^2(n^2-4) = 0 \end{aligned} \tag{5.46}$$

Finding the roots of this equation is not that trivial as it has been in the cases of $n = 0$ and $n = 1$. That is why we will describe the solution process in more detail. Take the first two additive terms in eqn (5.46) and factor their sum as shown

$$k(k-1)(k-2)(k-3) + 2k(k-1)(k-2) = k(k-1)^2(k-2)$$

while the sum of the third and the fourth terms can be simplified as

$$-(1+2n^2)k(k-1) + (1+2n^2)k = -(1+2n^2)k(k-2)$$

This converts (5.46) into

$$k(k-2)[(k-1)^2 - (1+2n^2)] + n^2(n^2-4) = 0$$

which can be rewritten as

$$k(k-2)[k(k-2) - 2n^2)] + n^2(n^2-4) = 0$$

or, removing the brackets, we have

$$\underline{[k(k-2)]^2 - 2n^2k(k-2) + n^4} - 4n^2 = 0$$

At this point of our development it is important to observe that the sum of the underlined terms in the above equation represent a complete square. This transforms the equation into

$$[k(k-2)-n^2]^2-4n^2=0$$

Viewing now the left-hand side of the above equation as a difference of squares, we factor it as

$$[k(k-2)-n^2-2n][k(k-2)-n^2+2n]=0$$

Hence, with the series of elegant transformations, we have managed to break the auxiliary equation in (5.46) onto two trivial quadratic equations. This brings four distinct real roots of the auxiliary equation as

$$k=n, \quad k=-n, \quad k=n+2, \quad \text{and} \quad k=2-n$$

Hence, the fundamental set of solutions of eqn (5.39) can be represented with the functions

$$r^n, \quad r^{-n}, \quad r^{n+2}, \quad \text{and} \quad r^{2-n}$$

This delivers the following expression

$$g_n(r,\rho)=-\frac{1}{8}\left\{\frac{r\rho}{n-1}\left[\left(\frac{r\rho}{a^2}\right)^{n-1}-\left(\frac{r}{\rho}\right)^{n-1}\right]+\frac{r^2+\rho^2}{n}\left[\left(\frac{r}{\rho}\right)^n-\left(\frac{r\rho}{a^2}\right)^n\right]\right.$$
$$\left.+\frac{r\rho}{n+1}\left[\left(\frac{r\rho}{a^2}\right)^{n+1}-\left(\frac{r}{\rho}\right)^{n+1}\right]\right\}, \quad r\le\rho \tag{5.47}$$

for the Green's function to the problem in eqns (5.39) and (5.40) for the general case of $n\ge 2$.

The reader probably noticed that the expressions for the Green's functions $g_0(r,\rho)$, $g_1(r,\rho)$, and $g_n(r,\rho)$ that we derived in eqns (5.43), (5.45), and (5.47) are valid only for $r\ge\rho$. But regardless of the index $n$, expressions for $g_0(r,\rho)$, $g_1(r,\rho)$, and $g_n(r,\rho)$, which are valid for $r\ge\rho$, can be obtained from the corresponding ones presented by eqns (5.43), (5.45), and (5.47) by interchanging of $r$ with $\rho$.

Tracing out our procedure developed earlier for rectangular-shaped plates, the influence function $G(r,\varphi;\rho,\psi)$ of a transverse point force for the clamped circular plate is expressed in terms of the Green's functions $g_0(r,\rho)$, $g_1(r,\rho)$, and $g_n(r,\rho)$ as

$$G(r,\varphi;\rho,\psi)=\frac{1}{2\pi D}\left[g_0(r,\rho)+2\sum_{n=1}^{\infty}g_n(r,\rho)\cos n(\varphi-\psi)\right] \tag{5.48}$$

Note that from the mathematics standpoint, the above represents the Green's function of the homogeneous boundary-value problem corresponding to that of eqns (5.35) and (5.36).

In what follows we are going to show that the series in (5.48) can be summed up completely. Notice that it does not matter which of the two branches of its coefficients (either the ones valid for $r \leq \rho$ or the other ones) is taken for the summation procedure. In our derivation, we take advantage of the branches presented in eqns (5.43), (5.45), and (5.47).

To sum up the series in (5.48), we somewhat regroup its terms. Since the coefficient $g_1(r, \rho)$ of the first term in the series part of $G(r, \varphi; \rho, \psi)$ is obtained in a form different of the rest of the coefficients $g_n(r, \rho)$, obtained for $n = 2, 3, \ldots,$ we isolate the entire first term

$$g_1(r, \rho)\cos(\varphi - \psi)$$

of the series and rewrite (5.48) as

$$G(r, \varphi; \rho, \psi) = \frac{1}{2\pi D}\Bigg[g_0(r, \rho) + 2g_1(r, \rho)\cos(\varphi - \psi) + 2\sum_{n=2}^{\infty} g_n(r, \rho)\cos n(\varphi - \psi)\Bigg] \tag{5.49}$$

Upon substituting the expressions for $g_0(r, \rho)$ and $g_1(r, \rho)$ presented in eqns (5.43) and (5.45) into this expansion, the first two terms in the brackets read as

$$\begin{aligned} g_0(r, \rho) &+ 2g_1(r, \rho)\cos(\varphi - \psi) \\ &= \frac{1}{8}\left[\frac{1}{a^2}(a^2 - \rho^2)(a^2 + r^2) + 2(r^2 + \rho^2)\ln\frac{\rho}{a}\right] \\ &\quad - \left[\frac{r(a^2 - \rho^2)}{8a^4\rho}(r^2(a^2 - \rho^2) + 2a^2\rho^2) + \frac{1}{2}r\rho\ln\frac{\rho}{a}\right]\cos(\varphi - \psi) \end{aligned}$$

Combining the logarithmic terms, we rewrite it as

$$\begin{aligned} g_0(r, \rho) &+ 2g_1(r, \rho)\cos(\varphi - \psi) \\ &= \frac{1}{8}\left\{\underline{\frac{1}{a^2}(a^2 - \rho^2)(a^2 + r^2)} + 2[r^2 - 2r\rho\cos(\varphi - \psi) + \rho^2]\ln\frac{\rho}{a}\right. \\ &\quad \left. \underline{\underline{- \frac{r(a^2 - \rho^2)}{a^4\rho}[r^2(a^2 - \rho^2) + 2a^2\rho^2]\cos(\varphi - \psi)}}\right\} \end{aligned} \tag{5.50}$$

Later this expression will be recalled when the series in eqn (5.49) is ready for the ultimate summation.

By substituting the expression for $g_n(r, \rho)$ from eqn (5.47) into the series part of eqn (5.49), we obtain

$$\begin{aligned} 2\sum_{n=2}^{\infty} & g_n(r, \rho)\cos n(\varphi - \psi) \\ &= -\frac{1}{4}\left\{r\rho\sum_{n=2}^{\infty}\frac{1}{n-1}\left[\left(\frac{r\rho}{a^2}\right)^{n-1} - \left(\frac{r}{\rho}\right)^{n-1}\right]\cos n(\varphi - \psi)\right. \end{aligned}$$

$$+ (r^2+\rho^2)\sum_{n=2}^{\infty}\frac{1}{n}\left[\left(\frac{r}{\rho}\right)^n - \left(\frac{r\rho}{a^2}\right)^n\right]\cos n(\varphi-\psi)$$

$$+ r\rho\sum_{n=2}^{\infty}\frac{1}{n+1}\left[\left(\frac{r\rho}{a^2}\right)^{n+1} - \left(\frac{r}{\rho}\right)^{n+1}\right]\cos n(\varphi-\psi)\Bigg\} \qquad (5.51)$$

Each of the series in this expression requires an individual treatment. To partially sum up the first of them, we change its summation index $n$ by making the substitution $k = n-1$. This yields

$$\sum_{n=2}^{\infty}\frac{1}{n-1}\left[\left(\frac{r\rho}{a^2}\right)^{n-1} - \left(\frac{r}{\rho}\right)^{n-1}\right]\cos n(\varphi-\psi)$$

$$= \sum_{k=1}^{\infty}\frac{1}{k}\left[\left(\frac{r\rho}{a^2}\right)^{k} - \left(\frac{r}{\rho}\right)^{k}\right]\cos (k+1)(\varphi-\psi)$$

$$= \sum_{k=1}^{\infty}\frac{1}{k}\left[\left(\frac{r\rho}{a^2}\right)^{k} - \left(\frac{r}{\rho}\right)^{k}\right]$$

$$\times [\cos k(\varphi-\psi)\cos(\varphi-\psi) - \sin k(\varphi-\psi)\sin(\varphi-\psi)]$$

$$= \cos(\varphi-\psi)\sum_{k=1}^{\infty}\frac{1}{k}\left[\left(\frac{r\rho}{a^2}\right)^{k} - \left(\frac{r}{\rho}\right)^{k}\right]\cos k(\varphi-\psi)$$

$$- \sin(\varphi-\psi)\sum_{k=1}^{\infty}\frac{1}{k}\left[\left(\frac{r\rho}{a^2}\right)^{k} - \left(\frac{r}{\rho}\right)^{k}\right]\sin k(\varphi-\psi)$$

To sum up the above cosine-series, we recall the summation formula that we derived in eqn (1.42) of Chapter 1, while the sine-series will be left in its current form. This yields, for the entire first term in eqn (5.51)

$$r\rho\sum_{n=2}^{\infty}\frac{1}{n-1}\left[\left(\frac{r\rho}{a^2}\right)^{n-1} - \left(\frac{r}{\rho}\right)^{n-1}\right]\cos n(\varphi-\psi)$$

$$= r\rho\cos(\varphi-\psi)\left[\frac{1}{2}\ln\left(1-2\frac{r}{\rho}\cos(\varphi-\psi)+\left(\frac{r}{\rho}\right)^2\right)\right.$$

$$\left.-\frac{1}{2}\ln\left(1-2\frac{r\rho}{a^2}\cos(\varphi-\psi)+\left(\frac{r\rho}{a^2}\right)^2\right)\right]$$

$$- r\rho\sin(\varphi-\psi)\sum_{k=1}^{\infty}\frac{1}{k}\left[\left(\frac{r\rho}{a^2}\right)^{k} - \left(\frac{r}{\rho}\right)^{k}\right]\sin k(\varphi-\psi) \qquad (5.52)$$

The sine-series in (5.52) can be also summed up. Indeed, it is evident that this can be done with the aid of another summation formula that was also derived in Chapter 1 (see eqn (1.43)). But for the sake of our further development, we, however, leave the sine-series in its current form.

To sum up the series in the second term of eqn (5.51), we rewrite it as

$$(r^2+\rho^2)\sum_{n=2}^{\infty}\frac{1}{n}\left[\left(\frac{r}{\rho}\right)^n-\left(\frac{r\rho}{a^2}\right)^n\right]\cos n(\varphi-\psi)$$
$$=(r^2+\rho^2)\sum_{n=1}^{\infty}\frac{1}{n}\left[\left(\frac{r}{\rho}\right)^n-\left(\frac{r\rho}{a^2}\right)^n\right]\cos n(\varphi-\psi)$$
$$+(r^2+\rho^2)\left[\left(\frac{r\rho}{a^2}\right)-\left(\frac{r}{\rho}\right)\right]\cos(\varphi-\psi)$$

and implement the summation formula of eqn (1.42). This allows us to obtain

$$(r^2+\rho^2)\sum_{n=2}^{\infty}\frac{1}{n}\left[\left(\frac{r}{\rho}\right)^n-\left(\frac{r\rho}{a^2}\right)^n\right]\cos n(\varphi-\psi)$$
$$=-(r^2+\rho^2)\left[\frac{1}{2}\ln\left(1-2\frac{r}{\rho}\cos(\varphi-\psi)+\left(\frac{r}{\rho}\right)^2\right)\right.$$
$$\left.-\frac{1}{2}\ln\left(1-2\frac{r\rho}{a^2}\cos(\varphi-\psi)+\left(\frac{r\rho}{a^2}\right)^2\right)\right]$$
$$+(r^2+\rho^2)\left[\left(\frac{r\rho}{a^2}\right)-\left(\frac{r}{\rho}\right)\right]\cos(\varphi-\psi)$$

For a partial summation of the series of the last term in eqn (5.51), we change its summation index $n$ by introducing $k=n+1$. This yields

$$r\rho\sum_{n=2}^{\infty}\frac{1}{n+1}\left[\left(\frac{r\rho}{a^2}\right)^{n+1}-\left(\frac{r}{\rho}\right)^{n+1}\right]\cos n(\varphi-\psi)$$
$$=r\rho\sum_{k=3}^{\infty}\frac{1}{k}\left[\left(\frac{r\rho}{a^2}\right)^k-\left(\frac{r}{\rho}\right)^k\right]\cos(k-1)(\varphi-\psi)$$
$$=r\rho\sum_{k=1}^{\infty}\frac{1}{k}\left[\left(\frac{r\rho}{a^2}\right)^k-\left(\frac{r}{\rho}\right)^k\right]\cos(k-1)(\varphi-\psi)$$
$$-r\rho\left[\left(\frac{r\rho}{a^2}\right)-\left(\frac{r}{\rho}\right)\right]-\frac{r\rho}{2}\left[\left(\frac{r\rho}{a^2}\right)^2-\left(\frac{r}{\rho}\right)^2\right]\cos(\varphi-\psi)$$
$$=r\rho\cos(\varphi-\psi)\sum_{k=1}^{\infty}\frac{1}{k}\left[\left(\frac{r\rho}{a^2}\right)^k-\left(\frac{r}{\rho}\right)^k\right]\cos k(\varphi-\psi)$$
$$+r\rho\sin(\varphi-\psi)\sum_{k=1}^{\infty}\frac{1}{k}\left[\left(\frac{r\rho}{a^2}\right)^k-\left(\frac{r}{\rho}\right)^k\right]\sin k(\varphi-\psi)$$
$$-r\rho\left[\left(\frac{r\rho}{a^2}\right)-\left(\frac{r}{\rho}\right)\right]-\frac{r\rho}{2}\left[\left(\frac{r\rho}{a^2}\right)^2-\left(\frac{r}{\rho}\right)^2\right]\cos(\varphi-\psi) \qquad (5.53)$$

Summing up the cosine-series in the above expression with the aid of the summation formula of eqn (1.42), and leaving the sine-series in its current form, the last term in eqn (5.51) is finally expressed as

$$
\begin{aligned}
& r\rho \sum_{n=2}^{\infty} \frac{1}{n+1}\left[\left(\frac{r\rho}{a^2}\right)^{n+1} - \left(\frac{r}{\rho}\right)^{n+1}\right] \cos n(\varphi - \psi) \\
& = r\rho \cos(\varphi - \psi)\left[\frac{1}{2}\ln\left(1 - 2\frac{r}{\rho}\cos(\varphi-\psi) + \left(\frac{r}{\rho}\right)^2\right)\right. \\
& \quad \left. - \frac{1}{2}\ln\left(1 - 2\frac{r\rho}{a^2}\cos(\varphi-\psi) + \left(\frac{r\rho}{a^2}\right)^2\right)\right] \\
& \quad + r\rho\sin(\varphi-\psi)\sum_{k=1}^{\infty}\frac{1}{k}\left[\left(\frac{r\rho}{a^2}\right)^k - \left(\frac{r}{\rho}\right)^k\right]\sin k(\varphi-\psi) \\
& \quad - \underline{r\rho\left[\left(\frac{r\rho}{a^2}\right) - \left(\frac{r}{\rho}\right)\right]} - \underline{\underline{\frac{r\rho}{2}\left[\left(\frac{r\rho}{a^2}\right)^2 - \left(\frac{r}{\rho}\right)^2\right]\cos(\varphi-\psi)}}
\end{aligned} \tag{5.54}
$$

At this point in our development, we substitute the expressions from eqns (5.52), (5.53), and (5.54) into eqn (5.51). In doing so, the two sine-series (one from each of eqns (5.52) and (5.54)) cancel out and all the logarithmic terms as well as the two double-underlined terms are accordingly combined. This yields, for the series term of the representation in eqn (5.49)

$$
\begin{aligned}
& 2\sum_{n=2}^{\infty} g_n(r,\rho)\cos n(\varphi-\psi) \\
& = -[r^2 - 2r\rho\cos(\varphi-\psi) + \rho^2]\ln\frac{\rho^2[a^4 - 2a^2 r\rho\cos(\varphi-\psi) + r^2\rho^2]}{a^4[r^2 - 2r\rho\cos(\varphi-\psi) + \rho^2]} \\
& \quad + \underline{r\rho\left[\left(\frac{r\rho}{a^2}\right) - \left(\frac{r}{\rho}\right)\right]} + \underline{\underline{\frac{r(a^2-\rho^2)}{a^4\rho}[r^2(a^2-\rho^2) + 2a^2\rho^2]\cos(\varphi-\psi)}}
\end{aligned}
$$

Upon substituting this expression, along with that of eqn (5.50), all in eqn (5.49), the two double-underlined terms cancel out, while the two logarithmic and the two simply-underlined terms are accordingly combined. This yields the final representation for the influence function $G(r, \varphi; \rho, \psi)$ for the clamped circular plate of radius $a$ in the form

$$
\begin{aligned}
& G(r, \varphi; \rho, \psi) \\
& = \frac{1}{16\pi D}\left\{\frac{1}{a^2}(a^2-\rho^2)(a^2-r^2)\right. \\
& \quad \left. - [r^2 - 2r\rho\cos(\varphi-\psi) + \rho^2]\ln\frac{a^4 - 2a^2 r\rho\cos(\varphi-\psi) + r^2\rho^2}{a^2[r^2 - 2r\rho\cos(\varphi-\psi) + \rho^2]}\right\}
\end{aligned} \tag{5.55}
$$

It is evident that this form is absolutely identical to the classical representation of the influence function for the clamped circular plate shown earlier in eqn (5.34). Indeed, the variables $r$ and $\rho$ are the moduli of the observation point $z$ and the force application point $\zeta$, respectively, while the expression

$$r^2 - 2r\rho\cos(\varphi - \psi) + \rho^2$$

represents a square of the distance $|z - \zeta|$ between $z$ and $\zeta$. As to the expression

$$a^4 - 2a^2r\rho\cos(\varphi - \psi) + r^2\rho^2$$

in the denominator of the logarithmic function, it can easily be shown that it represents a square of the modulus of $a^2 - z\bar{\zeta}$. ▲

**Example 2.6:** Consider a semi-circular plate of radius $a$, whose middle plane occupies the region $\Omega = \{(r, \varphi) : 0 < r < a,\ 0 < \varphi < \pi\}$, with the rectilinear edge being simply-supported while the curvilinear edge is clamped.

Omitting the detailed development that can be accomplished by simply tracing out the derivation utilized in the previous example for the clamped circular plate, we present just the final expression of the influence function of a point force for the semi-circular plate in the following closed form

$$\begin{aligned} G(r, \varphi; \rho, \psi) \\ = \frac{1}{16\pi D}\Bigg\{ & [r^2 - 2r\rho\cos(\varphi + \psi) + \rho^2]\ln\frac{a^4 - 2a^2r\rho\cos(\varphi + \psi) + r^2\rho^2}{a^2[r^2 - 2r\rho\cos(\varphi + \psi) + \rho^2]} \\ & - [r^2 - 2r\rho\cos(\varphi - \psi) + \rho^2]\ln\frac{a^4 - 2a^2r\rho\cos(\varphi - \psi) + r^2\rho^2}{a^2[r^2 - 2r\rho\cos(\varphi - \psi) + \rho^2]}\Bigg\} \end{aligned} \tag{5.56}$$

The reader is encouraged to go through the derivation procedure for the above representation in detail. ▲

The examples that have been completed so far bring a strong confidence in the power of the proposed technique that appears to be productive in a number of problems where influence functions are either not available at all or their existing representations do not meet the numerical implementation requirements. The next example is of just such a nature.

**Example 2.7:** Consider a simply-supported circular plate of radius $a$. Although the complex variable-based method for the construction of the influence function of a transverse point force for this problem was described in journal articles long ago (the information concerning this issue can be found in [45]), the influence function itself is not available in the existing handbooks.

The boundary conditions of a simple support are imposed on the edge $r = a$ in terms of the deflection function in the form

$$w(a, \varphi) = 0, \; M_r(a, \varphi) \equiv \left( \frac{\partial^2}{\partial r^2} + \frac{\sigma}{a}\left( \frac{\partial}{\partial r} + \frac{1}{a}\frac{\partial^2}{\partial \varphi^2} \right) \right) w(a, \varphi) = 0$$

assigning the deflection function itself and the radial bending moment to be zero on the edge $r = a$. Remember that the parameter $\sigma$ is the Poisson ratio of the material of which the plate is made.

The second of the above conditions can be simplified. Notice that, since the deflection $w(r, \varphi)$ is supposed to be identical zero along the edge $r = a$ of the plate, all the derivatives of $w(r, \varphi)$ with respect to the tangential variable $\varphi$ must also be zero. Hence, the second of the above two conditions reduces to

$$\frac{\partial^2 w(a, \varphi)}{\partial r^2} + \frac{\sigma}{a}\frac{\partial w(a, \varphi)}{\partial r} = 0$$

If we follow the procedure, a detailed description of which is available in Example 2.5, the influence function of the simply-supported circular plate is also obtained in the series form of eqn (5.48), whose coefficients, for $r \leq \rho$, are found in this case as

$$g_0(r, \rho) = \frac{1}{8}\left\{ \frac{a^2 - \rho^2}{a^2}\left[ (a^2 + r^2) + \frac{2}{1+\sigma}(a^2 - r^2) \right] + 2(r^2 + \rho^2) \ln\left( \frac{\rho}{a} \right) \right\},$$

$$g_1(r, \rho) = \frac{1}{8}\left\{ \frac{1+\sigma}{3+\sigma}[(r^2 + \rho^2) - a^2]\frac{r\rho}{a^2} - \frac{r^3}{2\rho}\left( 1 - \frac{1-\sigma}{3+\sigma}\frac{\rho^4}{a^4} \right) - 2r\rho \ln\left( \frac{\rho}{a} \right) \right\}$$

and

$$g_n(r, \rho) = -\frac{1}{8}\left\{ \frac{r\rho}{n-1}\left[ \left( \frac{r\rho}{a^2} \right)^{n-1} - \left( \frac{r}{\rho} \right)^{n-1} \right] + \frac{r\rho}{n+1}\left[ \left( \frac{r\rho}{a^2} \right)^{n+1} - \left( \frac{r}{\rho} \right)^{n+1} \right] \right.$$
$$\left. - \frac{r^2 + \rho^2}{n}\left[ \left( \frac{r\rho}{a^2} \right)^{n} - \left( \frac{r}{\rho} \right)^{n} \right] + \frac{1}{n+\omega}\frac{(r^2 - a^2)(a^2 - \rho^2)}{a^2}\left( \frac{r\rho}{a^2} \right)^{n} \right\}$$

where the parameter $\omega$ is introduced in $g_n(r, \rho)$ in terms of the Poisson ratio of the material as $\omega = (1 + \sigma)/2$.

Unfortunately, the series in eqn (5.48) cannot, in this case, be entirely summed up. This is so because of the last term in the above expression for $g_n(r, \rho)$, the one containing the parameter $\omega$. The partial summation, though, provides a quite compact representation for the influence function under consideration, that is

$$G(r, \varphi; \rho, \psi)$$
$$= \frac{1}{16\pi D}\left\{ \frac{(a^2 - \rho^2)(a^2 - r^2)}{a^2}\left[ \frac{3+\sigma}{1+\sigma} + 2\sum_{n=1}^{\infty} \frac{1}{n+\omega}\left( \frac{r\rho}{a^2} \right)^n \cos n(\varphi - \psi) \right] \right.$$
$$\left. - [r^2 - 2r\rho\cos(\varphi - \psi) + \rho^2] \ln \frac{a^4 - 2a^2 r\rho\cos(\varphi - \psi) + r^2\rho^2}{a^2[r^2 - 2r\rho\cos(\varphi - \psi) + \rho^2]} \right\} \quad (5.57)$$

This representation could be convenient if used for computing values of the influence function $G(r, \varphi; \rho, \psi)$ inside of the circle, because the series in the brackets uniformly converges if both the field point $(r, \varphi)$ and the force application point $(\rho, \psi)$ are interior for the circle, provided $r\rho/a^2 < 1$. Notice, however, that the rate of its convergence depends on the proximity of $r$ and $\rho$ to the plate's edge. Indeed, the convergence notably slows down if both the field and the force application point approach the edge of the plate (both $r$ and $\rho$ approach $a$).

In what follows we will show how the practicality of the representation in (5.57) radically improves by splitting off the slowly convergent component of the series and summing it up. In doing so, the coefficient of the series in eqn (5.57) is presented in the form

$$\frac{1}{n+\omega} = \frac{1}{n} - \frac{\omega}{n(n+\omega)}$$

yielding for the series itself

$$\begin{aligned}2\sum_{n=1}^{\infty}\frac{1}{n+\omega}\left(\frac{r\rho}{a^2}\right)^n \cos n(\varphi-\psi) &= 2\sum_{n=1}^{\infty}\left[\frac{1}{n} - \frac{\omega}{n(n+\omega)}\right]\left(\frac{r\rho}{a^2}\right)^n \cos n(\varphi-\psi)\\ &= 2\sum_{n=1}^{\infty}\frac{1}{n}\left(\frac{r\rho}{a^2}\right)^n \cos n(\varphi-\psi)\\ &\quad - 2\omega\sum_{n=1}^{\infty}\frac{1}{n(n+\omega)}\left(\frac{r\rho}{a^2}\right)^n \cos n(\varphi-\psi)\end{aligned}$$

The first of these series is summable with the aid of the standard summation formula from eqn (1.42). This yields

$$\begin{aligned}2\sum_{n=1}^{\infty}\frac{1}{n+\omega}\left(\frac{r\rho}{a^2}\right)^n \cos n(\varphi-\psi) &= -\ln\left(1 - 2\frac{r\rho}{a^2}\cos(\varphi-\psi) + \left(\frac{r\rho}{a^2}\right)^2\right)\\ &\quad - 2\omega\sum_{n=1}^{\infty}\frac{1}{n(n+\omega)}\left(\frac{r\rho}{a^2}\right)^n \cos n(\varphi-\psi)\end{aligned}$$

In view of the foregoing transformation, the expression for the influence function of a point force for the simply-supported circular plate, as of eqn (5.57), can be rewritten in the form

$$\begin{aligned}&G(r, \varphi; \rho, \psi)\\ &= \frac{1}{16\pi D}\left\{\frac{(a^2-\rho^2)(a^2-r^2)}{a^2}\left[\frac{3+\sigma}{1+\sigma} - \ln\left(1 - 2\frac{r\rho}{a^2}\cos(\varphi-\psi) + \left(\frac{r\rho}{a^2}\right)^2\right)\right.\right.\\ &\quad \left. - 2\omega\sum_{n=1}^{\infty}\frac{1}{n(n+\omega)}\left(\frac{r\rho}{a^2}\right)^n \cos n(\varphi-\psi)\right]\\ &\quad \left. - [r^2 - 2r\rho\cos(\varphi-\psi) + \rho^2]\ln\frac{a^4 - 2a^2 r\rho\cos(\varphi-\psi) + r^2\rho^2}{a^2[r^2 - 2r\rho\cos(\varphi-\psi) + \rho^2]}\right\} \qquad (5.58)\end{aligned}$$

It is evident that the series in the above representation converges at a much higher rate than that of eqn (5.57). Indeed, the convergence rate of the series in (5.58) is of the order of $1/n^2$, against $1/n$ for the series in (5.57).

This implies that in practical computing of values of $G(r, \varphi; \rho, \psi)$, the series in eqn (5.58) can be appropriately truncated. To help with finding the truncation parameter $N$

$$\sum_{n=1}^{N} \frac{1}{n(n+\omega)} \left(\frac{r\rho}{a^2}\right)^n \cos n(\varphi - \psi)$$

that provides us with a required accuracy level, we estimate the modulus of the series $N$-th remainder

$$\begin{aligned} |R_N(r, \varphi; \rho, \psi)| &= \left| \sum_{n=N+1}^{\infty} \frac{1}{n(n+\omega)} \left(\frac{r\rho}{a^2}\right)^n \cos n(\varphi - \psi) \right| \\ &\leq \sum_{n=N+1}^{\infty} \frac{1}{n(n+\omega)} \leq \sum_{n=N+1}^{\infty} \frac{1}{n^2} \\ &= \sum_{n=1}^{\infty} \frac{1}{n^2} - \sum_{n=1}^{N} \frac{1}{n^2} = \frac{\pi^2}{6} - \sum_{n=1}^{N} \frac{1}{n^2} \end{aligned}$$

Thus, the compact estimate of the remainder

$$|R_N(r, \varphi; \rho, \psi)| \leq \frac{\pi^2}{6} - \sum_{n=1}^{N} \frac{1}{n^2}$$

that we came up with, indeed gives us an effective tool to appropriately truncate the series in (5.58).

The expression in eqn (5.58) can be further transformed by introducing compact notation for the observation point $z = r(\cos\varphi + i \sin\varphi)$ and the force application point $\zeta = \rho(\cos\psi + i \sin\psi)$. After some elementary algebra, the logarithmic terms of eqn (5.58) are combined and rearranged. This finally yields

$$\begin{aligned} &G(r, \varphi; \rho, \psi) \\ &= \frac{1}{8\pi D} \left\{ |z - \zeta|^2 \ln \frac{|z - \zeta|}{a} - \frac{|a^2 - z\bar{\zeta}|^2}{a^2} \ln \frac{|a^2 - z\bar{\zeta}|}{a^2} \right. \\ &\quad \left. + \frac{(a^2 - \rho^2)(a^2 - r^2)}{2a^2} \left[ \frac{3+\sigma}{1+\sigma} - 2\omega \sum_{n=1}^{\infty} \frac{1}{n(n+\omega)} \left(\frac{r\rho}{a^2}\right)^n \cos n(\varphi - \psi) \right] \right\} \end{aligned} \tag{5.59}$$

Observe that the first logarithmic term in the above expression of the influence function contains the fundamental solution

$$|z - \zeta|^2 \ln|z - \zeta|$$

of the biharmonic equation. This term represents the singular component of the influence function. A clarification is required as to the word 'singular' which is conditionally applied to the above term. Indeed, the term itself is not singular, because its limit as $z$ approaches $\zeta$ is finite. It is actually zero

$$\lim_{z\to\zeta} |z-\zeta|^2 \ln|z-\zeta| = 0$$

which can easily be verified by applying the L'Hospital's rule to the above limit. But we use, nevertheless, the word 'singular' as to the first logarithmic term in (5.59) to highlight that stress-related components (bending moments and shear forces) associated with this term possess logarithmic and even higher order singularity. This stays in agreement with the known fact [65] that the bending moments and shear forces, generated in a Poisson–Kirchhoff plate by a transverse point force, are theoretically unbounded at the force application point. ▲

For the next example in this section, we take a look at a circular plate whose edge $r=a$ is not subject to the traditional boundary conditions of either simple support, or rigid clamp, or free edge conditions. The conditions to be considered are, nonetheless, of a notable importance in structural mechanics.

**Example 2.8:** Let the plate's edge be elastically clamped in the way that the slope of the deflection function is zero as $r=a$, while the shear force is directly proportional to the deflection. This is formalized as

$$\frac{\partial w(a,\varphi)}{\partial r} = 0, \quad D\frac{\partial}{\partial r}\left(\frac{\partial^2}{\partial r^2} + \frac{1}{r}\frac{\partial}{\partial r} + \frac{1}{r^2}\frac{\partial^2}{\partial \varphi^2}\right)w(a,\varphi) = Cw(a,\varphi) \quad (5.60)$$

where $D$ is the plate's flexural rigidity, while the parameter $C$ represents the coefficient of elastic edge support.

By removing the parenthesis, the operator in the left-hand side of the second equation in (5.60) transforms to

$$\frac{\partial^3}{\partial r^3} - \frac{1}{r^2}\frac{\partial}{\partial r} + \frac{1}{r}\frac{\partial^2}{\partial r^2} - \frac{2}{r^3}\frac{\partial^2}{\partial \varphi^2} + \frac{1}{r^2}\frac{\partial^3}{\partial r \partial \varphi^2}$$

In compliance with the first relation in (5.60), the second and the last additive terms in the above operator vanish and it can be rewritten as

$$\frac{\partial^3}{\partial r^3} + \frac{1}{r}\frac{\partial^2}{\partial r^2} - \frac{2}{r^3}\frac{\partial^2}{\partial \varphi^2}$$

This transforms the boundary conditions in (5.60) to

$$\frac{\partial w(a,\varphi)}{\partial r} = 0, \quad \left(\frac{\partial^3}{\partial r^3} + \frac{1}{a}\frac{\partial^2}{\partial r^2} - \frac{2}{a^3}\frac{\partial^2}{\partial \varphi^2}\right)w(a,\varphi) = k^* w(a,\varphi) \quad (5.61)$$

where the parameter $k^*$ is defined as $k^* = C/D$.

It is evident that the Green's function for the biharmonic equation on the circle of radius $a$, subject to the boundary conditions imposed by eqn (5.61), represents the influence function $G(r, \varphi; \rho, \psi)$ of a transverse point force for the plate under consideration.

Following our procedure, $G(r, \varphi; \rho, \psi)$ is obtained in a form of the expansion in eqn (5.49) whose coefficient $g_0(r, \rho)$ represents the Green's function to a boundary-value problem stated for equation (5.41) on the interval $[0, a]$. The boundary conditions at $r = a$ are in this case imposed as

$$\frac{dw_0(a)}{dr} = 0, \quad \frac{d^3w_0(a)}{dr^3} + \frac{1}{a}\frac{d^2w_0(a)}{dr^2} - k^* w_0(a) = 0 \tag{5.62}$$

The expression for $g_0(r, \rho)$ that is valid for $r \le \rho$, is found of the form

$$g_0(r, \rho) = \frac{1}{8}\left[\frac{1}{a^2}(a^2 + r^2)(a^2 - \rho^2) + 2(r^2 + \rho^2)\ln\frac{\rho}{a} + \frac{8}{k^* a}\right]$$

The coefficient $g_1(r, \rho)$ in the expansion of (5.49) represents the Green's function to the boundary-value problem

$$\frac{dw_1(a)}{dr} = 0, \quad \frac{d^3w_1(a)}{dr^3} + \frac{1}{a}\frac{d^2w_1(a)}{dr^2} + \frac{2 - k^* a^3}{a^3} w_1(a) = 0 \tag{5.63}$$

stated for the equation in (5.44). The expression for $g_1(r, \rho)$ that is valid for $r \le \rho$, is found as

$$g_1(r, \rho) = -\frac{1}{\Delta^*}\{k^* a^3[r(\rho^2 - a^2)(r^2(\rho^2 - a^2) - 2a^2\rho^2)]$$
$$+ 2ra^2[r^2(a^2 + 2\rho^2) + 2\rho^2(\rho^2 - 7a^2)]\} - \frac{r\rho}{4}\ln\frac{\rho}{a}$$

where $\Delta^* = 16a^4\rho(4 + k^* a^3)$.

The coefficient $g_n(r, \rho)$ in (5.49) represents the Green's function to the boundary-value problem

$$\frac{dw_n(a)}{dr} = 0, \quad \frac{d^3w_n(a)}{dr^3} + \frac{1}{a}\frac{d^2w_n(a)}{dr^2} + \frac{2n^2 - k^* a^3}{a^3} w_n(a) = 0 \tag{5.64}$$

stated for eqn (5.39). The expression for $g_n(r, \rho)$ valid for $r \le \rho$, was found as

$$g_n(r, \rho)$$
$$= -\frac{1}{\Delta^{**}}\left\{\frac{r^n}{n-1}[n(4(2 - n^2) + k^* a^3)a^{2-2n}\rho^n - (2n^2(n+1) + k^* a^3)\rho^{2-n}\right.$$
$$+ (n-1)(2n^2 - k^* a^3)a^{-2n}\rho^{n+2}] + [(n+1)(2n^2 - k^* a^3)a^{-2n}\rho^n$$
$$\left. + (2n^2(n+1) + k^* a^3)\rho^{-n} + nk^* a^3 a^{-2n-2}\rho^{n+2}]\frac{r^{n+2}}{n+1}\right\}$$

where the parameter $\Delta^{**}$ is introduced as $\Delta^{**} = 8n[2n^2(n+1) + k^* a^3]a^{2n-2}$.

Remember that the boundary-value problems that we have dealt with in obtaining the coefficients $g_0(r, \rho)$, $g_1(r, \rho)$, and $g_n(r, \rho)$ of the expansion (5.49), are stated for fourth order governing equations shown in (5.41), (5.44), and (5.39), respectively, on the interval $[0, a]$. This implies that the total number of four boundary conditions ought to be imposed for each problem. Two of those conditions are usually imposed at $r = a$ (see eqns (5.62)–(5.64)), while the other two ought to be imposed at $r = 0$. Due to the form of the governing equations, the boundedness conditions are assumed at $r = 0$ (see the conditions in eqn (5.40) of Section 5.2.2).

By substituting the coefficients $g_0(r, \rho)$, $g_1(r, \rho)$, and $g_n(r, \rho)$ just presented in the expansion of eqn (5.49), one obtains an expression for the influence function of a transverse point force for the elastically clamped circular plate. The expansion in (5.49) is in this case in a computer-friendly form. Indeed, it is convenient for computer implementations, because as analysis shows the series converges at the rate of $1/n^2$ for a finite value of $k^*$.

If, however, a limit of the series in (5.49) is taken as the parameter $k^*$ approaches infinity, then one obtains the series representation of the influence function for the clamped circular plate obtained earlier in Example 2.5, where the series convergences at the rate of $1/n$. It is evident that by letting $k^*$ approach infinity the boundary conditions in eqn (5.60) reduce to those in eqn (5.36). Taking the limit of the series in (5.49) as $k^*$ approaches infinity is not a trivial procedure and we leave it as one of the End Chapter Exercises. As a hint to that exercise we recommend the reader to revisit our algebra in Example 2.5 with changing the summation indices in the series. ▲

Completing the discussion in this section, note that, by learning the essentials of our approach, the reader can apply it to the construction of influence functions for thin plates considered within the scope other plate models. In Section 5.4 we will do so and focus on the Reissner plate model that accounts for the effect of transverse normal stress and transverse shear deformation.

## 5.3 Plates on Elastic Foundation

The routine developed in Section 5.2 will be applied herein to the construction of influence functions of a point force for Poisson–Kirchhoff plates resting on a simple (single parameter) elastic foundation [65].

**Example 3.1:** We begin with yet another classical example [65] of a simply-supported rectangular plate of uniform thickness $h$ made of a homogeneous isotropic elastic material whose properties are determined by the elasticity modulus $E$ and Poisson ratio $\sigma$. Let the plate undergo a distributed lateral load $q(x, y)$, the middle plane occupy the region $\Omega = \{(x, y) : 0 < x < a, 0 < y < b\}$, and the elastic coefficient of the foundation be denoted with $\lambda_0$. To obtain the influence function to this plate, we consider the boundary-value problem

$$\frac{\partial^4 w(x,y)}{\partial x^4}+2\frac{\partial^4 w(x,y)}{\partial x^2\partial y^2}+\frac{\partial^4 w(x,y)}{\partial y^4}+\lambda w(x,y)=-f(x,y) \tag{5.65}$$

$$w=\left.\frac{\partial^2 w}{\partial y^2}\right|_{y=0,b}=0,\quad w=\left.\frac{\partial^2 w}{\partial x^2}\right|_{x=0,a}=0 \tag{5.66}$$

where the parameter $\lambda > 0$ is defined in terms of the elastic coefficient $\lambda_0$ of the foundation and the plate's flexural rigidity $D = Eh^3/12(1-\sigma^2)$ as $\lambda = \lambda_0/D$. The right-hand side function in eqn (5.65) is defined in terms of the loading function as $f(x, y) = q(x, y)/D$.

Similarly to the derivation in Example 2.1 of Section 5.2, we apply the Navier approach to the above problem, in compliance with which the solution to (5.65) and (5.66) is expressed in a form of the double Fourier sine-series

$$w(x,y)=\sum_{m=1}^{\infty}\sum_{n=1}^{\infty} w_{mn}\sin\mu x\sin\nu y,\quad \mu=\frac{m\pi}{a},\quad \nu=\frac{n\pi}{b} \tag{5.67}$$

It is evident that the above representation for $w(x, y)$ satisfies all the boundary conditions imposed in (5.66).

We also express the right-hand side function $f(x, y)$ of eqn (5.65) in the identical double sine-series form

$$f(x,y)=\sum_{m=1}^{\infty}\sum_{n=1}^{\infty} f_{mn}\sin\mu x\sin\nu y \tag{5.68}$$

Substituting then expansions from (5.67) and (5.68) in (5.65) and combining the like terms in the left-hand side yields the equation

$$\sum_{m=1}^{\infty}\sum_{n=1}^{\infty}(\mu^4+2\mu^2\nu^2+\nu^4+\lambda)w_{mn}\sin\mu x\sin\nu y$$

$$=-\sum_{m=1}^{\infty}\sum_{n=1}^{\infty} f_{mn}\sin\mu x\sin\nu y$$

from which, equating the corresponding coefficients of the above series and performing a trivial algebra, one obtains

$$w_{mn}=-\frac{f_{mn}}{(\mu^2+\nu^2)^2+\lambda}$$

Substitution of this expression in (5.67) yields

$$w(x,y)=-\sum_{m=1}^{\infty}\sum_{n=1}^{\infty}\frac{f_{mn}}{(\mu^2+\nu^2)^2+\lambda}\sin\mu x\sin\nu y \tag{5.69}$$

Recall the Euler–Fourier formula from eqn (1.26) in Chapter 1 and adjust it to the Fourier double-series environment. This yields for the Fourier coefficients $f_{mn}$

of the right-hand side term in eqn (5.65)

$$f_{mn} = \frac{4}{ab} \int_0^a \int_0^b f(s, t) \sin \mu s \sin \nu t \, ds \, dt$$

Substitute this expression for $f_{mn}$ in (5.69) and assume that the summation and integration in it can be interchanged. This delivers the ultimate solution to the boundary-value problem of eqns (5.65) and (5.66) as

$$w(x, y) = -\frac{4}{ab} \int_0^a \int_0^b \sum_{m=1}^{\infty} \sum_{n=1}^{\infty} \frac{\sin \mu x \sin \mu s \sin \nu y \sin \nu t}{(\mu^2 + \nu^2)^2 + \lambda} f(s, t) \, ds \, dt$$

The above can be written in terms of the loading function $q(x, y)$ as

$$w(x, y) = -\frac{4}{abD} \int_0^a \int_0^b \sum_{m=1}^{\infty} \sum_{n=1}^{\infty} \frac{\sin \mu x \sin \mu s \sin \nu y \sin \nu t}{(\mu^2 + \nu^2)^2 + \lambda} q(s, t) \, ds \, dt \tag{5.70}$$

The change of order of the summation and integration that we just made is well justified in theory [60], but we will not go into details of the justification itself because it stays beyond of the scope of this text.

Thus, in light of the definition introduced earlier in this chapter (see eqn (5.3) in Section 5.1), we come to the conclusion that the kernel-function

$$G(x, y; s, t) = -\frac{4}{abD} \sum_{m=1}^{\infty} \sum_{n=1}^{\infty} \frac{\sin \mu x \sin \mu s \sin \nu y \sin \nu t}{(\mu^2 + \nu^2)^2 + \lambda} \tag{5.71}$$

in (5.70) represents the influence function of a transverse point force for the simply-supported rectangular plate resting on simple elastic foundation whose coefficient is $\lambda_0 = \lambda D$, with $(x, y)$ and $(s, t)$ representing the observation and the force application point, respectively.

It is evident that the above series converges at the same rate as the series in eqn (5.13) of Section 5.2. This implies that values of $G(x, y; s, t)$ can be accurately computed by a truncation of the series in (5.71). And, similarly to the situation with the simply-supported rectangular plate considered in Example 2.1 of Section 5.2, an accurate computation of the stress-related components in the plate resting on elastic foundation in the immediate vicinity of the force application point requires special attention.

At the same time, the influence function in eqn (5.71) allows accurate computation of all the components of the stress-strain state in the simply-supported rectangular plate resting on elastic foundation if the plate undergoes a transverse distributed load. Indeed, the integration in eqn (5.70) increases the convergence rate of the series. The reader will find some illustrations of this statement in the End Chapter Exercises. ▲

**Example 3.2:** Let a plate, whose middle plane occupies the region $\Omega = \{(x, y) : 0 < x < \infty, 0 < y < b\}$, rest on the elastic foundation. Let also the edges $y = 0$,

$y = b$ and $x = 0$ be simply-supported. Evidently, this setting is simulated by the boundary-value problem

$$w = \left.\frac{\partial^2 w}{\partial y^2}\right|_{y=0,b} = 0, \qquad w = \left.\frac{\partial^2 w}{\partial x^2}\right|_{x=0} = 0 \tag{5.72}$$

for equation (5.65) in $\Omega$. Note that in addition to the above conditions the plate's state is supposed to be bounded as $x$ approaches infinity.

We express the solution $w(x, y)$ to the boundary-value problem in eqns (5.65) and (5.72) in a form of the Fourier sine-series

$$w(x, y) = \sum_{n=1}^{\infty} w_n(x) \sin \nu y, \qquad \nu = \frac{n\pi}{b} \tag{5.73}$$

and express the right-hand side function $f(x, y)$ of eqn (5.65) in the identical sine-series form

$$f(x, y) = \sum_{n=1}^{\infty} f_n(x) \sin \nu y \tag{5.74}$$

Once the expansions for $w(x, y)$ and $f(x, y)$ are substituted in (5.65), and the corresponding coefficients of the two Fourier sine-series that arise on the right and on the left of the equal sign are set equal, the following set ($n = 1, 2, 3, \ldots$) of boundary-value problems

$$\frac{d^4 w_n(x)}{dx^4} - 2\nu^2 \frac{d^2 w_n(x)}{dx^2} + (\nu^4 + \lambda) w_n(x) = -f_n(x), \quad x \in (0, \infty) \tag{5.75}$$

$$w_n(0) = \frac{d^2 w_n(0)}{dx^2} = 0, \quad |w_n(\infty)| < \infty, \quad \left|\frac{dw_n(\infty)}{dx}\right| < \infty \tag{5.76}$$

arises in the coefficients $w_n(x)$ of the series from eqn (5.73).

A fundamental set of solutions to the homogeneous equation corresponding to (5.75) can be found from its characteristic equation

$$k^4 - 2\nu^2 k^2 + (\nu^4 + \lambda) = 0$$

which is biquadratic in nature provided that its solution set is represented by the four complex numbers

$$k_j = \pm\sqrt{\nu^2 \pm i\sqrt{\lambda}}, \quad j = \overline{1, 4} \tag{5.77}$$

expressed in terms of the parameters $\nu$ and $\lambda$. The radicals in the above and in what follows are understood as principal (arithmetic) values.

To separate real and imaginary parts of $k_j$, we express, in compliance with Section 1.3, the radicand of (5.77) in trigonometric form as

$$\nu^2 \pm i\sqrt{\lambda} = \sqrt{\nu^4+\lambda}\left[\cos\left(\arctan\left(\frac{\sqrt{\lambda}}{\nu^2}\right)\right) \pm i\,\sin\left(\arctan\left(\frac{\sqrt{\lambda}}{\nu^2}\right)\right)\right]$$

This implies that

$$k_j = \pm\sqrt[4]{\nu^4+\lambda}\left[\cos\left(\frac{1}{2}\arctan\left(\frac{\sqrt{\lambda}}{\nu^2}\right)\right) \pm i\,\sin\left(\frac{1}{2}\arctan\left(\frac{\sqrt{\lambda}}{\nu^2}\right)\right)\right] \tag{5.78}$$

which can be simplified by recalling the half-angle identities from trigonometry

$$\cos\left(\frac{1}{2}\arctan\left(\frac{\sqrt{\lambda}}{\nu^2}\right)\right) = \frac{\sqrt{2}}{2}\sqrt{1+\cos\left(\arctan\left(\frac{\sqrt{\lambda}}{\nu^2}\right)\right)}$$
$$= \frac{\sqrt{2}}{2}\sqrt{1+\frac{\nu^2}{\sqrt{\nu^4+\lambda}}} = \frac{\sqrt{2}}{2}\sqrt{\frac{\sqrt{\nu^4+\lambda}+\nu^2}{\sqrt{\nu^4+\lambda}}}$$

and

$$\sin\left(\frac{1}{2}\arctan\left(\frac{\sqrt{\lambda}}{\nu^2}\right)\right) = \frac{\sqrt{2}}{2}\sqrt{1-\cos\left(\arctan\left(\frac{\sqrt{\lambda}}{\nu^2}\right)\right)}$$
$$= \frac{\sqrt{2}}{2}\sqrt{1-\frac{\nu^2}{\sqrt{\nu^4+\lambda}}} = \frac{\sqrt{2}}{2}\sqrt{\frac{\sqrt{\nu^4+\lambda}-\nu^2}{\sqrt{\nu^4+\lambda}}}$$

Upon substituting these in (5.78), we finally obtain the following four complex numbers for the solution set of the characteristic equation

$$k_j = \pm\frac{\sqrt{2}}{2}\left(\sqrt{\sqrt{\nu^4+\lambda}+\nu^2} \pm i\sqrt{\sqrt{\nu^4+\lambda}-\nu^2}\right)$$

Thus, a fundamental set of solutions to the homogeneous equation corresponding to (5.75) can be presented by the following four linearly independent functions

$$\begin{aligned} w_{n,1}(x) &= e^{\alpha x}\cos\beta x, \quad & w_{n,2}(x) &= e^{\alpha x}\sin\beta x \\ w_{n,3}(x) &= e^{-\alpha x}\cos\beta x, \quad & w_{n,4}(x) &= e^{-\alpha x}\sin\beta x \end{aligned} \tag{5.79}$$

where the parameters $\alpha$ and $\beta$ are defined as

$$\alpha = \frac{\sqrt{2}}{2}\sqrt{\sqrt{\nu^4+\lambda}+\nu^2} \quad \text{and} \quad \beta = \frac{\sqrt{2}}{2}\sqrt{\sqrt{\nu^4+\lambda}-\nu^2}$$

Based on the fundamental set of solutions in (5.79), we solve the boundary-value problem in (5.75) and (5.76) by the method of variation of parameters. This implies that the general solution to equation (5.75) can be written as the linear combination

$$w_n(x) = C_1(x)w_{n,1}(x) + C_2(x)w_{n,2}(x) + C_3(x)w_{n,3}(x) + C_4(x)w_{n,4}(x) \tag{5.80}$$

of the components in eqn (5.79). Tracing out the procedure, one obtains the following system of linear algebraic equations

$$\begin{pmatrix} w_{n,1}(x) & w_{n,2}(x) & w_{n,3}(x) & w_{n,4}(x) \\ w'_{n,1}(x) & w'_{n,2}(x) & w'_{n,3}(x) & w'_{n,4}(x) \\ w''_{n,1}(x) & w''_{n,2}(x) & w''_{n,3}(x) & w''_{n,4}(x) \\ w'''_{n,1}(x) & w'''_{n,2}(x) & w'''_{n,3}(x) & w'''_{n,4}(x) \end{pmatrix} \times \begin{pmatrix} C'_1(x) \\ C'_2(x) \\ C'_3(x) \\ C'_4(x) \end{pmatrix} = \begin{pmatrix} 0 \\ 0 \\ 0 \\ -f_n(x) \end{pmatrix}$$

in the derivatives of the parameters $C_1(x)$, $C_2(x)$, $C_3(x)$, and $C_4(x)$. The solution of this system is found in a compact form as

$$C'_1(x) = \frac{\alpha \sin \beta x + \beta \cos \beta x}{4\alpha\beta(\alpha^2 + \beta^2)e^{\alpha x}} f_n(x), \quad C'_2(x) = -\frac{\alpha \cos \beta x - \beta \sin \beta x}{4\alpha\beta(\alpha^2 + \beta^2)e^{\alpha x}} f_n(x)$$

$$C'_3(x) = \frac{\alpha \sin \beta x - \beta \cos \beta x}{4\alpha\beta(\alpha^2 + \beta^2)e^{-\alpha x}} f_n(x), \quad C'_4(x) = -\frac{\alpha \cos \beta x + \beta \sin \beta x}{4\alpha\beta(\alpha^2 + \beta^2)e^{-\alpha x}} f_n(x)$$

We integrate the above relations to find the functions $C_1(x)$, $C_2(x)$, $C_3(x)$, and $C_4(x)$ themselves as

$$C_1(x) = \int_0^x \frac{\alpha \sin \beta s + \beta \cos \beta s}{4\alpha\beta(\alpha^2 + \beta^2)e^{\alpha s}} f_n(s)\, ds + M_1$$

$$C_2(x) = -\int_0^x \frac{\alpha \cos \beta s - \beta \sin \beta s}{4\alpha\beta(\alpha^2 + \beta^2)e^{\alpha s}} f_n(s)\, ds + M_2$$

$$C_3(x) = \int_0^x \frac{\alpha \sin \beta s - \beta \cos \beta s}{4\alpha\beta(\alpha^2 + \beta^2)e^{-\alpha s}} f_n(s)\, ds + M_3$$

and

$$C_4(x) = -\int_0^x \frac{\alpha \cos \beta s + \beta \sin \beta s}{4\alpha\beta(\alpha^2 + \beta^2)e^{-\alpha s}} f_n(s)\, ds + M_4$$

Upon substituting these in (5.80), we obtain $w_n(x)$ in the form

$$w_n(x) = \left[\int_0^x \frac{\alpha \sin \beta s + \beta \cos \beta s}{4\alpha\beta(\alpha^2 + \beta^2)e^{\alpha s}} f_n(s)\, ds + M_1\right] e^{\alpha x} \cos \beta x$$
$$+ \left[-\int_0^x \frac{\alpha \cos \beta s - \beta \sin \beta s}{4\alpha\beta(\alpha^2 + \beta^2)e^{\alpha s}} f_n(s)\, ds + M_2\right] e^{\alpha x} \sin \beta x$$

$$+\left[\int_0^x \frac{\alpha \sin \beta s - \beta \cos \beta s}{4\alpha\beta(\alpha^2+\beta^2)e^{\alpha s}} f_n(s)\,ds + M_3\right] e^{-\alpha x} \cos \beta x$$

$$+\left[-\int_0^x \frac{\alpha \cos \beta s + \beta \sin \beta s}{4\alpha\beta(\alpha^2+\beta^2)e^{\alpha s}} f_n(s)\,ds + M_4\right] e^{-\alpha x} \sin \beta x \qquad (5.81)$$

It is evident that, since the plate's state is bounded as $x$ approaches infinity, the factors

$$\int_0^x \frac{\alpha \sin \beta s + \beta \cos \beta s}{4\alpha\beta(\alpha^2+\beta^2)e^{\alpha s}} f_n(s)\,ds + M_1$$

and

$$-\int_0^x \frac{\alpha \cos \beta s - \beta \sin \beta s}{4\alpha\beta(\alpha^2+\beta^2)e^{\alpha s}} f_n(s)\,ds + M_2$$

of the unbounded functions $e^{\alpha x} \cos \beta x$ and $e^{\alpha x} \sin \beta x$ in (5.81) ought to be set equal to zero when $x$ approaches infinity. This implies

$$M_1 = -\int_0^\infty \frac{\alpha \sin \beta s + \beta \cos \beta s}{4\alpha\beta(\alpha^2+\beta^2)e^{\alpha s}} f_n(s)\,ds$$

and

$$M_2 = \int_0^\infty \frac{\alpha \cos \beta s - \beta \sin \beta s}{4\alpha\beta(\alpha^2+\beta^2)e^{\alpha s}} f_n(s)\,ds$$

To obtain the constants $M_3$ and $M_4$, we substitute the above expressions for $M_1$ and $M_2$ in (5.81) and regroup it as

$$w_n(x) = \int_0^x \frac{\beta \cos \beta(x-s) \sinh \beta(x-s) - \alpha \sin \beta(x-s) \cosh \beta(x-s)}{2\alpha\beta(\alpha^2+\beta^2)} f_n(s)\,ds$$
$$+ \int_0^\infty \frac{\alpha \sin \beta(x-s) - \beta \cos \beta(x-s)}{4\alpha\beta(\alpha^2+\beta^2)e^{\alpha(s-x)}} f_n(s)\,ds$$
$$+ M_3 e^{-\alpha x} \cos \beta x + M_4 e^{-\alpha x} \sin \beta x \qquad (5.82)$$

From the first condition in (5.76) it follows that

$$M_3 = \int_0^\infty \frac{\alpha \sin \beta s + \beta \cos \beta s}{4\alpha\beta(\alpha^2+\beta^2)e^{\alpha s}} f_n(s)\,ds$$

while the second condition in (5.76) implies

$$M_4 = \int_0^\infty \frac{\alpha \cos \beta s - \beta \sin \beta s}{4\alpha\beta(\alpha^2+\beta^2)e^{\alpha s}} f_n(s)\,ds$$

Upon substituting these expressions for $M_3$ and $M_4$ in (5.82), the solution to the boundary-value problem in eqns (5.75) and (5.76) is finally found as

$$w_n(x) = \int_0^x \frac{\beta \cos \beta(x-s) \sinh \beta(x-s) - \alpha \sin \beta(x-s) \cosh \beta(x-s)}{2\alpha\beta(\alpha^2+\beta^2)} f_n(s)\, ds$$
$$+ \int_0^\infty \frac{1}{\Delta}\{e^{-\alpha(x+s)}[\alpha \sin \beta(x+s) + \beta \cos \beta(x+s)]$$
$$- e^{-\alpha|x-s|}[\alpha \sin \beta|x-s| + \beta \cos \beta(x-s)]\} f_n(s)\, ds$$

where $\Delta = 4\alpha\beta(\alpha^2+\beta^2)$. This representation can be rewritten as a single integral

$$w_n(x) = \int_0^\infty g_n(x, s) f_n(s)\, ds$$

where the kernel function $g_n(x, s)$ is defined as

$$g_n(x, s) = \frac{1}{\Delta}\{e^{-\alpha(x+s)}[\alpha \sin \beta(x+s) + \beta \cos \beta(x+s)]$$
$$- e^{-\alpha|x-s|}[\alpha \sin \beta|x-s| + \beta \cos \beta(x-s)]\} \tag{5.83}$$

In the series expansion of eqn (5.74), the Fourier coefficients $f_n(s)$ of the right-hand side term $f(x, y)$ in eqn (5.65) can be written by means of the Euler–Fourier formula as

$$f_n(s) = \frac{2}{b} \int_0^b f(s, t) \sin \nu t\, dt$$

providing for $w_n(x)$ the integral representation

$$w_n(x) = \frac{2}{b} \int_0^b \int_0^\infty g_n(x, s) \sin \nu t\, f(s, t)\, ds\, dt$$

Upon substituting this in eqn (5.73) and changing the order of the summation and the integration, the solution of the boundary-value problem posed by eqns (5.65) and (5.72) is ultimately found as

$$w(x, y) = \int_0^b \int_0^\infty \left( \frac{2}{b} \sum_{n=1}^{\infty} g_n(x, s) \sin \nu y \sin \nu t \right) f(s, t)\, ds\, dt \tag{5.84}$$

In light of the definition introduced earlier in Section 5.1, the kernel-function

$$G(x, y; s, t) = \frac{2}{bD} \sum_{n=1}^{\infty} g_n(x, s) \sin \nu y \sin \nu t \tag{5.85}$$

in eqn (5.84) represents the influence function of a transverse point force for the semi-strip-shaped plate resting on elastic foundation, with all three edges of the plate simply-supported.

## 5.4 Reissner Plates

In this section, the influence function method is extended to the mathematical formulation of thin plate problems based on the Reissner theory (see [56]), which accounts for the effect of transverse normal stress and transverse shear deformation and admits any standard physically feasible boundary conditions (clamped, simply-supported, elastically supported, or free edge). The technique, described in the previous sections of this book, is here employed to analytically construct influence matrices that can be used for numerical solutions of actual plate problems. A validation example is presented revealing computational properties of such an approach.

Consider a thin plate occupying a region $\Omega$ whose boundary is a piecewise smooth curve $\Gamma$. Suppose the plate is subjected to a transverse distributed load $q = q(x, y)$ per unit area. Assume, in addition, that the plate has a uniform thickness $h$ and is composed of a homogeneous isotropic elastic material having the modulus of elasticity $E$ and Poisson ratio $\sigma$.

Let the equilibrium state of the plate be described in compliance with the Reissner theory [56] with the following system of equations

$$\frac{\partial^2\beta_x}{\partial x^2}+\frac{1-\sigma}{2}\frac{\partial^2\beta_x}{\partial y^2}+\frac{1+\sigma}{2}\frac{\partial^2\beta_y}{\partial x\partial y}-\frac{5(1-\sigma)}{h^2}\left(\beta_x+\frac{\partial w}{\partial x}\right)$$
$$=-\frac{1}{D}\frac{\partial}{\partial x}\left(\frac{\sigma h^2 q}{10(1-\sigma)}\right)$$
$$\frac{\partial^2\beta_y}{\partial y^2}+\frac{1-\sigma}{2}\frac{\partial^2\beta_y}{\partial x^2}+\frac{1+\sigma}{2}\frac{\partial^2\beta_x}{\partial x\partial y}-\frac{5(1-\sigma)}{h^2}\left(\beta_y+\frac{\partial w}{\partial y}\right)$$
$$=-\frac{1}{D}\frac{\partial}{\partial y}\left(\frac{\sigma h^2 q}{10(1-\sigma)}\right)$$
$$\frac{\partial^2 w}{\partial x^2}+\frac{\partial^2 w}{\partial y^2}+\frac{\partial\beta_x}{\partial x}+\frac{\partial\beta_y}{\partial y}=-\frac{12(1+\sigma)}{5Eh}q \tag{5.86}$$

in the out-of-plane deflection $w = w(x, y)$ and rotations $\beta_x = \beta_x(x, y)$ and $\beta_y = \beta_y(x, y)$ of a point $(x, y)$ of the middle plane in the plate. The stress resultants can be written in terms of the basic unknowns of the above system in the form

$$M_x = D\left(\frac{\partial\beta_x}{\partial x}+\sigma\frac{\partial\beta_y}{\partial y}\right)+\frac{\sigma h^2 q}{10(1-\sigma)}$$
$$M_y = D\left(\frac{\partial\beta_y}{\partial y}+\sigma\frac{\partial\beta_x}{\partial x}\right)+\frac{\sigma h^2 q}{10(1-\sigma)}$$
$$M_{xy} = D\frac{1-\sigma}{2}\left(\frac{\partial\beta_x}{\partial y}+\frac{\partial\beta_y}{\partial x}\right) \tag{5.87}$$

$$V_x = \frac{5Eh}{12(1+\sigma)}\left(\beta_x + \frac{\partial w}{\partial x}\right), \quad V_y = \frac{5Eh}{12(1+\sigma)}\left(\beta_y + \frac{\partial w}{\partial y}\right) \tag{5.88}$$

where $M_x$, $M_y$, and $M_{xy}$ represent the bending and twisting moments, respectively, $V_x$ and $V_y$ are the shear forces, and $D = Eh^3/(12(1-\sigma^2))$ represents the flexural rigidity of the plate.

Notice that for the first time in this text, we are involved with a formulation that includes a system of partial differential equations of higher order. The total order of the system in eqn (5.86) is six. Thus, to complete the problem formulation, three linearly independent boundary conditions are to be imposed at each point on the contour $\Gamma$.

Assume the plate occupies a rectangular region $\Omega_r = \{(x, y) : 0 < x < a, 0 < y < b\}$ and let the edges $y = 0$ and $y = b$ be simply-supported yielding the following formulation of the boundary conditions along them

$$w|_{y=0,b} = 0, \quad \beta_x|_{y=o,b} = 0, \quad M_y|_{y=0,b} = 0 \tag{5.89}$$

Any combination of the conventional boundary conditions (either simply-supported, or clamped, or elastically supported, or free edge) is allowed on the edges $x = 0$ and $x = a$.

We are now in a position to construct the influence matrix of a transverse concentrated unit force applied at an arbitrary point in the plate. In doing so, we represent the loading function $q(x, y)$ by means of the Fourier series

$$q(x, y) = \sum_{n=1}^{\infty} q_n(x) \sin \nu y, \quad \nu = \frac{n\pi}{b} \tag{5.90}$$

The solution vector $\mathbf{U}(x, y) = (w(x, y), \beta_x(x, y), \beta_y(x, y))^T$ of the boundary-value problem just formulated is expanded in the following manner

$$\mathbf{U}(x, y) = \sum_{n=0}^{\infty} Q_n(y)\mathbf{U}_n(x) \tag{5.91}$$

where we denote

$$Q_n(y) = \begin{pmatrix} \sin \nu y & 0 & 0 \\ 0 & \sin \nu y & 0 \\ 0 & 0 & \cos \nu y \end{pmatrix}, \quad \mathbf{U}_n(x) = \begin{pmatrix} w_n(x) \\ \beta_{xn}(x) \\ \beta_{yn}(x) \end{pmatrix}$$

Notice that the expansion in eqn (5.91) satisfies the boundary conditions of eqn (5.89). The components of the vector $\mathbf{U}_n(x)$ are, subsequently, to satisfy the following system of ordinary differential equations

$$\frac{d^2\beta_{xn}}{dx^2} - \frac{1-\sigma}{2}\nu^2\beta_{xn} - \frac{1+\sigma}{2}\nu\frac{d\beta_{yn}}{dx} - \frac{5(1-\sigma)}{h^2}\left(\beta_{xn} + \frac{dw_n}{dx}\right)$$

$$= -\frac{6\sigma(1+\sigma)}{5Eh}\frac{dq_n}{dx}$$

$$\frac{d^2\beta_{yn}}{dx^2} - \frac{2\nu^2}{1-\sigma}\beta_{yn} + \frac{1+\sigma}{1-\sigma}\nu\frac{d\beta_{xn}}{dx} - \frac{10}{h^2}(\beta_{yn} + \nu w_n) = -\frac{12\sigma(1+\sigma)}{5Eh(1-\sigma)}\nu q_n$$

$$\frac{d^2 w_n}{dx^2} - \nu^2 w_n + \frac{d\beta_{xn}}{dx} - \nu\beta_{yn} = -\frac{12(1+\sigma)}{5Eh}q_n \tag{5.92}$$

This system can be subject to any of the following combinations of boundary conditions at $x = 0$ and $x = a$:

$$w_n = 0, \quad \frac{d\beta_{xn}}{dx} = 0, \quad \beta_{yn} = 0 \tag{5.93}$$

(as associated with a simply-supported edge in the original statement);

$$w_n = 0, \quad \beta_{xn} = 0, \quad \beta_{yn} = 0 \tag{5.94}$$

(as associated with a clamped edge);

$$\beta_{xn} + \frac{dw_n}{dx} = 0, \quad \nu\beta_{xn} + \frac{d\beta_{yn}}{dx} = 0, \quad \sigma\frac{d\beta_{xn}}{dx} - \nu\beta_{yn} = 0 \tag{5.95}$$

(as associated with a free edge).

Clearly, the system in eqn (5.92) is linear and has constant coefficients. Hence, for the corresponding homogeneous ($q_n(x) \equiv 0$) system, the fundamental set of solutions (which is a set of six linearly independent vector-functions) can be found with the aid of standard procedures. Omitting the elementary but quite cumbersome algebra, we present the fundamental set of solutions in the form

$$\mathbf{U}_n^{(1)}(x) = \begin{pmatrix} e^{\nu x} \\ -\nu e^{\nu x} \\ -\nu e^{\nu x} \end{pmatrix}, \quad \mathbf{U}_n^{(2)}(x) = \begin{pmatrix} xe^{\nu x} \\ -(\nu x + \omega + 1)e^{\nu x} \\ -(\nu x + \omega)e^{\nu x} \end{pmatrix}$$

$$\mathbf{U}_n^{(3)}(x) = \begin{pmatrix} e^{-\nu x} \\ \nu e^{-\nu x} \\ -\nu e^{-\nu x} \end{pmatrix}, \quad \mathbf{U}_n^{(4)}(x) = \begin{pmatrix} xe^{-\nu x} \\ (\nu x - \omega - 1)e^{-\nu x} \\ -(\nu x - \omega)e^{-\nu x} \end{pmatrix}$$

$$\mathbf{U}_n^{(5)}(x) = \begin{pmatrix} 0 \\ e^{px} \\ p_1 e^{px} \end{pmatrix}, \quad \mathbf{U}_n^{(6)}(x) = \begin{pmatrix} 0 \\ e^{-px} \\ -p_1 e^{-px} \end{pmatrix} \tag{5.96}$$

The parameters $\omega$, $p$, and $p_1$ in the above expressions are defined as

$$\omega = \frac{2\nu^2 h^2}{5(1-\sigma)}, \quad p = \sqrt{\nu^2 + \frac{10}{h^2}}, \quad p_1 = \frac{p}{\nu}$$

In compliance with Lagrange's method of variation of parameters, the general solution of the system in eqn (5.92) can be sought in the form

$$\mathbf{U}_n(x) = P_n(x)\mathbf{C}_n(x) \tag{5.97}$$

where

$$P_n(x) = \left(\mathbf{U}_n^{(j)}(x)\right), \quad j = 1, \ldots, 6$$

is a rectangular matrix of order $3 \times 6$ whose columns are the vectors from eqn (5.96), while $\mathbf{C}_n(x)$ is a column-vector of dimension six, whose entries are undetermined functions.

Proceeding further with Lagrange's method, one obtains the system of six linear algebraic equations

$$P_n^*(x)\mathbf{C}_n'(x) = \mathbf{F}_n^*(x) \tag{5.98}$$

in the components of the derivative $\mathbf{C}_n'(x)$ of the vector $\mathbf{C}_n(x)$.

Structure of this system can be viewed by the horizontal partition of its coefficient matrix $P_n^*(x)$ into two equidimensional sub-matrices. Its upper $3 \times 6$ sub-matrix represents matrix $P_n(x)$ from eqn (5.97), while its lower $3 \times 6$ sub-matrix represents the derivative of $P_n(x)$. Partitioning analogously the right-hand side vector $\mathbf{F}_n^*(x)$, we view its upper sub-vector as a three-dimensional zero-vector, while its lower sub-vector is viewed as the right-hand side vector

$$\mathbf{F}_n(x) = \left(-\frac{6\sigma(1+\sigma)}{5Eh}\frac{dq_n}{dx}, -\frac{12\sigma(1+\sigma)}{5Eh(1-\sigma)}\nu q_n, -\frac{12(1+\sigma)}{5Eh}q_n\right)^T$$

of the system in eqn (5.92). That is

$$P_n^*(x) = \begin{pmatrix} P_n(x) \\ P_n'(x) \end{pmatrix}, \quad \mathbf{F}_n^*(x) = \begin{pmatrix} 0 \\ \mathbf{F}_n(x) \end{pmatrix}$$

It is evident that the coefficient matrix of the system in eqn (5.98) represents the Wronskian of the fundamental set of solutions of the homogeneous system corresponding to that of eqn (5.92). The system in eqn (5.98) has, consequently, a unique solution that can be written in terms of the inverse of $P_n^*(x)$ as

$$\mathbf{C}_n'(x) = (P_n^*(x))^{-1}\mathbf{F}_n^*(x)$$

Recalling the vector's $\mathbf{F}_n^*(x)$ structure (its first three components are zeros), the above equation can be rewritten in terms of the vector $\mathbf{F}_n(x)$ as

$$\mathbf{C}_n'(x) = R_n(x)\mathbf{F}_n(x) \tag{5.99}$$

where $R_n(x)$ represents a matrix of order $6 \times 3$, which is a sub-matrix of $(P_n^*(x))^{-1}$ consisting of the latter's fourth, fifth, and sixth columns.

To obtain an explicit expression for $\mathbf{C}_n(x)$, we integrate the relation in eqn (5.99). This yields

$$\mathbf{C}_n(x) = \int_0^x R_n(s)\mathbf{F}_n(s)\,ds + \mathbf{D}_n$$

Upon substituting this expression for $\mathbf{C}_n(x)$ into eqn (5.97) and taking the factor $P_n(x)$ under the integral sign, one obtains

$$\mathbf{U}_n(x) = \int_0^x S_n(x, s)\mathbf{F}_n(s)\,ds + P_n(x)\mathbf{D}_n \tag{5.100}$$

where the kernel-matrix $S_n(x, s) = P_n(x)R_n(s)$ is a square matrix of order $3 \times 3$, whose entries $S_{ij}^n(x, s)$ are found in the form

$$S_{11}^n(x, s) = \frac{1}{\omega\nu}[(\omega + 1)\sinh\nu(x - s) - \nu(x - s)\cosh\nu(x - s)]$$

$$S_{12}^n(x, s) = -\frac{1}{2\nu}(x - s)\sinh\nu(x - s)$$

$$S_{13}^n(x, s) = \frac{1}{\omega(p^2 - \nu^2)}[\nu(x - s)\cosh\nu(x - s) - \sinh\nu(x - s)]$$

$$S_{21}^n(x, s) = \frac{1}{\omega}\nu(x - s)\sinh\nu(x - s)$$

$$S_{22}^n(x, s) = \frac{1}{2p\nu}[p(\omega + 1)\sinh\nu(x - s) + p\nu(x - s)\cosh\nu(x - s) - \omega\nu\sinh p(x - s)]$$

$$S_{23}^n(x, s) = \frac{\nu}{\omega(p^2 - \nu^2)}[\omega\cosh\nu(x - s) - \nu(x - s)\sinh\nu(x - s) - \omega\cosh\nu(x - s)]$$

$$S_{31}^n(x, s) = \frac{1}{\omega}[\nu(x - s)\cosh\nu(x - s) - \sinh\nu(x - s)]$$

$$S_{32}^n(x, s) = \frac{1}{2\nu}[\omega\cosh\nu(x - s) + \nu(x - s)\sinh\nu(x - s) - \omega\cosh p(x - s)]$$

$$S_{33}^n(x, s) = \frac{1}{\omega(p^2 - \nu^2)}[p\omega\sinh(p(x - s)) - \nu^2(x - s)\cosh\nu(x - s) - \nu(\omega - 1)\sinh\nu(x - s)]$$

In the development that follows, we take advantage of a special feature of the above entries of $S_n(x, s)$. The point is that, being functions of the two variables $x$ and $s$, they actually represent functions of a single variable $x - s$, that is $S_{ij}^n(x, s) = S_{ij}^n(x - s)$. Notice also that the relation $S_{ij}^n(x, x) = 0$ holds for all of the entries of $S_n(x, s)$. This feature is taken into account later when the differentiation of the vector $\mathbf{U}_n(x)$ is performed while satisfying the boundary conditions.

To make the subsequent development as compact as possible, we introduce the operator form

$$B_{1n}[w_n, \beta_{xn}, \beta_{yn}] = 0, \quad B_{2n}[w_n, \beta_{xn}, \beta_{yn}] = 0$$
$$B_{3n}[w_n, \beta_{xn}, \beta_{yn}] = 0 \tag{5.101}$$

for the boundary conditions at $x = 0$ and $x = a$, imposed by eqns (5.93)–(5.95).

Satisfying the boundary conditions by the components of the vector $\mathbf{U}_n(x)$ from eqn (5.100), one obtains a system of six linear algebraic equations

$$T_n\mathbf{D}_n = \mathbf{K}_n \tag{5.102}$$

in the components of the vector $\mathbf{D}_n$ that appears in eqn (5.100). The first, second, and third rows of the coefficient matrix $T_n$ of the above system are defined as $B_{1n}[P_n(0)]$, $B_{2n}[P_n(0)]$, and $B_{3n}[P_n(0)]$, respectively, while the fourth, fifth, and sixth rows are defined as $B_{1n}[P_n(a)]$, $B_{2n}[P_n(a)]$, and $B_{3n}[P_n(a)]$, respectively. The first, second, and third entries of the right-hand side vector $\mathbf{K}_n$ are zero, while its fourth, fifth, and sixth entries are defined as

$$\int_0^a Z_n(a, s)\mathbf{F}_n(s)\,ds$$

where $Z_n(x, s)$ is a square matrix of dimension $3 \times 3$, whose rows are defined as $-B_{1n}[S_n(x, s)]$, $-B_{2n}[S_n(x, s)]$, and $-B_{3n}[S_n(x, s)]$, respectively.

In compliance with the described structure of the coefficient matrix and the right-hand side vector $\mathbf{K}_n$ of the system in eqn (5.102), the solution to the latter can be written in the form

$$\mathbf{D}_n = \int_0^a N_n(s)\mathbf{F}_n(s)\,ds \tag{5.103}$$

with $N_n(s)$ being expressed in terms of the inverse $T_n^{-1}$ of the coefficient matrix of the system in eqn (5.102) as

$$N_n(s) = T_n^{-1} \times H_n(a, s)$$

where $H_n(a, s)$ is a matrix of order $6 \times 3$, whose first three rows are zero, while its remaining $3 \times 3$ sub-matrix represents the matrix $Z_n(a, s)$ recently introduced.

By substituting the expression for $\mathbf{D}_n$ derived in eqn (5.103) into eqn (5.100), one obtains

$$\mathbf{U}_n(x) = \int_0^x S_n(x, s)\mathbf{F}_n(s)\, ds + \int_0^a P_n(x)N_n(s)\mathbf{F}_n(s)\, ds$$

This can readily be put in the form of a single integral

$$\mathbf{U}_n(x) = \int_0^a g_n(x, s)\mathbf{F}_n(s)\, ds \tag{5.104}$$

where $g_n(x, s)$ is defined in pieces as

$$g_n(x, s) = \begin{cases} P_n(x)N_n(s) + S_n(x, s), & x \geq s \\ P_n(x)N_n(s), & x \leq s \end{cases}$$

This represents the Green's matrix of the homogeneous boundary-value problem corresponding to that posed by eqns (5.92) and (5.101).

Substituting the expression for $\mathbf{U}_n(x)$ from eqn (5.104) into the expansion in eqn (5.91), and then proceeding in conformity with the relation from eqn (5.90), we finally obtain the following integral representation

$$\mathbf{U}(x, y) = \iint_{\Omega_r} \left[\frac{2}{b}\sum_{n=1}^{\infty} Q_n(y)g_n(x, s)Q_n(t)\right]\mathbf{F}(s, t)\, d\Omega_r(s, t) \tag{5.105}$$

for the solution vector $\mathbf{U}(x, y)$ of the original boundary-value problem for the system in eqn (5.86). Here $\mathbf{F}(s, t)$ represents a vector whose components are the right-hand terms of the system in eqn (5.86). Thus, in accordance with the definition introduced in the opening part of this chapter, the kernel-matrix

$$G(x, y; s, t) = \frac{2}{b}\sum_{n=1}^{\infty} Q_n(y)g_n(x, s)Q_n(t) \tag{5.106}$$

of the representation in eqn (5.105) is the influence matrix of a transverse concentrated unit force applied at an arbitrary point in the rectangular $a \times b$ plate considered in the scope of the Reissner's theory.

One finds below the entries of $G(x, y; s, t)$ obtained for the plate, all edges of which are simply-supported. We present the branches $G^+_{ij} = G^+_{ij}(x, y; s, t)$ valid for $x \leq s$ in the form

$$G^+_{11} = \frac{2}{b}\sum_{n=1}^{\infty}[\varphi^n_{11}(x)S^n_{11}(a, s) + \varphi^n_{12}(x)\overline{S}^n_{21}(a, s) + \varphi^n_{13}(x)S^n_{31}(a, s)] \sin \nu y \sin \nu t$$

$$G^+_{12} = \frac{2}{b}\sum_{n=1}^{\infty}[\varphi^n_{11}(x)S^n_{12}(a, s) + \varphi^n_{12}(x)\overline{S}^n_{22}(a, s) + \varphi^n_{13}(x)S^n_{32}(a, s)] \sin \nu y \sin \nu t$$

$$G_{13}^{+} = \frac{4}{b(1-\sigma)} \sum_{n=1}^{\infty} [\varphi_{11}^{n}(x) S_{13}^{n}(a, s) + \varphi_{12}^{n}(x) \overline{S}_{23}^{n}(a, s) + \varphi_{13}^{n}(x) S_{33}^{n}(a, s)]$$
$$\times \sin \nu y \cos \nu t$$

and

$$G_{21}^{+} = \frac{2}{b} \sum_{n=1}^{\infty} [\varphi_{12}^{n}(x) S_{11}^{n}(a, s) + \varphi_{22}^{n}(x) \overline{S}_{21}^{n}(a, s) + \varphi_{23}^{n}(x) S_{31}^{n}(a, s)] \sin \nu y \sin \nu t$$

$$G_{22}^{+} = \frac{2}{b} \sum_{n=1}^{\infty} [\varphi_{12}^{n}(x) S_{12}^{n}(a, s) + \varphi_{22}^{n}(x) \overline{S}_{22}^{n}(a, s) + \varphi_{23}^{n}(x) S_{32}^{n}(a, s)] \sin \nu y \sin \nu t$$

$$G_{23}^{+} = \frac{4}{b(1-\sigma)} \sum_{n=1}^{\infty} [\varphi_{21}^{n}(x) S_{13}^{n}(a, s) + \varphi_{22}^{n}(x) \overline{S}_{23}^{n}(a, s) + \varphi_{23}^{n}(x) S_{33}^{n}(a, s)]$$
$$\times \sin \nu y \cos \nu t$$

and

$$G_{31}^{+} = \frac{2}{b} \sum_{n=1}^{\infty} [\varphi_{31}^{n}(x) S_{11}^{n}(a, s) + \varphi_{32}^{n}(x) \overline{S}_{21}^{n}(a, s) + \varphi_{33}^{n}(x) S_{31}^{n}(a, s)]$$
$$\times \cos \nu y \sin \nu t$$

$$G_{32}^{+} = \frac{2}{b} \sum_{n=1}^{\infty} [\varphi_{31}^{n}(x) S_{12}^{n}(a, s) + \varphi_{32}^{n}(x) \overline{S}_{22}^{n}(a, s) + \varphi_{33}^{n}(x) S_{32}^{n}(a, s)]$$
$$\times \cos(\nu y) \sin(\nu t)$$

$$G_{33}^{+} = \frac{4}{b(1-\sigma)} \sum_{n=1}^{\infty} [\varphi_{31}^{n}(x) S_{13}^{n}(a, s) + \varphi_{32}^{n}(x) \overline{S}_{23}^{n}(a, s) + \varphi_{33}^{n}(x) S_{33}^{n}(a, s)]$$
$$\times \cos \nu y \cos \nu t$$

where we use notations as follows

$$\overline{S}_{21}^{n}(a, s) = (S_{21}^{n}(x, s))_{,x}|_{x=a}, \quad \overline{S}_{22}^{n}(a, s) = (S_{22}^{n}(x, s))_{,x}|_{x=a}$$
$$\overline{S}_{23}^{n}(a, s) = (S_{23}^{n}(x, s))_{,x}|_{x=a}$$

and

$$\varphi_{11}^{n}(x) = \frac{\sinh \nu x}{\sinh \nu a}, \quad \varphi_{12}^{n}(x) = \frac{a \cosh \nu a \sinh \nu x - x \cosh \nu x \sinh \nu a}{2\nu \sinh^2 \nu a}$$
$$\varphi_{13}^{n}(x) = \frac{x \cosh \nu x \sinh \nu a - a \cosh \nu a \sinh \nu x}{2\nu \sinh^2 \nu a}$$

$$\varphi_{21}^n(x) = \nu\left(\frac{\nu \cosh px}{p \sinh pa} - \frac{\cosh \nu x}{\sinh \nu a}\right)$$

$$\varphi_{22}^n(x) = \frac{1}{2}\left(\frac{x \sinh \nu x}{\sinh \nu a} + \frac{(\omega+1)\sinh \nu a - \nu a \cosh \nu a}{\nu \sinh^2 \nu a(\cosh \nu x)^{-1}} - \frac{\omega \cosh px}{p \sinh pa}\right)$$

$$\varphi_{23}^n(x) = \frac{1}{2}\left(\frac{(\omega+2)\cosh px}{p_1 \sinh pa} - \frac{\nu x \sinh \nu x}{\sinh \nu a} - \frac{(\omega+1)\sinh \nu a - \nu a \cosh \nu a}{\sinh^2 \nu a(\cosh \nu x)^{-1}}\right)$$

$$\varphi_{31}^n(x) = \nu\left(\frac{\sinh px}{\sinh pa} - \frac{\sinh \nu x}{\sinh \nu a}\right)$$

$$\varphi_{32}^n(x) = \frac{1}{2}\left(\frac{x \cosh \nu x}{\sinh \nu a} + \frac{\omega \sinh \nu a - \nu a \cosh \nu a}{\nu a \sinh^2 \nu a(\sinh \nu x)^{-1}} - \frac{\omega \sinh px}{\nu \sinh pa}\right)$$

$$\varphi_{33}^n(x) = \frac{1}{2}\left(\frac{(\omega+2)\sinh px}{\sinh pa} - \frac{\omega \sinh \nu a - \nu a \cosh \nu a}{\sinh^2 \nu a(\sinh \nu x)^{-1}} - \frac{x \cosh \nu x}{\sinh \nu a}\right)$$

The branches $G_{ij}^- = G_{ij}^-(x, y; s, t)$ of $G(x, y; s, t)$, valid for $x \geq s$, can readily be obtained from $G_{ij}^+$ in compliance with Betti's reciprocal work theorem, whose implementation to this case provides

$$G_{11}^-(x, y; s, t) = G_{11}^+(s, t; x, y)$$

$$\frac{1}{D}G_{12}^-(x, y; s, t) = \frac{12(1+\sigma)}{5Eh}G_{21}^+(s, t; x, y)$$

$$\frac{1}{D}G_{13}^-(x, y; s, t) = \frac{12(1+\sigma)}{5Eh}G_{31}^+(s, t; x, y)$$

$$G_{22}^-(x, y; s, t) = G_{22}^+(s, t; x, y)$$

and

$$G_{23}^-(x, y; s, t) = G_{32}^+(s, t; x, y), \quad G_{33}^-(x, y; s, t) = G_{33}^+(s, t; x, y)$$

In the discussion that follows, it is necessary to accurately compute the domain integrals of the entries of the influence matrix just derived as well as the domain integrals of the partial derivatives of these entries. Note that some of the entries possess singularity. That is, not all of the series from eqn (5.106) converge uniformly. Thus, the differentiation and integration mentioned above cannot be accomplished in a straightforward manner. For that reason, before proceeding any further with the development of the computational algorithm, we will clarify the integral and differential properties of the series from eqn (5.106).

In doing so, we stress that for any fixed position of the observation point $(x, y)$ within the basic region $\Omega_r$, the series in eqn (5.106) converges uniformly to a corresponding entry $G_{ij}(x, y; s, t)$ of the influence matrix everywhere in $\Omega_r$ except perhaps at the point $(s, t) = (x, y)$. The practical convergence takes place as long as $(s, t) \in \Omega_r^* = \Omega_r \setminus B(x, y)$, where $B(x, y)$ is in the immediate vicinity of the observation point $(x, y)$.

A similar property also remains valid for the partial derivatives of the first order of the series of eqn (5.106). They converge uniformly to the corresponding derivatives $\partial G_{ij}/\partial x$ or $\partial G_{ij}/\partial y$ as $(s, t) \in \Omega_r^*$. In other words, it is possible to differentiate the series of eqn (5.106) with respect to either $x$ or $y$ in a term-by-term fashion inside the region $\Omega_r^*$.

The improper domain integrals of the form

$$\iint_{\Omega_0} G_{ij}(x, y; s, t)\, d\Omega_0(s, t), \qquad \iint_{\Omega_0} \frac{\partial}{\partial x} G_{ij}(x, y; s, t)\, d\Omega_0(s, t)$$

and

$$\iint_{\Omega_0} \frac{\partial}{\partial y} G_{ij}(x, y; s, t)\, d\Omega_0(s, t)$$

converge absolutely over a subregion $\Omega_0$ of $\Omega_r$. Thus, the series of eqn (5.106) can be integrated in a term-by-term manner over a subregion of the basic region $\Omega_r$, again resulting in a uniformly convergent series.

Based on the influence matrix recently derived, one can readily rewrite the integral representation of the solution vector $\mathbf{U}(x, y)$ from eqn (5.105) in the expanded form

$$\begin{pmatrix} w(P) \\ \beta_x(P) \\ \beta_y(P) \end{pmatrix} = \int_0^a \int_0^b \begin{pmatrix} G_{11}(P; Q) & G_{12}(P; Q) & G_{13}(P; Q) \\ G_{11}(P; Q) & G_{12}(P; Q) & G_{13}(P; Q) \\ G_{11}(P; Q) & G_{12}(P; Q) & G_{13}(P; Q) \end{pmatrix} \begin{pmatrix} F_1(Q) \\ F_2(Q) \\ F_3(Q) \end{pmatrix} dQ$$

where $F_j(Q)$, $j = 1, 2, 3$, represent the components of the loading vector of the right-hand side of the system in eqn (5.86), while $P$ and $Q$ are the observation point $(x, y)$ and the source point $(s, t)$, respectively.

To compute the approximate values $w^h(x, y)$, $\beta_x^h(x, y)$, and $\beta_y^h(x, y)$ of the components of $\mathbf{U}(x, y)$, we partition the basic region $\Omega_r$ into a set of elementary rectangles $\Omega_m$, $m = 1, \dots, M$ by using the scheme as shown

$$\Omega_r = \bigcup_{m=1}^{M} \Omega_m, \quad \Omega_m = \{(x, y) \in \Omega_r,\, x_m^1 \le x \le x_m^2,\, y_m^1 \le y \le y_m^2\} \tag{5.107}$$

and then apply the following cubature formulae

$$w^h(x, y) = \sum_{m=1}^{M} \left\{ \sum_{j=1}^{3} F_j(x_m, y_m) \int_{x_m^1}^{x_m^2} \int_{y_m^1}^{y_m^2} G_{1j}(x, y; s, t)\, dt\, ds \right\}$$

$$\beta_x^h(x, y) = \sum_{m=1}^{M} \left\{ \sum_{j=1}^{3} F_j(x_m, y_m) \int_{x_m^1}^{x_m^2} \int_{y_m^1}^{y_m^2} G_{2j}(x, y; s, t)\, dt\, ds \right\}$$

and

$$\beta_y^h(x, y) = \sum_{m=1}^{M} \left\{ \sum_{j=1}^{3} F_j(x_m, y_m) \int_{x_m^1}^{x_m^2} \int_{y_m^1}^{y_m^2} G_{3j}(x, y; s, t)\, dt\, ds \right\} \tag{5.108}$$

based on the mean value theorem of integration. Here

$$x_m = \tfrac{1}{2}(x_m^1 + x_m^2), \quad y_m = \tfrac{1}{2}(y_m^1 + y_m^2)$$

Clearly, approximate expressions for the first order partial derivatives of the functions from eqn (5.108) can be obtained analytically by straightforward differentiation. For the derivative $\partial w^h/\partial x$, for instance, one obtains

$$\frac{\partial w^h(x, y)}{\partial x} = \sum_{m=1}^{M} \left\{ \sum_{j=1}^{3} F_j(x_m, y_m) \int_{x_m^1}^{x_m^2} \int_{y_m^1}^{y_m^2} \frac{\partial G_{1j}(x, y; s, t)}{\partial x}\, dt\, ds \right\} \quad (5.109)$$

All other derivatives needed for the evaluation of the stress resultants from eqns (5.87) and (5.88) can be written in a similar manner.

As we have already mentioned, the integrals over the elementary rectangles $\Omega_m$ in eqns (5.108) and (5.109) converge absolutely, providing a suitable basis for approximate computation of the components of $\mathbf{U}(x, y)$ and its derivatives.

One can readily derive the error estimates for the approximations obtained in eqns (5.108) and (5.109). Indeed, let us assume that the functions $F_j(s, t)$, $j = 1, 2, 3$ satisfy the Lipschitz condition [60]

$$|F_j(s_2, t_2) - F_j(s_1, t_1)| \le \alpha_m \sqrt{(s_2 - s_1)^2 + (t_2 - t_1)^2}$$

in each of the elementary rectangles $\Omega_m$. Taking into account that the entries $G_{ij}(x, y; s, t)$ of the influence matrix are absolutely integrable functions of $s$ and $t$ over $\Omega_r$, we obtain, for any fixed position of the observation point $(x, y) \in \Omega_r$

$$|w(x, y) - w^h(x, y)| \le \max_m(\alpha_m D_m) \int_{x_m^1}^{x_m^2} \int_{y_m^1}^{y_m^2} \sum_{j=1}^{3} |G_{1j}(x, y; s, t)|\, dt\, ds$$

$$|\beta_x(x, y) - \beta_x^h(x, y)| \le \max_m(\alpha_m D_m) \int_{x_m^1}^{x_m^2} \int_{y_m^1}^{y_m^2} \sum_{j=1}^{3} |G_{2j}(x, y; s, t)|\, dt\, ds$$

and

$$|\beta_y(x, y) - \beta_y^h(x, y)| \le \max_m(\alpha_m D_m) \int_{x_m^1}^{x_m^2} \int_{y_m^1}^{y_m^2} \sum_{j=1}^{3} |G_{3j}(x, y; s, t)|\, dt\, ds \quad (5.110)$$

where with $D_m$ we denote the length of the biggest side of the elementary rectangle $\Omega_m$.

It is evident that the error estimates for the first order partial derivatives of the components of the vector $\mathbf{U}^h(x, y)$ can be derived in a similar manner. That is, the error estimate for the stress components can also be evaluated.

From the estimates derived in eqn (5.110), it appears that the approximate solution vector $\mathbf{U}^h(x, y)$, whose components are given in eqn (5.108), converges to the true solution of the original boundary-value problem for the system in eqn (5.86) as the partition parameter $M$ in eqn (5.107) approaches infinity

$$\lim_{M\to\infty}(\mathbf{U}(x, y) - \mathbf{U}^h(x, y)) = 0$$

This conclusion is based on the following observation. The limit of the first factor in the right-hand side of the inequalities in eqn (5.110) equals zero, because $D_m \to 0$ as $M$ approaches infinity (this is true at least for the uniform partition specified by eqn (5.107)), while the limit of the second factor (double integral) is bounded from above, because of the absolute integrability of the entries of the influence matrix over $\Omega_r$.

Speaking of the computational procedure, the mathematical basis of which has just been described, it is worth noting that we here completely avoid numerical differentiation while computing stress components. Consequently, values of the moment and shear resultants are actually computed, within this study, as accurately as those of the deflection and rotation functions. This observation is reiterated in the validation example that is later presented.

We complete the discussion in this section by solving a validation example for a Reissner plate. That is going to be a test problem whose exact solution is known. A simply-supported square plate of a uniform thickness $h$ is considered. Let the plate's middle plane occupy the region $\Omega_{sq} = \{(x, y) : 0 < x, y < a\}$ and the plate is subjected to a transverse load

$$q(x, y) = q_0 \sin\frac{\pi x}{a} \sin\frac{\pi y}{a}$$

Boundary conditions of simple support can be written in terms of the deflection, rotations, and bending moments. Clearly, these components of the stress-strain state vanish along the simply supported edges. This yields

$$\begin{aligned} &w|_{x=0,a} = 0, \quad M_x|_{x=0,a} = 0, \quad \beta_y|_{x=0,a} = 0 \\ &w|_{y=0,a} = 0, \quad M_y|_{y=0,a} = 0, \quad \beta_x|_{y=0,a} = 0 \end{aligned} \tag{5.111}$$

It can easily be verified by inspection that the deflection function $w(x, y)$ taken in the form of the following trigonometric function

$$w(x, y) = \frac{q_0}{D}\frac{(2-\sigma)\chi^2 + \delta_1}{4\delta_1\chi^4} \sin\frac{\pi x}{a} \sin\frac{\pi y}{a}$$

along with the rotation functions $\beta_x(x, y)$ and $\beta_y(x, y)$ taken as

$$\beta_x(x, y) = \frac{q_0}{D}\frac{\delta_2\chi^2 - 1}{4\chi^3} \cos\frac{\pi x}{a} \sin\frac{\pi y}{a}$$

and

$$\beta_y(x, y) = \frac{q_0}{D}\frac{\delta_2\chi^2 - 1}{4\chi^3} \sin\frac{\pi x}{a} \cos\frac{\pi y}{a}$$

represent in this case the components of the true solution vector $\mathbf{U}(x, y)$ to the boundary-value problem posed by eqns (5.86) and (5.111). The parameters $\chi$, $\delta_1$

and $\delta_2$ in the above expressions for $w(x, y)$, $\beta_x(x, y)$ and $\beta_y(x, y)$ are defined as

$$\chi = \frac{\pi}{a}, \quad \delta_1 = \frac{5(1-\sigma)}{h^2}, \quad \delta_2 = \frac{\sigma}{\delta_1}$$

In actually computing the components of the vector $\mathbf{U}^h(x, y)$ for this test problem, we uniformly partitioned the basic region $\Omega_{sq}$ into $M = m_x \times m_y$ elementary rectangles. The uniformly convergent series, which represent the domain integrals over the elementary rectangles $\Omega_m$ (see eqns (5.107)–(5.109)), have been truncated to attain the level of accuracy of $10^{-5}$. The physical and geometric parameters in the statement are $E = 0.21 \times 10^6 MPa$, $\sigma = 0.3$, $a = 1.0m$, and $h = 0.025m$.

Some data that are obtained within this study are displayed in Table 5.1. The maximal values of relative error are shown for various components of the stress-strain state (the deflection, rotations, bending moments, and shear forces), versus the dimension $M = m_x \times m_y$ of the partition used.

**Table 5.1:** Relative errors of the approximate solution for the test problem posed by eqns (5.86) and (5.111)

| Partition, | Relative error, % | | | |
|---|---|---|---|---|
| $M = m_x \times m_y$ | Deflection | Rotations | Moments | Shear forces |
| $5 \times 5$ | 3.16 | 3.36 | 3.42 | 3.46 |
| $8 \times 8$ | 1.24 | 1.26 | 1.34 | 1.39 |
| $10 \times 10$ | 0.78 | 0.80 | 0.82 | 0.85 |
| $12 \times 12$ | 0.65 | 0.66 | 0.68 | 0.71 |

Two evident observations follow from the data of Table 5.1. First, the data show a high degree of practical convergence of the computational procedure developed in this study. Indeed, for the partition of $5 \times 5$ (the partition parameter $M = 25$), the relative error slightly varies from column to column but stays within the range of 3.5%. The relative error then drops down to the value of about 0.7% for the partition of $12 \times 12$ ($M = 144$). Hence, the accuracy level for all of the components is nearly directly proportional to the partition parameter $M$.

The second observation is even more impressive. From the data presented, it is evident that for a fixed partition, the displacements and stresses have been computed with an equal level of accuracy. This appearance is not accidental, because it is typical for computational implementations based on the influence function method, representing one of its most distinguishable and superior features.

## 5.5 Thin Shells of Revolution

Beam and plate problems do not represent the only area of structural mechanics where the method of influence functions appears to be productive. Another class of problems will be brought to the reader's attention in this section. The technique that was used in the previous section for the construction of Green's matrices for systems of partial differential equations will be implemented to problems simulating the static equilibrium of thin elastic shells of revolution.

It is worth making an interesting historical observation regarding the application of the Green's function method to shell problems. The point is that static equilibrium of shells represents a class of problems in structural mechanics to which the Green's (influence) matrix method had been applied much earlier than to other problems in the field. Indeed, pioneering works [29] touching upon the application of this method to the equilibrium of shells had been published nearly four decades ago.

### 5.5.1 Construction of influence matrices

We focus first on a general case procedure for the construction of influence matrices of a point force for a shell of revolution. Consider geometrically linear elastic equilibrium of a thin shell whose middle surface represents a surface of revolution closed in the longitudinal direction, with $x \in (0, l)$ and $\varphi \in (0, 2\pi)$ representing its meridian (latitudinal) and longitudinal coordinates, respectively. A system of partial differential equations modeling the equilibrium state is written in displacements

$$\Lambda\left(\frac{\partial}{\partial x}, \frac{\partial}{\partial \varphi}, x\right)\mathbf{W}(x, \varphi) = -\mathbf{F}(x, \varphi) \tag{5.112}$$

where

$$\mathbf{W}(x, \varphi) = \begin{pmatrix} u(x, \varphi) \\ v(x, \varphi) \\ w(x, \varphi) \end{pmatrix} \quad \text{and} \quad \mathbf{F}(x, \varphi) = \begin{pmatrix} X(x, \varphi) \\ Y(x, \varphi) \\ Z(x, \varphi) \end{pmatrix}$$

are the displacement and the load vectors, respectively, with $u(x, \varphi)$ and $v(x, \varphi)$ representing the components of the displacement vector in the axial $x$, and circumferential $\varphi$ direction, respectively, while $w(x, \varphi)$ represents the normal to the middle surface deflection. $X(x, \varphi)$, $Y(x, \varphi)$, and $Z(x, \varphi)$ represent the components of the loading vector in the corresponding directions. The coefficients of the entries $\Lambda_{ij}$ of the matrix-operator $\Lambda$

$$\Lambda\left(\frac{\partial}{\partial x}, \frac{\partial}{\partial \varphi}, x\right) = \left(\Lambda_{ij}\left(\frac{\partial}{\partial x}, \frac{\partial}{\partial \varphi}, x\right)\right)_{3\times 3}$$

in eqn (5.112) represent functions of the meridian coordinate $x$. In accordance with the total order of the system in eqn (5.112), which is equal to eight, the boundary conditions on the edges $x = 0$ and $x = l$ are written in the form

$$B_0\left(\frac{\partial}{\partial x}, \frac{\partial}{\partial \varphi}\right)\mathbf{W}(0, \varphi) = 0, \quad B_l\left(\frac{\partial}{\partial x}, \frac{\partial}{\partial \varphi}\right)\mathbf{W}(l, \varphi) = 0 \tag{5.113}$$

where $B_0$ and $B_l$ represent $4 \times 3$ matrix-operators. That is, four linearly independent boundary conditions are imposed at each edge of the shell.

To construct the influence matrix of the homogeneous boundary-value problem corresponding to that posed by eqns (5.112) and (5.113), we expand the vectors $\mathbf{W}(x, \varphi)$ and $\mathbf{F}(x, \varphi)$ in the following trigonometric series with respect to the variable $\varphi$

$$\mathbf{W}(x, \varphi) = \sum_{n=0}^{\infty} Q_n(\varphi)\mathbf{W}_n(x), \quad \mathbf{F}(x, \varphi) = \sum_{n=0}^{\infty} Q_n(\varphi)\mathbf{F}_n(x) \tag{5.114}$$

where the transformation matrix $Q_n(\varphi)$ is given as

$$Q_n(\varphi) = \begin{pmatrix} \cos n\varphi & 0 & 0 \\ 0 & \sin n\varphi & 0 \\ 0 & 0 & \cos n\varphi \end{pmatrix}$$

The expansions in eqn (5.114) imply that the components of the vector-functions $\mathbf{W}(x, \varphi)$ and $\mathbf{F}(x, \varphi)$ are $2\pi$-periodic with respect to the longitudinal coordinate $\varphi$.

The vectors $\mathbf{W}_n(x)$ and $\mathbf{F}_n(x)$ in eqn (5.114) are expressed as

$$\mathbf{W}_n(x) = \begin{pmatrix} u_n(x) \\ v_n(x) \\ w_n(x) \end{pmatrix}, \quad \mathbf{F}_n(x) = \begin{pmatrix} X_n(x) \\ Y_n(x) \\ Z_n(x) \end{pmatrix}$$

Upon substituting the expansions from eqn (5.114) into eqn (5.112), we obtain

$$\Lambda\left(\frac{\partial}{\partial x}, \frac{\partial}{\partial \varphi}, x\right)\left[\sum_{n=0}^{\infty} Q_n(\varphi)\mathbf{W}_n(x)\right] = -\sum_{n=0}^{\infty} Q_n(\varphi)\mathbf{F}_n(x)$$

or

$$\sum_{n=0}^{\infty} Q_n(\varphi)\left[\Lambda_n\left(\frac{d}{dx}, x\right)\mathbf{W}_n(x)\right] = -\sum_{n=0}^{\infty} Q_n(\varphi)\mathbf{F}_n(x)$$

From this equation, it follows that the vectors $\mathbf{W}_n(x)$ should satisfy the following system of ordinary differential equations

$$\Lambda_n\left(\frac{d}{dx}, x\right)\mathbf{W}_n(x) = \mathbf{F}_n(x), \quad (n = 0, 1, 2, \ldots) \tag{5.115}$$

Since the displacement vector $\mathbf{W}(x, \varphi)$ is expressed in terms of the series in eqn (5.114), the boundary conditions from eqn (5.113) result in

$$\sum_{n=0}^{\infty} Q_n(\varphi)\left[B_{0n}\left(\frac{d}{dx}\right)\mathbf{W}_n(0)\right]=0$$

$$\sum_{n=0}^{\infty} Q_n(\varphi)\left[B_{ln}\left(\frac{d}{dx}\right)\mathbf{W}_n(l)\right]=0$$

or

$$B_{0n}\left(\frac{d}{dx}\right)\mathbf{W}_n(0)=0, \quad B_{ln}\left(\frac{d}{dx}\right)\mathbf{W}_n(l)=0 \tag{5.116}$$

Thus, the original boundary-value problem stated by eqns (5.112) and (5.113) has reduced to a set ($n = 0, 1, 2, \ldots$) of systems of linear ordinary differential equations (5.115) subject to the boundary conditions in eqn (5.116). The system in eqn (5.115) has variable (generally speaking) coefficients. Notice that only cylindrical and conical shells yield systems with constant coefficients. The total order of the system in (5.115) is eight. As it is known [15, 23, 60, 68], systems with variable coefficients do not allow analytical solution. Hence, their fundamental sets of solutions that are required for the construction of Green's matrices can be computed by using numerical methods.

In doing so, the system in eqn (5.115) is converted to the normal form

$$\frac{dy_i(x)}{dx}=\sum_{j=1}^{8}\alpha_{ij}(x)y_j(x)+f_i(x), \quad (i=\overline{1,8}) \tag{5.117}$$

where the unknown functions $y_i(x)$ are defined in terms of the components of the vector $\mathbf{W}_n(x)$ as

$$\begin{aligned} &y_1(x)=u_n(x), \quad y_2(x)=\frac{du_n(x)}{dx}\\ &y_3(x)=v_n(x), \quad y_4(x)=\frac{dv_n(x)}{dx}\\ &y_5(x)=w_n(x), \quad y_6(x)=\frac{dw_n(x)}{dx}\\ &y_7(x)=\frac{d^2w_n(x)}{dx^2}, \quad y_8(x)=\frac{d^3w_n(x)}{dx^3} \end{aligned} \tag{5.118}$$

The coefficients $\alpha_{ij}(x)$ of the system in eqn (5.117) and its right-hand side $f_i(x)$ are defined by the operator $\Lambda_n$ and the right-hand side vector $\mathbf{F}_n(x)$, respectively, of the system in eqn (5.115).

To show how the boundary conditions in eqn (5.116) can be expressed in terms of the newly introduced functions $y_i(x)$, we consider a particular case of the edge conditions. Let, for example, the edges $x = 0$ and $x = l$ of the shell be clamped and simply-supported, respectively. This yields, for the matrices $B_{0n}$ and $B_{ln}$

$$B_{0n}\left(\frac{d}{dx}\right) \equiv \begin{pmatrix} I & 0 & 0 \\ 0 & I & 0 \\ 0 & 0 & I \\ 0 & 0 & d/dx \end{pmatrix}, \quad B_{ln}\left(\frac{d}{dx}\right) \equiv \begin{pmatrix} I & 0 & 0 \\ 0 & I & 0 \\ 0 & 0 & I \\ 0 & 0 & d^2/dx^2 \end{pmatrix}$$

Thus, in light of the relations from eqn (5.118), the boundary conditions stated by eqn (5.116) can, in this case, be reformulated in terms of $y_i(x)$ as

$$\begin{aligned} y_i(0) &= 0, \quad \text{for } i = 1, 3, 5, 6 \\ y_i(l) &= 0, \quad \text{for } i = 1, 3, 5, 7 \end{aligned} \tag{5.119}$$

To obtain the Green's matrix for the homogeneous system

$$\frac{dy_i(x)}{dx} = \sum_{j=1}^{8} \alpha_{ij}(x)y_j(x), \quad (i = \overline{1, 8}) \tag{5.120}$$

associated with that found in eqn (5.117) subject to the boundary conditions in eqn (5.119), one is required to have the fundamental set of solutions for the above system. That is the set ($j = 1, 2, \ldots, 8$) of its eight linearly independent vector-solutions $\{y_{ij}(x)\}$. These can be computed with the aid of standard procedures for the numerical solution of Cauchy problems for systems of ordinary differential equations.

Note that for cylindrical and conical shells fundamental sets of solutions can be obtained analytically, since the corresponding systems of differential equations have constant coefficients and allow, therefore, analytic solution. In Section 5.5.2 the reader can find important comments as to possible forms of fundamental sets of solutions for a cylindrical shell.

In compliance with Lagrange's method of variation of parameters, the general solution of the system in eqn (5.117) can be written in terms of the fundamental set of solutions $y_{ij}(x)$ as

$$y_i(x) = \sum_{j=1}^{8} y_{ij}(x)C_j(x), \quad (i = \overline{1, 8}) \tag{5.121}$$

where $C_j(x)$ represent unknown functions whose derivatives $C_j'(x)$ have to satisfy the following system of linear algebraic equations

$$\sum_{j=1}^{8} y_{ij}(x)C_j'(x) = f_i(x), \quad (i = \overline{1, 8}) \tag{5.122}$$

Recall that $f_i(x)$ are the right-hand terms of the system in eqn (5.117).

Write the solution to the system from eqn (5.122) in the form

$$C_i'(x) = \sum_{j=1}^{8} y_{ij}^{-1}(x) f_j(x), \quad (i = \overline{1, 8})$$

with $y_{ij}^{-1}(x)$ representing entries of the inverse of the coefficient matrix of the system in eqn (5.122).

Upon integrating the above expressions for $C_i'(x)$, one obtains the following integral representations for $C_i(x)$ themselves

$$C_i(x) = \int_0^x \sum_{j=1}^{8} y_{ij}^{-1}(s) f_j(s)\, ds + D_i, \quad (i = \overline{1, 8})$$

where $D_i$ are arbitrary coefficients. This allows the general solution of the system in eqn (5.117) to be written in the form

$$y_i(x) = \int_0^x \sum_{j=1}^{8} T_{ij}(x, s) f_j(s)\, ds + \sum_{j=1}^{8} y_{ij}(x) D_j \tag{5.123}$$

where

$$T_{ij}(x, s) = \sum_{k=1}^{8} y_{ik}(x) y_{kj}^{-1}(s)$$

By virtue of the boundary conditions imposed by eqn (5.119), one obtains, for the coefficients $D_i$, the following system of linear algebraic equations

$$\sum_{j=1}^{8} r_{ij} D_j = S_i, \quad (i = \overline{1, 8}) \tag{5.124}$$

The entries $r_{ij}$ of the first four rows of the $8 \times 8$ coefficient matrix $R$ of this system are defined as

$$r_{ij} = y_{mj}(0), \quad (i = \overline{1, 4},\ m = 1, 3, 5, 6)$$

The entries of the last four rows of that matrix are defined as

$$r_{ij} = y_{mj}(l), \quad (i = \overline{5, 8},\ m = 1, 3, 5, 7)$$

The components $S_i$ of the right-hand side vector of the system in eqn (5.124) are expressed as

$$S_i = \int_0^l \sum_{j=1}^{8} T_{ij}^*(l, s) f_j(s)\, ds, \quad (i = \overline{1, 8}) \tag{5.125}$$

with $T^*_{ij}(l, s)$ being defined by

$$T^*_{ij}(x, s) = \begin{cases} 0, & (i = \overline{1, 4}) \\ -T_{ij}(l, s), & (i = \overline{5, 8}) \end{cases}$$

where $T_{ij}(l, s)$ represent the boundary-values of the functions $T_{ij}(x, s)$ defined earlier through $y_{ik}(x)$ and $y^{-1}_{kj}(s)$ in the equation that immediately follows that of (5.123).

Thus, the solution of the system in eqn (5.124) can be found in terms of the inverse of its coefficient matrix $R$ as

$$D_i = \sum_{j=1}^{8} r^{-1}_{ij} S_j, \quad (i = \overline{1, 8})$$

where $r^{-1}_{ij}$ represents the entries of the inverse $R^{-1}$ of $R$.

Substituting the components $S_j$ from eqn (5.125) into the above relation, one obtains for $D_i$

$$D_i = \int_0^l \sum_{j=1}^{8} P_{ij}(l, s) f_j(s)\, ds$$

where

$$P_{ij}(l, s) = \sum_{k=1}^{8} r^{-1}_{ik} T^*_{kj}(l, s)$$

Substitution of the above expression for $D_i$ into eqn (5.123) yields, for the general solution of the system from eqn (5.117)

$$y_i(x) = \int_0^x \sum_{j=1}^{8} T_{ij}(x, s) f_j(s)\, ds + \int_0^l \sum_{j=1}^{8} H_{ij}(x, s) f_j(s)\, ds \tag{5.126}$$

where

$$H_{ij}(x, s) = \sum_{j=1}^{8} y_{ik}(x) P_{kj}(l, s)$$

The expression for $y_i(x)$ in eqn (5.126) can finally be written in a compact integral form

$$y_i(x) = \int_0^l \sum_{j=1}^{8} G_{ij}(x, s) f_j(s)\, ds, \quad i = \overline{1, 8} \tag{5.127}$$

in which

$$G_{ij}(x, s) = \begin{cases} H_{ij}(x, s), & x \leq s \\ H_{ij}(x, s) + T_{ij}(x, s), & x \geq s \end{cases} \tag{5.128}$$

Thus, the solution of the boundary-value problem posed by eqns (5.117) and (5.119) is finally found in the form of the definite integral in eqn (5.127).

Subsequently, in view of the relations from eqn (5.118), the vector

$$\mathbf{W}_n(x) = \begin{pmatrix} u_n(x) \\ v_n(x) \\ w_n(x) \end{pmatrix}$$

that represents the solution to the problem stated by eqns (5.115) and (5.116), can be written as

$$\mathbf{W}_n(x) = \int_0^l G^n(x, s)\mathbf{F}_n(s)\, ds$$

or, in an extended form

$$\begin{pmatrix} u_n(x) \\ v_n(x) \\ w_n(x) \end{pmatrix} = \int_0^l \begin{pmatrix} G_{12}(x, s) & G_{14}(x, s) & G_{18}(x, s) \\ G_{32}(x, s) & G_{34}(x, s) & G_{38}(x, s) \\ G_{52}(x, s) & G_{54}(x, s) & G_{58}(x, s) \end{pmatrix} \begin{pmatrix} X_n(s) \\ Y_n(s) \\ Z_n(s) \end{pmatrix} ds \quad (5.129)$$

The entries $G_{ij}(x, s)$ (with $i = 1, 3, 5$ and $j = 2, 4, 8$) of the kernel-matrix $G^n(x, s)$ of the integral representation in eqn (5.129) can be found in the $8 \times 8$ kernel-matrix from eqn (5.127). They are shown in eqn (5.128).

Upon substituting expressions for the components of $\mathbf{W}_n(x)$ from eqn (5.129) into the first of the expansions of eqn (5.114) and then applying the Euler–Fourier formula

$$\mathbf{F}_n(s) = \frac{\varepsilon_n}{2\pi} \int_0^{2\pi} Q_n(\psi)\mathbf{F}(s, \psi)\, d\psi, \quad \varepsilon_n = \begin{cases} 1, & n = 0 \\ 2, & n > 0 \end{cases}$$

for the coefficients of the second of the expansions from eqn (5.114), one obtains the following integral representation

$$\mathbf{W}(x, \varphi) = \int_0^l \int_0^{2\pi} G(x, \varphi; s, \psi)\mathbf{F}(s, \psi)\, ds\, d\psi \quad (5.130)$$

of the solution to the original boundary-value problem posed by eqns (5.112) and (5.113).

Hence, in light of the definition introduced in the prefatory part of this chapter, it follows that the kernel-matrix

$$G(x, \varphi; s, \psi) = \sum_{n=0}^{\infty} \frac{\varepsilon_n}{2\pi} Q_n(\varphi)G^n(x, s)Q_n(\psi)$$

of the integral in eqn (5.130) represents the influence matrix of the homogeneous problem corresponding to that posed by eqns (5.112) and (5.113).

What has just been presented is the algorithm for constructing influence matrices for problems modeling static equilibrium of thin elastic shells of revolution. Recall that the variability of the coefficients of the governing system (5.112) is taken care of by numerically solving a set of linearly independent Cauchy problems for the system in eqn (5.120). In regard to the numerical solution of such problems, it is worth noting that, due to the *stiffness* [3, 19] of the original equation (5.112), caused by small leading coefficients, the effectiveness of a numerical solution significantly depends upon the shell's length. For "long" shells, special care is required to obtain accurate results.

### 5.5.2 Circular cylindrical shell

Appearance of the system in (5.112), simulating elastic equilibrium of a particular shell of revolution, depends upon the shape of its meridian $\varphi = const$. To find out how the system in (5.112) may look and to give the reader a clear sense of the form of the influence matrix of a point force and a fundamental set of solutions for the homogeneous system corresponding to that in eqn (5.115), consider a circular cylindrical shell of radius $a$, which is subject to a transverse distributed load $Z = Z(x, \varphi)$.

The system in (5.112) can be written [28, 65] for a cylindrical shell as

$$\begin{gathered}
\frac{\partial^2 u}{\partial x^2} + \frac{1-\sigma}{2a^2}\frac{\partial^2 u}{\partial \varphi^2} + \frac{1+\sigma}{2a}\frac{\partial^2 v}{\partial x \partial \varphi} - \frac{\sigma}{a}\frac{\partial w}{\partial x} = 0 \\
\frac{1+\sigma}{2}\frac{\partial^2 u}{\partial x \partial \varphi} + a\frac{1-\sigma}{2}\frac{\partial^2 v}{\partial x^2} + \frac{1}{a}\frac{\partial^2 v}{\partial \varphi^2} - \frac{1}{a}\frac{\partial w}{\partial \varphi} = 0 \\
\sigma\frac{\partial u}{\partial x} + \frac{1}{a}\frac{\partial v}{\partial \varphi} - \frac{w}{a} - \frac{h^2}{12}\left(a\frac{\partial^4 w}{\partial x^4} + \frac{2}{a}\frac{\partial^4 w}{\partial x^2 \partial \varphi^2} + \frac{1}{a^3}\frac{\partial^4 w}{\partial \varphi^4}\right) = -\frac{a(1-\sigma^2)}{Eh}Z
\end{gathered} \tag{5.131}$$

where $E$, $\sigma$, and $h$ represent the elasticity modulus, the Poisson ratio of the material of which the shell is made, and the thickness of the shell.

**Example 5.1:** We begin with the construction of the influence function of a transverse point force for a section of a cylindrical shell whose middle surface occupies the region $\Omega = \{0 < x < l, 0 < \varphi < \varphi_0\}$ and the edges $x = 0$, $x = l$, $\varphi = 0$ and $\varphi = \varphi_0$ are simply supported.

To understand how different types of boundary conditions for the system in (5.131) can be formulated in terms of the components $u = u(x, \varphi)$, $v = v(x, \varphi)$ and $w = w(x, \varphi)$ of the displacement vector, we recall [65] the following expressions

$$N_x = \frac{Eh}{1-\sigma^2}\left[\frac{\partial u}{\partial x} + \frac{\sigma}{a}\left(\frac{\partial v}{\partial \varphi} - w\right)\right]$$

$$N_\varphi = \frac{Eh}{1-\sigma^2}\left[\sigma\frac{\partial u}{\partial x} + \frac{1}{a}\left(\frac{\partial v}{\partial \varphi} - w\right)\right]$$

$$M_x = -\frac{Eh^3}{12(1-\sigma^2)}\left(\frac{\partial^2 w}{\partial x^2} + \frac{\sigma}{a^2}\frac{\partial^2 w}{\partial \varphi^2}\right)$$

$$M_\varphi = -\frac{Eh^3}{12(1-\sigma^2)}\left(\sigma\frac{\partial^2 w}{\partial x^2} + \frac{1}{a^2}\frac{\partial^2 w}{\partial \varphi^2}\right)$$

for the normal forces $N_x$, $N_\varphi$ and the bending moments $M_x$, $M_\varphi$ acting in the middle surface of the shell.

The boundary conditions of simple support on the contour of $\Omega$ imply [65] that the deflection $w$, the normal force $N_x$, and the bending moment $M_x$ vanish along the edges $x = 0$ and $x = l$

$$w(x, \varphi)|_{x=0,l} = 0, \quad N_x(x, \varphi)|_{x=0,l} = 0, \quad M_x(x, \varphi)|_{x=0,l} = 0 \tag{5.132}$$

while the deflection $w$, the normal force $N_\varphi$, and the bending moment $M_\varphi$ vanish along the edges $\varphi = 0$ and $\varphi = \varphi_0$

$$w(x, \varphi)|_{\varphi=0,\varphi_0} = 0, \quad N_x(x, \varphi)|_{\varphi=0,\varphi_0} = 0, \quad M_x(x, \varphi)|_{\varphi=0,\varphi_0} = 0 \tag{5.133}$$

From Section 5.5.1 it follows that, if

$$G(x, \varphi; s, \psi) = (G_{ij}(x, \varphi; s, \psi))_{i,j=\overline{1,3}}$$

represents the Green's matrix to the homogeneous boundary-value problem corresponding to that of eqns (5.131)–(5.133), then the components of the solution vector to (5.131)–(5.133) can be written in terms of $G(x, \varphi; s, \psi)$ and the right-hand side vector of (5.131) as

$$\begin{pmatrix} u(x, \varphi) \\ v(x, \varphi) \\ w(x, \varphi) \end{pmatrix} = \int_0^{\varphi_0}\int_0^l G(x, \varphi; s, \psi) \begin{pmatrix} 0 \\ 0 \\ \dfrac{a(1-\sigma^2)}{Eh} Z(s, \psi) \end{pmatrix} ds\, d\psi \tag{5.134}$$

Since the first and the second components in the right-hand side vector in (5.131) are zero, the components of the solution vector of the problem in eqns (5.131)–(5.133) are defined in terms of the entries of the third column only of $G(x, \varphi; s, \psi)$. Indeed, the matrix integral representation in (5.134) transforms into three scalar relations as

$$\begin{aligned} u(x, \varphi) &= \int_0^{\varphi_0}\int_0^l G_{13}(x, \varphi; s, \psi)\frac{a(1-\sigma^2)}{Eh} Z(s, \psi)\, ds\, d\psi \\ v(x, \varphi) &= \int_0^{\varphi_0}\int_0^l G_{23}(x, \varphi; s, \psi)\frac{a(1-\sigma^2)}{Eh} Z(s, \psi)\, ds\, d\psi \\ u(x, \varphi) &= \int_0^{\varphi_0}\int_0^l G_{33}(x, \varphi; s, \psi)\frac{a(1-\sigma^2)}{Eh} Z(s, \psi)\, ds\, d\psi \end{aligned} \tag{5.135}$$

Due to the specificity of the edge conditions, we can expand the components $u = u(x, \varphi)$, $v = v(x, \varphi)$ and $w = w(x, \varphi)$ of the displacement vector, and the loading function $Z = Z(x, \varphi)$ in the series

$$u(x, \varphi) = \sum_{m=1}^{\infty}\sum_{n=1}^{\infty} u_{mn} \cos \mu x \sin \nu\varphi$$
$$v(x, \varphi) = \sum_{m=1}^{\infty}\sum_{n=1}^{\infty} v_{mn} \sin \mu x \cos \nu\varphi \tag{5.136}$$
$$w(x, \varphi) = \sum_{m=1}^{\infty}\sum_{n=1}^{\infty} w_{mn} \sin \mu x \sin \nu\varphi$$

and

$$Z(x, \varphi) = \sum_{m=1}^{\infty}\sum_{n=1}^{\infty} Z_{mn} \sin \mu x \sin \nu\varphi \tag{5.137}$$

where the parameters $\mu$ and $\nu$ are defined as

$$\mu = \frac{m\pi}{l} \quad \text{and} \quad \nu = \frac{n\pi}{\varphi_0}$$

It is evident that, if the components of the deflection vector are expressed as in (5.136), then the simple support boundary conditions in eqns (5.132) and (5.133) are satisfied.

Substituting the expansions from (5.136) and (5.137) in the system of eqn (5.131), we obtain the system of linear algebraic equations

$$\left(\mu^2 + \frac{1-\sigma}{2a^2}\nu^2\right)u_{mn} + \frac{1+\sigma}{2a}\mu\nu v_{mn} + \frac{\sigma}{a}\mu w_{mn} = 0$$
$$\frac{1+\sigma}{2}\mu\nu u_{mn} + \left(a\frac{1-\sigma}{2}\mu^2 + \frac{1}{a}\nu^2\right)v_{mn} + \frac{1}{a}\nu w_{mn} = 0 \tag{5.138}$$
$$\sigma\mu u_{mn} + \frac{1}{a}\nu v_{mn} + \left[\frac{1}{a} + \frac{h^2}{12}\left(a\mu^4 + \frac{2}{a}\mu^2\nu^2 + \frac{1}{a^3}\nu^4\right)\right]w_{mn} = \frac{a(1-\sigma^2)}{Eh}Z_{mn}$$

in the coefficients $u_{mn}$, $v_{mn}$, and $w_{mn}$ of the expansions in eqn (5.136).

The above system can be reduced to a more compact form. This can be done by multiplying the first equation in (5.138) by the factor of $a^2$, while the other two equations are multiplied through by $a$. This reduces the above system to

$$\left(\tilde{\mu}^2 + \frac{1-\sigma}{2}\nu^2\right)u_{mn} + \frac{1+\sigma}{2}\tilde{\mu}\nu v_{mn} + \sigma\tilde{\mu} w_{mn} = 0$$
$$\frac{1+\sigma}{2}\tilde{\mu}\nu u_{mn} + \left(\frac{1-\sigma}{2}\tilde{\mu}^2 + \nu^2\right)v_{mn} + \nu w_{mn} = 0 \tag{5.139}$$
$$\sigma\tilde{\mu} u_{mn} + \nu v_{mn} + \left[1 + \frac{h^2}{12a^2}(\tilde{\mu}^2 + \nu^2)^2\right]w_{mn} = \frac{a^2(1-\sigma^2)}{Eh}Z_{mn}$$

where the parameter $\widetilde{\mu}$ is defined as $\widetilde{\mu} = \mu a$. It can be found that the determinant $\Delta$ of the coefficient matrix

$$\begin{pmatrix} \widetilde{\mu}^2 + \dfrac{1-\sigma}{2}\nu^2 & \dfrac{1+\sigma}{2}\widetilde{\mu}\nu & \sigma\widetilde{\mu} \\ \dfrac{1+\sigma}{2}\widetilde{\mu}\nu & \dfrac{1-\sigma}{2}\widetilde{\mu}^2 + \nu^2 & \nu \\ \sigma\widetilde{\mu} & \nu & 1 + \dfrac{h^2}{12a^2}(\widetilde{\mu}^2+\nu^2)^2 \end{pmatrix}$$

of the system in (5.139) is expressed as

$$\Delta = 12\widetilde{\mu}^4 a^2(1-\sigma^2) + h^2(\widetilde{\mu}^2+\nu^2)^4$$

It is evident that the above represents a non-zero quantity. Indeed, the expression for $\Delta$ is a sum of two non-negative terms $12\widetilde{\mu}^4 a^2(1-\sigma^2)$ and $h^2(\widetilde{\mu}^2+\nu^2)^4$. This implies that the system in (5.139) has a unique solution which is ultimately found in the form

$$\begin{aligned} u_{mn} &= -\frac{12\widetilde{\mu}a^4(1-\sigma^2)}{Eh\Delta}[\widetilde{\mu}^2(1+\sigma) - (\widetilde{\mu}^2+\nu^2)]Z_{mn} \\ v_{mn} &= -\frac{12\nu a^4(1-\sigma^2)}{Eh\Delta}[\widetilde{\mu}^2(1+\sigma) + (\widetilde{\mu}^2+\nu^2)]Z_{mn} \\ w_{mn} &= \frac{12a^4(1-\sigma^2)}{Eh\Delta}(\widetilde{\mu}^2+\nu^2)^2 Z_{mn} \end{aligned}$$

Substituting the above expressions for $u_{mn}$, $v_{mn}$, and $w_{mn}$ in (5.136), we rewrite the components $u(x,\varphi)$, $v(x,\varphi)$ and $w(x,\varphi)$ of the displacement vector for the section of the cylindrical shell with simply-supported edges, subject to a transverse load $Z(x,\varphi)$ as

$$\begin{aligned} u(x,\varphi) &= -\sum_{m=1}^{\infty}\sum_{n=1}^{\infty} 12\widetilde{\mu}a^4(1-\sigma^2) \\ &\quad \times \frac{[\widetilde{\mu}^2(1+\sigma) - (\widetilde{\mu}^2+\nu^2)]}{Eh\Delta} Z_{mn}\cos\mu x \sin\nu\varphi \\ v(x,\varphi) &= -\sum_{m=1}^{\infty}\sum_{n=1}^{\infty} 12\nu a^4(1-\sigma^2) \\ &\quad \times \frac{[\widetilde{\mu}^2(1+\sigma) + (\widetilde{\mu}^2+\nu^2)]}{Eh\Delta} Z_{mn}\sin\mu x \cos\nu\varphi \\ w(x,\varphi) &= \sum_{m=1}^{\infty}\sum_{n=1}^{\infty} 12a^4(1-\sigma^2)\frac{(\widetilde{\mu}^2+\nu^2)^2}{Eh\Delta} Z_{mn}\sin\mu x \sin\nu\varphi \end{aligned} \tag{5.140}$$

The coefficients $Z_{mn}$ of the trigonometric series in (5.137) can be expressed [60] in terms of the loading function $Z(x, \varphi)$ as

$$Z_{mn} = \frac{4}{l\varphi_0} \int_0^{\varphi_0} \int_0^l Z(s, \psi) \sin \mu s \sin \nu\psi \, ds \, d\psi$$

This yields for (5.140)

$$u(x, \varphi) = - \int_0^{\varphi_0} \int_0^l \sum_{m=1}^{\infty} \sum_{n=1}^{\infty} \frac{48\widetilde{\mu} a^3}{l\varphi_0 \Delta} [\widetilde{\mu}^2(1+\sigma) - (\widetilde{\mu}^2 + \nu^2)]$$
$$\times \cos \mu x \sin \nu\varphi \sin \mu s \sin \nu\psi \frac{a(1-\sigma^2)}{Eh} Z(s, \psi) \, ds \, d\psi$$
$$v(x, \varphi) = - \int_0^{\varphi_0} \int_0^l \sum_{m=1}^{\infty} \sum_{n=1}^{\infty} \frac{48\nu a^3}{l\varphi_0 \Delta} [\widetilde{\mu}^2(1+\sigma) + (\widetilde{\mu}^2 + \nu^2)]$$
$$\times \sin \mu x \cos \nu\varphi \sin \mu s \sin \nu\psi \frac{a(1-\sigma^2)}{Eh} Z(s, \psi) \, ds \, d\psi \quad (5.141)$$
$$w(x, \varphi) = \int_0^{\varphi_0} \int_0^l \sum_{m=1}^{\infty} \sum_{n=1}^{\infty} \frac{48a^3}{l\varphi_0 \Delta} (\widetilde{\mu}^2 + \nu^2)^2$$
$$\times \sin \mu x \sin \nu\varphi \sin \mu s \sin \nu\psi \frac{a(1-\sigma^2)}{Eh} Z(s, \psi) \, ds \, d\psi$$

Thus, by virtue of eqn (5.135), we conclude that the entries of the third column in the Green's matrix for the simply supported section $\Omega$ of the cylindrical shell are obtained in the form

$$G_{13}(x, \varphi; s, \psi) = -\frac{48a^3}{l\varphi_0} \sum_{m=1}^{\infty} \sum_{n=1}^{\infty} \widetilde{\mu}[\widetilde{\mu}^2(1+\sigma) - (\widetilde{\mu}^2 + \nu^2)]$$
$$\times \frac{\cos \mu x \sin \nu\varphi \sin \mu s \sin \nu\psi}{12\widetilde{\mu}^4 a^2(1-\sigma^2) + h^2(\widetilde{\mu}^2 + \nu^2)^4}$$
$$G_{23}(x, \varphi; s, \psi) = -\frac{48a^3}{l\varphi_0} \sum_{m=1}^{\infty} \sum_{n=1}^{\infty} \nu[\widetilde{\mu}^2(1+\sigma) + (\widetilde{\mu}^2 + \nu^2)]$$
$$\times \frac{\sin \mu x \cos \nu\varphi \sin \mu s \sin \nu\psi}{12\widetilde{\mu}^4 a^2(1-\sigma^2) + h^2(\widetilde{\mu}^2 + \nu^2)^4} \quad (5.142)$$
$$G_{33}(x, \varphi; s, \psi) = \frac{48a^3}{l\varphi_0} \sum_{m=1}^{\infty} \sum_{n=1}^{\infty} (\widetilde{\mu}^2 + \nu^2)^2$$
$$\times \frac{\sin \mu x \sin \nu\varphi \sin \mu s \sin \nu\psi}{12\widetilde{\mu}^4 a^2(1-\sigma^2) + h^2(\widetilde{\mu}^2 + \nu^2)^4}$$

The above series representations determine the *influence matrix of a transverse point force* for the simply supported section of a cylindrical shell, where

$G_{13}(x, \varphi; s, \psi)$, $G_{23}(x, \varphi; s, \psi)$ and $G_{33}(x, \varphi; s, \psi)$ are components of the displacement vector of a point $(x, \varphi)$ in the middle surface of the shell caused by a transverse to the middle surface unit point force applied at $(s, \psi)$. ▲

In the following two examples, we demonstrate how the influence matrix just obtained can be employed to determine components of the displacement vector for the simply supported section $\Omega$ of a cylindrical shell loaded with different distributed transverse load.

**Example 5.2:** Let the edges of the section $\{0 < x < l, 0 < \varphi < \varphi_0\}$ of a cylindrical shell of radius $a$ be simply-supported. Let also the section be loaded with a uniform transverse load $Z_0$.

Replacing the load function $Z(s, \psi)$ in (5.141) with the constant $Z_0$ and performing the definite integration with respect to $s$ and $\psi$, we obtain the components of the displacement vector as

$$u(x, \varphi) = -\frac{48a^4(1-\sigma^2)Z_0}{Ehl\varphi_0} \sum_{m=1}^{\infty} \sum_{n=1}^{\infty} [\tilde{\mu}^2(1+\sigma) - (\tilde{\mu}^2 + \nu^2)] \times \frac{(\cos \mu l - 1)(\cos \nu\varphi_0 - 1) \cos \mu x \sin \nu\varphi}{\nu[12\tilde{\mu}^4 a^2(1-\sigma^2) + h^2(\tilde{\mu}^2+\nu^2)^4]}$$

$$v(x, \varphi) = -\frac{48a^4(1-\sigma^2)Z_0}{Ehl\varphi_0} \sum_{m=1}^{\infty} \sum_{n=1}^{\infty} [\tilde{\mu}^2(1+\sigma) + (\tilde{\mu}^2 + \nu^2)] \times \frac{(\cos \mu l - 1)(\cos \nu\varphi_0 - 1) \sin \mu x \cos \nu\varphi}{\tilde{\mu}[12\tilde{\mu}^4 a^2(1-\sigma^2) + h^2(\tilde{\mu}^2+\nu^2)^4]}$$

$$w(x, \varphi) = \frac{48a^4(1-\sigma^2)Z_0}{Ehl\varphi_0} \sum_{m=1}^{\infty} \sum_{n=1}^{\infty} (\tilde{\mu}^2 + \nu^2)^2 \times \frac{(\cos \mu l - 1)(\cos \nu\varphi_0 - 1) \sin \mu x \sin \nu\varphi}{\tilde{\mu}\nu[12\tilde{\mu}^4 a^2(1-\sigma^2) + h^2(\tilde{\mu}^2+\nu^2)^4]}$$

Using the above representations for the components of the displacement vector, one can obtain stress-related components of the stress-strain state of the shell. For the bending moment $M_x$, for example,

$$M_x(x, \varphi) = -\frac{Eh^3}{12(1-\sigma^2)} \left( \frac{\partial^2 w}{\partial x^2} + \frac{\sigma}{a^2} \frac{\partial^2 w}{\partial \varphi^2} \right)$$

after differentiation and an elementary algebra, one arrives at

$$M_x(x, \varphi) = \frac{4a^2h^2 Z_0}{l\varphi_0} \sum_{m=1}^{\infty} \sum_{n=1}^{\infty} (\tilde{\mu}^2 + \nu^2)^2 (\tilde{\mu}^2 + \sigma\nu^2) \times \frac{(\cos \mu l - 1)(\cos \nu\varphi_0 - 1) \sin \mu x \sin \nu\varphi}{\tilde{\mu}\nu[12\tilde{\mu}^4 a^2(1-\sigma^2) + h^2(\tilde{\mu}^2+\nu^2)^4]}$$

▲

**Example 5.3:** Consider the section $\{0 < x < l, 0 < \varphi < \varphi_0\}$ of a cylindrical shell of radius $a$ with simply-supported edges loaded with a linear load $Z(x, \varphi) = Z_0\varphi + Z_1$.

The integration in (5.141) with respect to $s$ and $\psi$, where $Z(s, \psi) = Z_0\psi + Z_1$, yields the components of the displacement vector in the form

$$u(x, \varphi) = -\frac{48a^4(1-\sigma^2)}{Ehl\varphi_0} \sum_{m=1}^{\infty} \sum_{n=1}^{\infty} [\widetilde{\mu}^2(1+\sigma) - (\widetilde{\mu}^2 + \nu^2)]$$
$$\times \frac{1}{\nu^2}[Z_0(\nu\varphi_0 \cos \nu\varphi_0 - \sin \nu\varphi_0) + \nu Z_1(\cos \nu\varphi_0 - 1)]$$
$$\times \frac{(\cos \mu l - 1) \cos \mu x \sin \nu\varphi}{[12\widetilde{\mu}^4 a^2(1-\sigma^2) + h^2(\widetilde{\mu}^2 + \nu^2)^4]}$$

$$v(x, \varphi) = -\frac{48a^4(1-\sigma^2)}{Ehl\varphi_0} \sum_{m=1}^{\infty} \sum_{n=1}^{\infty} [\widetilde{\mu}^2(1+\sigma) + (\widetilde{\mu}^2 + \nu^2)]$$
$$\times \frac{1}{\widetilde{\mu}\nu}[Z_0(\nu\varphi_0 \cos \nu\varphi_0 - \sin \nu\varphi_0) + \nu Z_1(\cos \nu\varphi_0 - 1)]$$
$$\times \frac{(\cos \mu l - 1) \sin \mu x \cos \nu\varphi}{[12\widetilde{\mu}^4 a^2(1-\sigma^2) + h^2(\widetilde{\mu}^2 + \nu^2)^4]}$$

$$w(x, \varphi) = \frac{48a^4(1-\sigma^2)Z_0}{Ehl\varphi_0} \sum_{m=1}^{\infty} \sum_{n=1}^{\infty} (\widetilde{\mu}^2 + \nu^2)^2$$
$$\times \frac{1}{\widetilde{\mu}\nu^2}[Z_0(\nu\varphi_0 \cos \nu\varphi_0 - \sin \nu\varphi_0) + \nu Z_1(\cos \nu\varphi_0 - 1)]$$
$$\times \frac{(\cos \mu l - 1) \sin \mu x \sin \nu\varphi}{[12\widetilde{\mu}^4 a^2(1-\sigma^2) + h^2(\widetilde{\mu}^2 + \nu^2)^4]}$$ ▲

In the development that follows, the emphasis is on a fundamental set of solutions that is required for the construction of the influence matrix for the cylindrical shell undergoing an axially symmetric load. Expanding the components $u = u(x, \varphi)$, $v = v(x, \varphi)$ and $w = w(x, \varphi)$ of the deflection vector, and the loading function $Z = Z(x, \varphi)$ in the series of eqn (5.114), and substituting these expansions in the system (5.131), one arrives at a system of ordinary differential equations, which appears, in this case, as

$$\frac{d^2u_n}{dx^2} - \frac{1-\sigma}{2a^2}n^2 u_n + \frac{1+\sigma}{2a}n\frac{dv_n}{dx} - \frac{\sigma}{a}\frac{dw_n}{dx} = 0$$
$$-\frac{1+\sigma}{2}n\frac{du_n}{dx} + a\frac{1-\sigma}{2}\frac{d^2v_n}{dx^2} - \frac{n^2}{a}v_n + \frac{n}{a}w_n = 0 \tag{5.143}$$
$$\sigma\frac{du_n}{dx} + \frac{n}{a}v_n - \frac{w_n}{a} - \frac{h^2}{12}\left(a\frac{d^4w_n}{dx^4} - \frac{2n^2}{a}\frac{d^2w}{dx^2} + \frac{n^4}{a^2}w_n\right) = -\frac{aZ_n(1-\sigma^2)}{Eh}$$

where $n = 0, 1, 2, \ldots$. Notice that this system has constant coefficients.

A fundamental set of solutions to the homogeneous system corresponding to (5.143) can be obtained by using standard procedures. Note, however, that the system is cumbersome and a manual use of standard procedures could be way too time-consuming. A computer algebra-based software like *Maple V* or *Mathematica* could be recommended to facilitate the work. We will go into some detail of that for the case of $n = 0$, for which the system in (5.143) degenerates into

$$\frac{d^2u_0}{dx^2} - \frac{\sigma}{a}\frac{dw_0}{dx} = 0$$

$$\sigma\frac{du_0}{dx} - \frac{w_0}{a} - \frac{ah^2}{12}\frac{d^4w_0}{dx^4} = -\frac{aZ_0(1-\sigma^2)}{Eh} \tag{5.144}$$

This system simulates the axially symmetric stress and strain state of the cylindrical shell. Direct application of the *Maple V* program yields the following general solution

$$u_0(x) = C_1 + C_2\frac{x}{\sigma a} + C_3\frac{i\sigma}{a\beta}e^{-i\beta x} - C_4\frac{\sigma}{a\beta}e^{-\beta x} - C_5\frac{i\sigma}{a\beta}e^{i\beta x} + C_6\frac{\sigma}{a\beta}e^{\beta x}$$

and

$$w_0(x) = C_2 + C_3e^{-i\beta x} + C_4e^{-\beta x} + C_5e^{i\beta x} + C_6e^{\beta x}$$

for the homogeneous system corresponding to that in eqn (5.144). The parameter $\beta$ is defined as

$$\beta = \frac{\sqrt[4]{12a^2h^2(\sigma^2-1)}}{ah} \tag{5.145}$$

This implies that the set of vector-functions

$$\begin{pmatrix}1\\0\end{pmatrix}, \begin{pmatrix}\dfrac{x}{\sigma a}\\1\end{pmatrix}, \begin{pmatrix}\dfrac{i\sigma}{a\beta}e^{-i\beta x}\\e^{-i\beta x}\end{pmatrix}$$

$$\begin{pmatrix}-\dfrac{\sigma}{a\beta}e^{-\beta x}\\e^{-\beta x}\end{pmatrix}, \begin{pmatrix}-\dfrac{i\sigma}{a\beta}e^{i\beta x}\\e^{i\beta x}\end{pmatrix} \text{ and } \begin{pmatrix}\dfrac{\sigma}{a\beta}e^{\beta x}\\e^{\beta x}\end{pmatrix} \tag{5.146}$$

can be selected as a fundamental set of solutions to the homogeneous system corresponding to (5.144).

Since the entries of the vectors in (5.146) are complex-valued functions, it is inconvenient to use them to practically construct the Green's matrix to a boundary-value problem for the governing system in eqn (5.144). This drawback can be avoided by separating real and imaginary parts in the entries of the vector-functions in (5.146). The latter transforms subsequently into

$$\begin{pmatrix} 1 \\ 0 \end{pmatrix}, \quad \begin{pmatrix} \dfrac{x}{\sigma a} \\ 1 \end{pmatrix}, \quad \begin{pmatrix} \dfrac{\sigma}{a\beta} \sin \beta x \\ \cos \beta x \end{pmatrix}$$

$$\begin{pmatrix} -\dfrac{\sigma}{a\beta} \cos \beta x \\ \sin \beta x \end{pmatrix}, \quad \begin{pmatrix} -\dfrac{\sigma}{a\beta} e^{-\beta x} \\ e^{-\beta x} \end{pmatrix} \quad \text{and} \quad \begin{pmatrix} \dfrac{\sigma}{a\beta} e^{\beta x} \\ e^{\beta x} \end{pmatrix} \tag{5.147}$$

This set looks formally more attractive than (5.146) because it does not contain the imaginary one explicitly. But a close analysis reveals, however, a drawback in (5.147) similar to that of (5.146). Indeed, according to the theory of elasticity [12, 21, 27, 58], the upper bound for the Poisson ratio $\sigma$ of an elastic material is 0.5, which makes the factor $(\sigma^2 - 1)$ in (5.145) negative. This implies that the radicand in (5.145) appears to be negative making the parameter $\beta$ complex-valued. Hence, the fundamental set of solutions in eqn (5.147) is also impractical.

To refrain from the drawback taking place in (5.146) and (5.147), we look for another fundamental set of solutions to the homogeneous system corresponding to that in (5.144). In doing so, use the De Moivre formula expressing the radical of $\sqrt[4]{-1}$ as

$$\sqrt[4]{-1} = \cos \frac{(2k+1)\pi}{4} + i \sin \frac{(2k+1)\pi}{4}, \qquad k = 0, 1, 2, 3$$

and pick up its principal value ($k = 0$)

$$\sqrt[4]{-1} = \frac{\sqrt{2}}{2}(1+i)$$

which transforms the parameter $\beta$ in (5.145) to

$$\beta = (1+i)\frac{\sqrt[4]{3a^2h^2(1-\sigma^2)}}{ah}$$

This allows us to ultimately obtain a fundamental set of solutions to the system in eqn (5.144) in the form

$$\begin{pmatrix} 1 \\ 0 \end{pmatrix}, \quad \begin{pmatrix} \dfrac{x}{\sigma a} \\ 1 \end{pmatrix}, \quad \begin{pmatrix} \dfrac{\sigma}{2a\omega} e^{\omega x} \overline{S}(x) \\ e^{\omega x} \cos \omega x \end{pmatrix}$$

$$\begin{pmatrix} -\dfrac{\sigma}{2a\omega} e^{-\omega x} \underline{S}(x) \\ e^{-\omega x} \cos \omega x \end{pmatrix}, \quad \begin{pmatrix} \dfrac{\sigma}{2a\omega} e^{\omega x} \underline{S}(x) \\ -e^{\omega x} \sin \omega x \end{pmatrix}, \quad \begin{pmatrix} \dfrac{\sigma}{2a\omega} e^{-\omega x} \overline{S}(x) \\ -e^{-\omega x} \cos \omega x \end{pmatrix} \tag{5.148}$$

where $\omega$ represents a real-valued parameter

$$\omega = \frac{\sqrt[4]{3a^2h^2(1-\sigma^2)}}{ah}$$

and the functions $\overline{S}(x)$ and $\underline{S}(x)$ are defined as

$$\overline{S}(x) = \cos \omega x + \sin \omega x \quad \text{and} \quad \underline{S}(x) = \cos \omega x - \sin \omega x$$

The set in (5.148) appears to be helpful in obtaining influence matrices of a concentrated axially symmetric load for the cylindrical shell. The reader is encouraged to implement this set as the fundamental set of solutions in solving Exercises 5.9 and 5.10.

## 5.6 End Chapter Exercises

**5.1** Construct the influence function of a transverse point concentrated unit force for the infinite strip-shaped Poisson–Kirchhoff plate occupying the region $\Omega = \{(x, y) : -\infty < x < \infty, 0 < y < b\}$, with simply-supported edges.

**5.2** Construct the influence functions for the rectangular Poisson–Kirchhoff plate occupying the region $\Omega = \{(x, y) : 0 < x < a, 0 < y < b\}$, with the following edge conditions:

**a)** $w(0, y) = \partial^2 w(0, y)/\partial x^2 = 0$, $\partial^2 w(a, y)/\partial x^2 = \partial^3 w(a, y)/\partial x^3 = 0$,
$w(x, 0) = \partial^2 w(x, 0)/\partial y^2 = 0$, $w(x, b) = \partial^2 w(x, b)/\partial y^2 = 0$
(three edges are simply-supported, while one is free);

**b)** $w(0, y) = \partial w(0, y)/\partial x = 0$, $w(a, y) = \partial w(a, y)/\partial x = 0$,
$w(x, 0) = \partial^2 w(x, 0)/\partial y^2 = 0$, $w(x, b) = \partial^2 w(x, b)/\partial y^2 = 0$
(two opposite edges are simply-supported, while the other two are clamped);

**c)** $w(0, y) = \partial^2 w(0, y)/\partial x^2 = 0$, $w(a, y) = \partial w(a, y)/\partial x = 0$,
$w(x, 0) = \partial^2 w(x, 0)/\partial y^2 = 0$, $w(x, b) = \partial^2 w(x, b)/\partial y^2 = 0$
(three edges are simply-supported, while one is clamped).

**5.3** Determine deflection $w(x, y)$, and bending moments $M_x(x, y)$ and $M_y(x, y)$ in a simply-supported rectangular $\{(x, y) : 0 < x < a, 0 < y < b\}$ Poisson–Kirchhoff plate subject to a uniform transverse load of magnitude $Q_0$ applied to a rectangular region $\{(x, y) : a_1 < x < a_2, b_1 < y < b_2\}$, with $a_2 < a$ and $b_2 < b$.

**5.4** Determine deflection $w(x, y)$ of a simply-supported rectangular $\{(x, y) : 0 < x < a, 0 < y < b\}$ Poisson–Kirchhoff plate loaded with:

**a)** $q(x, y) = Q_0$ applied to $\{(x, y) : 0 < x < a, 0 < y < b/2\}$;

**b)** $q(x, y) = Q_0 xy$ applied to $\{(x, y) : 0 < x < a/2, 0 < y < b\}$;

**c)** $q(x, y) = Q_0 x(a - x)y(b - y)$ applied to
$\{(x, y) : 0 < x < a, 0 < y < b\}$.

**5.5** Determine deflection $w(x, y)$ of a simply-supported rectangular $\{(x, y) : 0 < x < a, 0 < y < b\}$ Poisson–Kirchhoff plate resting on elastic foundation ($\lambda$) and loaded with:

**a)** $q(x, y) = Q_0$ applied to $\{(x, y) : 0 < x < a/2, 0 < y < b\}$;

**b)** $q(x, y) = Q_0$ applied to $\{(x, y) : 0 < x < a/2, 0 < y < b/2\}$;

**c)** $q(x, y) = Q_0 xy$ applied to $\{(x, y) : 0 < x < a, 0 < y < b\}$.

**5.6** Construct the influence function of a transverse point concentrated unit force for the circular Poisson–Kirchhoff plate whose edge $r = a$ is elastically supported.

**5.7** Construct the influence matrix of a transverse concentrated unit force for the infinite strip-shaped Reissner plate occupying the region $\Omega = \{(x, y) : -\infty < x < \infty, 0 < y < b\}$, with simply-supported edges.

**5.8** Find components of the displacement vector for the section $\Omega = \{(x, \varphi) : 0 < x < l, 0 < \varphi < \varphi_0\}$ of a cylindrical shell of radius $a$ loaded with the uniform $Z_0$ transverse load acting on the half-section $\{0 < x < l/2, 0 < \varphi < \varphi_0\}$.

**5.9** Construct the influence matrix of a point force for an axially symmetric state of the semi-infinite cylindrical shell whose edge is clamped.

**5.10** Construct the influence matrix of a point force for an axially symmetric state of the semi-infinite cylindrical shell whose edge is simply supported.

# Answers to End Chapter Exercises

## Chapter 1

**1.1:**

**a)** No (linearly dependent); **b)** No (linearly dependent);

**c)** Yes (linearly independent); **d)** Yes (linearly independent);

**e)** Yes (linearly independent); **f)** No (linearly dependent);

**g)** Yes (linearly independent); **h)** Yes (linearly independent).

**1.2:**

**a)** $\{1, e^{-4x}\}$; **b)** $\{e^{2x}, e^{-4x}\}$;

**c)** $\{e^{-\frac{3x}{2}} \cos \frac{x\sqrt{11}}{2}, e^{-\frac{3x}{2}} \sin \frac{x\sqrt{11}}{2}\}$; **d)** $\{1, e^{-3x}, e^{3x}\}$;

**e)** $\{1, e^{-3x}, e^{5x}\}$; **f)** $\{e^x, e^{(1+\sqrt{3})x}, e^{(1-\sqrt{3})x}\}$;

**g)** $\{e^x, e^{-x}, \sin x, \cos x\}$; **(h)** $\{e^{\sqrt{3}x}, e^{-\sqrt{3}x}, \sin 2x, \cos 2x\}$.

**1.3:**

**a)** $y_p = 4x(\ln x - 1)$, (the method of variation of parameters);

**b)** $y_p = -\frac{3}{8} \cos 2x$, (either the method of undetermined coefficients or the method of variation of parameters can be applied);

**c)** $y_p = \frac{2}{25} e^{-3x}$, (either the method of undetermined coefficients or the method of variation of parameters can be applied);

**d)** $y_p = \frac{29}{6280} \sin 3x + \frac{27}{6280} \cos 3x - \frac{15}{104} \sin x - \frac{3}{104} \cos x$, (either the method of variation of parameters or the method of undetermined coefficients can be applied; in the latter case, a trigonometric transformation is required before the method is used);

**e)** $y_p = \left(\frac{21}{2} e^{-x} + \frac{3}{10} x^2 + \frac{4}{25} x - \frac{113}{125}\right) \cos x + \left(\frac{3}{2} e^{-x} + \frac{1}{10} x^2 - \frac{22}{25} x - \frac{7}{250}\right) \sin x$, (transposition should be used before either the method of undetermined coefficients or the method of variation of parameters is applied);

**f)** $y_p = -\frac{1}{24} x^4 + \frac{1}{8} x^3 - x^2$, (either of the two methods is applicable; if the method of undetermined coefficients is applied, then a fourth-degree polynomial is the guess for $y_p$);

**g)** $y_p = \frac{1}{12}x^4 - \frac{2}{3}x^3 - x^2$, (either of the two methods is applicable).

**1.4:**

**a)** Yes; **b)** Yes; **c)** Yes; **d)** Yes; **e)** Yes; **f)** Yes;

**g)** No, $y = C$, where $C$ is an arbitrary constant, is also a solution;

**h)** No, $y = Cx$, where $C$ is an arbitrary constant, is also a solution.

**1.5:**

**a)** $$e^{2x} = \sinh 2\left\{\frac{1}{2} + 2\sum_{n=1}^{\infty} \frac{(-1)^n}{4+n^2\pi^2}(2\cos n\pi x - n\pi \sin n\pi x)\right\};$$

**b)** $$\cos^8 x = \frac{35}{128} + \frac{7}{16}\cos 2x + \frac{7}{32}\cos 4x + \frac{1}{16}\cos 6x + \frac{1}{128}\cos 8x;$$

**c)** $$f(x) = \frac{\pi+2}{4} + \frac{1}{\pi}\sum_{n=1}^{\infty}\left[\frac{(-1)^n - 1}{n^2}\cos nx - \frac{(\pi-1)(-1)^n + 1}{n}\sin nx\right].$$

**1.6:**

**a)** $$e^x - 1 = \frac{1}{\pi}\left[(e^\pi - \pi - 1) + 2\sum_{n=1}^{\infty}\frac{(-1)^n e^\pi - 1}{1+n^2}\cos nx\right];$$

$$e^x - 1 = -\frac{2}{\pi}\sum_{n=1}^{\infty}\frac{(-1)^n(n^2e^\pi - n^2 - 1) + 1}{1+n^2}\sin nx.$$

**b)** $$x - x^2 = -\frac{1}{3} - \frac{4}{\pi^2}\sum_{n=1}^{\infty}\frac{3(-1)^n + 1}{n^2}\cos\frac{n\pi x}{2};$$

$$x - x^2 = \frac{4}{\pi^3}\sum_{n=1}^{\infty}\frac{(-1)^n(n^2\pi^2 - 4) + 4}{n^3}\sin\frac{n\pi x}{2}.$$

**1.7:**

**a)** $\sqrt[3]{8-6i} = \sqrt[3]{10}\exp\left(i\left(\frac{1}{3}\arctan\frac{4}{3} + \frac{\pi}{6}(3+4k)\right)\right)$, $k = 0, 1, 2$;

**b)** $\sqrt{-5-4i} = \sqrt[4]{41}\exp\left(i\left(\frac{1}{2}\arctan\frac{4}{5} + \frac{\pi}{2}(1+2k)\right)\right)$, $k = 0, 1$;

**c)** $\sqrt[4]{-4+5i} = \sqrt[8]{41}\exp\left(i\left(\frac{1}{4}\arctan\frac{4}{5} + \frac{\pi}{8}(1+4k)\right)\right)$, $k = 0, 1, 2, 3$.

## Chapter 2

**2.1:**

**a)** $$g(x,s) = \begin{cases} -x, & \text{for } x \le s \\ -s, & \text{for } x \ge s \end{cases}$$

**b)** $$g(x,s) = \frac{1}{1+ha}\begin{cases} x[h(s-a) - 1], & \text{for } x \le s \\ s[h(x-a) - 1], & \text{for } x \ge s \end{cases}$$

In the case of $h = 0$, the boundary conditions in this problem reduce to the ones in part **a)**, which reduces, consequently, the above Green's function to that of part **a)**.

**c)** $g(x, s) = \dfrac{1}{H}\begin{cases} (1 + h_1 x)[h_2(s - a) - 1], & \text{for } x \leq s \\ (1 + h_1 s)[h_2(x - a) - 1], & \text{for } x \geq s \end{cases}$
where $H = h_2 + h_1(1 + h_2 a)$. If $h_1 = 0$, then the present Green's function reduces to that of Example 1.3 of Section 1.1.

**d)** $g(x, s) = \dfrac{1}{M}\begin{cases} \ln[(ms + p)/(ma + p)] \ln[(mx + p)/p], & x \leq s \\ \ln[(mx + p)/(ma + p)] \ln[(ms + p)/p], & x \geq s \end{cases}$

where $M = m \ln[(ma + p)/p]$

**e)** $g(x, s) = \dfrac{1}{\beta(e^{-\beta a} - 1)}\begin{cases} (e^{-\beta x} - 1)(e^{-\beta a} - e^{-\beta s}), & \text{for } x \leq s \\ (e^{-\beta s} - 1)(e^{-\beta a} - e^{-\beta x}), & \text{for } x \geq s \end{cases}$

**f)** $g(x, s) = \dfrac{1}{\beta}\begin{cases} e^{-\beta x} - 1, & \text{for } x \leq s \\ e^{-\beta s} - 1, & \text{for } x \geq s \end{cases}$

**g)** $g(x, s) = \dfrac{1}{k \sin(ka)}\begin{cases} \sin k(a - s) \sin kx, & \text{for } x \leq s \\ \sin k(a - x) \sin ks, & \text{for } x \geq s \end{cases}$

**h)** $g(x, s) = \dfrac{1}{6}\begin{cases} x^2(x - 3s), & \text{for } x \leq s \\ s^2(s - 3x), & \text{for } x \geq s \end{cases}$

**2.2:**

**a)** Yes; **b)** Yes;

**c)** No, because the condition of self-adjointness in eqn (2.54) is not met;

**d)** No, because the condition in eqn (2.54) is not met;

**e)** Yes.

**2.3:**

**a)** The integrating factor $e^{-2x}$ reduces the given equation to the following self-adjoint form

$$e^{-2x}y''(x) - 2e^{-2x}y'(x) + 4e^{-2x}y = 0;$$

**b)** The integrating factor $e^{x^2/2}$ reduces the given equation to the following self-adjoint form

$$e^{x^2/2}y''(x) + xe^{x^2/2}y'(x) - x^2e^{x^2/2}y(x) = 0;$$

**c)** The integrating factor $x^{-3}$ reduces the given equation to the following self-adjoint form

$$x^{-1}y''(x) - x^{-2}y'(x) + x^{-3}y(x) = 0;$$

**d)** The integrating factor $x^{-1}$ reduces the given equation to the following self-adjoint form

$$xy''(x) + y'(x) - x^{-1}y(x) = 0.$$

**2.4:**

**a)** Yes; **b)** Yes; **c)** Yes; **d)** Yes; **e)** Yes; **f)** No.

**2.5:**

By applying the integrating factor $e^{3x}$, the original nonself-adjoint boundary-value problem, whose Green's function

$$g(x, s) = \frac{1}{7}\begin{cases} e^{-(5x+2s)} - e^{2(x-s)}, & \text{for } x \le s \\ e^{-(5x+2s)} - e^{-5(x-s)}, & \text{for } x \ge s \end{cases}$$

appears to be non-symmetric, reduces to the following self-adjoint form

$$e^{3x}y''(x) + 3e^{3x}y'(x) - 10e^{3x}y(x) = 0, \quad y(0) = 0, \quad y(\infty) < \infty,$$

whose Green's function

$$g(x, s) = \frac{1}{7}\begin{cases} e^{-2(x+s)} - e^{5x-2s}, & \text{for } x \le s \\ e^{-2(x+s)} - e^{5s-2x}, & \text{for } x \ge s \end{cases}$$

is symmetric.

**2.6:**

The Green's function is found as

$$g(x, s) = \frac{1}{k \cos k}\begin{cases} \sin kx \cos k(s-1), & \text{for } x \le s \\ \sin ks \cos k(x-1), & \text{for } x \ge s \end{cases}$$

The construction procedure, which uses a *standard* fundamental set of solutions (in this case it is composed of the functions: $y_1 = \sin kx$ and $y_2 = \cos kx$), appears to be more cumbersome compared to the procedure, which uses the *special* fundamental set of solutions: $y_1(x) = \sin kx$ and $y_2(x) = \cos k(x-1)$. This occurs because the latter procedure does not imply a direct satisfaction of the boundary conditions as a part of the construction of the Green's function. The boundary conditions are taken care of in advance when the *special* fundamental set of solutions is obtained.

**2.7:**

**a)** $g(x,s) = \begin{cases} -\ln(s/a), & \text{for } x \le s \\ -\ln(x/a), & \text{for } x \ge s \end{cases}$

**b)** $g(x,s) = \dfrac{1}{k(e^{ka} + \lambda e^{-ka})} \begin{cases} (e^{kx} + \lambda e^{-kx}) \sinh k(s-a), & \text{for } x \le s \\ (e^{ks} + \lambda e^{-ks}) \sinh k(x-a), & \text{for } x \ge s \end{cases}$

where $\lambda = (k-h)/(k+h)$.

**c)** $g(x,s) = \dfrac{1}{6} \begin{cases} x^2(s-1)^2[s(2x-3)+x], & \text{for } x \le s \\ s^2(x-1)^2[x(2s-3)+s], & \text{for } x \ge s \end{cases}$

**d)** $g(x,s) = \dfrac{1}{6} \begin{cases} x(1-s)[(1-s)^2+(x^2-1)], & \text{for } x \le s \\ s(1-x)[(1-x)^2+(s^2-1)], & \text{for } x \ge s \end{cases}$

**2.8:**

**a)** $$y(x) = \frac{1}{5}e^{2x} - \frac{1}{5}[h\sin(b-a)+\cos(b-a)]^{-1} \times \{e^{2a}[h\sin(b-x)+\cos(b-x)] + (h+2)e^{2b}\sin(x-a)\};$$

**b)** $$y(x) = [(h+1)e^{x+2b} + (h-1)e^{x+2a}]^{-1}\{(hb+1)(e^{2x+b} - e^{2a+b}) - (h+1)(xe^{x+2b} - xe^{x+2a} - ae^{a+2b}) - (h-1)ae^{2x+a}\};$$

**c)** $$y(x) = \frac{1}{2(b-a)}[(b-1-x)e^{a-x}\sin a - (a-1-x)e^{b-x}\sin b] - \frac{1}{2}\cos x;$$

**d)** $y(x) = \frac{1}{360}x^2(x-1)^2(x^2+2x+3);$

**e)** $y(x) = \frac{1}{2}(x-3)e^{x-1} + \frac{1}{2}(5x+3)e^{-(x+1)} - 2(x+1)e^{-x} + 2.$

**2.10:**

**a)** The entries $g_{ij}(x,s)$, $(i,j=1,2)$ of the matrix of Green's type are defined in the form:

$$g_{11}(x,s) = \frac{1}{H}\begin{cases} (1-\lambda ks)(x+a), & \text{for } -a \le x \le s \le 0 \\ (1-\lambda kx)(s+a), & \text{for } -a \le s \le x \le 0 \end{cases}$$

$$g_{12}(x,s) = \frac{\lambda k(x+a)}{H}e^{-ks}, \quad \text{for } -a \le x \le 0,\ 0 \le s < \infty$$

$$g_{21}(x,s) = \frac{\lambda k(s+a)}{H}e^{-kx}, \quad \text{for } -a \le s \le 0,\ 0 \le x < \infty$$

$$g_{22}(x,s) = \frac{1}{2H}\begin{cases} He^{k(x-s)} - (1-\lambda ka)e^{-k(x+s)}, & 0 \le x \le s < \infty \\ He^{k(s-x)} - (1-\lambda ka)e^{-k(s+x)}, & 0 \le s \le x < \infty \end{cases}$$

where $H = 1 + \lambda ka$.

**b)** The entries $g_{ij}(x,s)$, $(i, j = 1, 2)$ of the matrix of Green's type are defined in the form:

$$g_{11}(x,s) = \frac{1}{k\Delta^*}\begin{cases} \sinh k(x+a)(\cosh ks - \lambda \sinh ks), \\ \qquad \text{for } -a \le x \le s \le 0 \\ \sinh k(s+a)(\cosh kx - \lambda \sinh kx), \\ \qquad \text{for } -a \le s \le x \le 0 \end{cases}$$

$$g_{12}(x,s) = \frac{\lambda}{k\Delta^*}e^{-ks}\sinh k(x+a), \quad \text{for } -a \le x \le 0 \le s < \infty$$

$$g_{21}(x,s) = \frac{1}{k\Delta^*}e^{-kx}\sinh(k(s+a)), \quad \text{for } -a \le s \le 0 \le x < \infty$$

$$g_{22}(x,s) = \frac{1}{k\Delta^*}\begin{cases} e^{-ks}(\lambda \sinh ka \cosh kx + \cosh ka \sinh kx), \\ \qquad x \le s \\ e^{-kx}(\lambda \sinh ka \cosh ks + \cosh ka \sinh ks), \\ \qquad s \le x \end{cases}$$

where $\Delta^* = \lambda \sinh ka + \cosh ka$.

**c)** The entries $g_{ij}(x,s)$, $(i, j = 1, 2)$ of the matrix of Green's type are defined in the form:

$$g_{11}(x,s) = \frac{1}{\Delta^*}\begin{cases} \sin k(x+a)(\lambda \sin ks \cos ka - \sin ka \cos ks), \\ \qquad x \le s \\ \sin k(s+a)(\lambda \sin kx \cos ka - \sin ka \cos kx), \\ \qquad s \le x \end{cases}$$

$$g_{12}(x,s) = \frac{\lambda}{\Delta^*}\sin k(x+a)\sin k(s-a), \quad -a \le x \le 0 \le s \le a$$

$$g_{21}(x,s) = \frac{1}{\Delta^*}\sin k(x-a)\sin k(s+a), \quad -a \le s \le 0 \le x \le a$$

$$g_{22}(x,s) = \frac{1}{\Delta^*}\begin{cases} \sin k(s-a)(\lambda \sin ka \cos kx + \sin kx \cos ka), \\ \qquad x \le s \\ \sin k(x-a)(\lambda \sin ka \cos ks + \sin ks \cos ka), \\ \qquad s \le x \end{cases}$$

where $\Delta^* = k(1+\lambda)\sin ka \cos ka$.

**d)** The diagonal entries $g_{ii}(x, s)$ of the matrix of Green's type are defined as follows:

$$g_{ii}(x, s) = \frac{1}{H_0}\begin{cases} e^{-ks}[h_1 \cosh kx + h_2 \sinh kx + h_3 \sinh kx], \\ \qquad x \le s \\ e^{-kx}[h_1 \cosh ks + h_2 \sinh ks + h_3 \sinh ks], \\ \qquad s \le x \end{cases}$$

where $i = 1, 2, 3$ and $H_0 = k(h_1 + h_2 + h_3)$. The peripheral entries $g_{ij}(x, s)$, $(i \neq j)$ of this matrix are defined in the form:

$$g_{ij}(x, s) = -H_0^{-1} h_j e^{-k(x+s)}, \quad i \neq j, \quad 0 \le x, s < \infty.$$

**2.11:** $y_1(x) = \dfrac{1}{\pi} H h_1(x-1) - \dfrac{1}{\pi^2}\sin \pi x; \quad y_2(x) = \dfrac{1}{\pi} H h_1(x-1);$

$$y_3(x) = \frac{1}{\pi} H h_1(x-1), \quad \text{where } H = (h_1 + h_2 + h_3)^{-1}.$$

## Chapter 3

**3.1:** $g(x, s) = \dfrac{1}{6a}\begin{cases} x(a-s)[(a-s)^2 + (x^2 - a^2)], & x \le s \\ s(a-x)[(a-x)^2 + (s^2 - a^2)], & s \le x \end{cases}$

**3.3:** $g(x, s) = \dfrac{1}{12a^3}\begin{cases} x(a-s)^2[s(a^2 - x^2) - 2a(x^2 - as)], & x \le s \\ s(a-x)^2[x(a^2 - s^2) - 2a(s^2 - ax)], & s \le x \end{cases}$

**3.4:** $g(x, s) = \dfrac{1}{\Delta}\begin{cases} k_a(a-s)\{(x-a) \\ \quad + k_0 ax[(x^2 - a^2) + (s-a)^2]\} - k_0 xs, & x \le s \\ k_a(a-x)\{(s-a) \\ \quad + k_0 as[(s^2 - a^2) + (x-a)^2]\} - k_0 xs, & x \ge s \end{cases}$

where $\Delta = 6a^2 k_0 k_a$, $k_0 = k_0^*/6EI$ and $k_a = k_a^*/6EI$.

**3.5:** **a)** It is evident that the boundary conditions of elastic support reduce to the free edge conditions, if the coefficient of support goes to zero. Thus, the limit of the influence function derived in Exercise 3.4, as either $k_0^*$ or $k_a^*$ (or both) approach zero, is undefined. This reflects an obvious physical interpretation of the influence function. Indeed, the corresponding boundary conditions reduce, in such cases, to the free edge conditions. And the influence function is undefined for a beam, one edge of which is elastically supported, while the other one is free (or both edges are free).

**b)** The boundary condition of elastic support reduces to the condition of simple support if the coefficient of support tends to infinity. Thus, if both the coefficients $k_0^*$ and $k_a^*$ approach infinity, then the influence function derived in Exercise 3.4 reduces to that derived in Exercise 3.1 for a simply supported beam.

**c)** If $k_0^*$ approaches infinity while $k_a^*$ remains finite, then the influence function derived in Exercise 3.4 reduces to that derived in Exercise 3.6 for a beam whose left-hand edge is simply supported, while the right-hand edge is elastically supported.

**3.6:** $$g(x,s)=\frac{1}{6a^2}\begin{cases} x\{a(a-s)[(x^2-a^2)+(s-a)^2]-s/k_a\}, & x\le s\\ s\{a(a-x)[(s^2-a^2)+(x-a)^2]-x/k_a\}, & x\ge s\end{cases}$$

where $k_a=k_a^*/6EI$.

This influence function can be constructed by either the method of variation of parameters or by the modification of the standard method. It is worth noting that it can also be obtained as a particular case of the influence function derived in Exercise 3.4 for the beam, both edges of which are elastically supported. Indeed, the boundary conditions of elastic support can be reduced to the conditions of simple support

$$w(0)=0,\qquad \frac{d^2w(0)}{dx^2}=0$$

when the coefficient $k_0$ approaches infinity. Hence, to derive the influence function in the present exercise, one takes the limit of the influence function in Exercise 3.4, as $k_0$ approaches infinity. The influence function for a simply supported beam

$$g(x,s)=\frac{1}{6a}\begin{cases} x(a-s)[(x^2-a^2)+(s-a)^2], & x\le s\\ s(a-x)[(s^2-a^2)+(x-a)^2], & x\ge s\end{cases}$$

constructed earlier in Exercise 3.1, can, in turn, be obtained from the influence function in the present exercise by taking the limit of it as the coefficient $k_a^*$ approaches infinity. As $k_a^*$ approaches zero, the influence function in the present exercise is undefined.

**3.7:** $$g(x,s)=\frac{1}{6}\begin{cases} x(3s^2+x^2-6as), & x\le s\\ s(3x^2+s^2-6ax), & s\le x\end{cases}$$

**3.8:** $$g(x,s)=\frac{1}{12a}\begin{cases} x^2(3s^2+2ax-6as), & x\le s\\ s^2(3x^2+2as-6ax), & s\le x\end{cases}$$

**3.9:**

**a)** $$w(x) = \frac{q_0}{92160EI} x a^3 (57a^2 - 103x^2), \quad M(x) = -\frac{103}{15360} q_0 a^3 x,$$

$$Q(x) = -\frac{103}{15360} q_0 a^3, \quad \text{for } 0 \le x \le a/4;$$

$$w(x) = \frac{q_0}{1474560EI}(12288x^5 a - 11520x^4 a^2 + 3472x^3 a^3 - 4096x^6 - 1200x^2 a^4 - 7a^6 + 1056xa^5),$$

$$M(x) = -\frac{q_0}{15360}[1280x^2(x-a)^2 + a^2(160x^2 - 217ax + 25a^2)],$$

$$Q(x) = -\frac{q_0}{15360}(5120x^3 - 7680ax^2 + 2880a^2 x - 217a^3),$$

for $a/4 \le x \le 3a/4$;

$$w(x) = \frac{q_0}{92160EI} a^3 (217x - 46a)(x-a)^2,$$

$$M(x) = \frac{217 q_0 a^3}{46080}(3x - 2a),$$

$$Q(x) = \frac{217}{15360} q_0 a^3, \quad \text{for } 3a/4 \le x \le a.$$

**b)** $$w(x) = -\frac{x}{162EI}[P_1(6a^2 - 14x^2) + P_2(5a^2 - 4x^2)],$$

$$M(x) = \frac{2}{27} x (7P_1 + 2P_2), \quad Q(x) = \frac{2}{27}(7P_1 + 2P_2),$$

for $0 \le x \le a/3$;

$$w(x) = -\frac{1}{162EI}[P_1(a-x)^2(13x - a) + P_2 x (5a^2 - 4x^2)],$$

$$M(x) = \frac{1}{27}[P_1(9a - 13x) + 4P_2 x], \quad Q(x) = -\frac{1}{27}(13P_1 - 4P_2),$$

for $a/3 \le x \le 2a/3$;

$$w(x) = -\frac{(a-x)^2}{162EI}[P_1(13x - a) + P_2(23x - 8a)],$$

$$M(x) = \frac{1}{27}[P_1(9a - 13x) + P_2(18a - 23x)],$$

$$Q(x) = -\frac{1}{27}(13P_1 + 23P_2), \quad \text{for } 2a/3 \le x \le a.$$

**c)** $$w(x) = -\frac{x}{36aEI}[8x^2 M_1 + (5x^2 - 3a^2)M_2],$$

$$M(x) = -\frac{x}{6a}(8M_1 + 5M_2),$$

$$Q(x) = -\frac{1}{6a}(8M_1 + 5M_2), \quad \text{for } 0 \le x \le a/3;$$

$$w(x) = -\frac{1}{36aEI}[2(6a^2x - 9x^2a - a^3 + 4x^3)M_1 + x(5x^2 - 3a^2)M_2],$$

$$M(x) = -\frac{1}{6a}[5M_2x + 2M_1(4x - 3a)],$$

$$Q(x) = -\frac{1}{6a}(8M_1 + 5M_2), \quad \text{for } a/3 \le x \le 2a/3;$$

$$w(x) = -\frac{(a-x)^2}{36aEI}[2(4x - a)M_1 + (5x - 8a)M_2],$$

$$M(x) = -\frac{1}{6a}[2M_1(4x - 3a) + M_2(5x - 6a)],$$

$$Q(x) = -\frac{1}{6a}(8M_1 + 5M_2), \quad \text{for } 2a/3 \le x \le a.$$

**3.10:** The parameter $k_a$ is introduced as $k_a = k_a^*/6EI$.

**a)**

$$w(x) = \frac{q_0 x}{360EI}\left(x^5 - 3ax^4 + 5a^3x^2 - 3a^5 - \frac{5a^2}{k_a}\right),$$

$$M(x) = \frac{q_0}{12}x(x - a)(x^2 - ax - a^2),$$

$$Q(x) = \frac{q_0}{12}(2x - a)(2x^2 - 2ax - a^2).$$

**b)**

$$w(x) = \frac{x}{96a^2EI}\left\{M_1\left[a(16x^2 + 11a^2) + \frac{16}{k_a}\right] + 4M_2\left[a(4x^2 - a^2) + \frac{4}{k_a}\right]\right\},$$

$$M(x) = \frac{x}{a}(M_1 + M_2), \quad Q(x) = \frac{1}{a}(M_1 + M_2), \quad \text{for } 0 \le x \le a/4;$$

$$w(x) = \frac{1}{96a^2EI}\left\{M_1\left[a(a-x)(13a^2 - 16(a-x)^2) + \frac{16x}{k_a}\right] + 4M_2x\left[a(4x^2 - a^2) + \frac{4}{k_a}\right]\right\},$$

$$M(x) = \frac{1}{a}[M_1(x - a) + M_2x],$$

$$Q(x) = \frac{1}{a}(M_1 + M_2), \quad \text{for } a/4 \le x \le a/2;$$

$$w(x) = \frac{1}{96a^2EI}\left\{M_1\left[a(a-x)(13a^2 - 16(a-x)^2) + \frac{16x}{k_a}\right]\right.$$

$$\left. + 4M_2\left[a(a-x)(a^2 - 4(a-x)^2) + \frac{4x}{k_a}\right]\right\},$$

$$M(x) = \frac{x-a}{a}(M_1 + M_2),$$

$$Q(x) = \frac{1}{a}(M_1 + M_2), \quad \text{for } a/2 \le x \le a.$$

**c)** $$w(x) = \frac{x}{162aEI}\left\{P_1\left[2a(5a^2 - 9x^2) + \frac{9}{k_a}\right]\right.$$

$$\left. + P_2\left[a(8a^2 - 9x^2) + \frac{18}{k_a}\right]\right\},$$

$$M(x) = -\frac{x}{3}(2P_1 + P_2),$$

$$Q(x) = -\frac{1}{3}(2P_1 + P_2), \quad \text{for } 0 \le x \le a/3;$$

$$w(x) = \frac{1}{162aEI}\left\{P_1\left[a(x-a)(9(x-a)^2 - 8a^2) + \frac{9x}{k_a}\right]\right.$$

$$\left. + P_2x\left[a(8a^2 - 9x^2) + \frac{18}{k_a}\right]\right\},$$

$$M(x) = \frac{1}{3}[(x-a)P_1 + xP_2],$$

$$Q(x) = \frac{1}{3}(P_1 - P_2), \quad \text{for } a/3 \le x \le 2a/3;$$

$$w(x) = \frac{1}{162aEI}\left\{P_1\left[a(x-a)(9(x-a)^2 - 8a^2) + \frac{9x}{k_a}\right]\right.$$

$$\left. + 2P_2\left[a(x-a)(9(x-a)^2 - 5a^2) + \frac{9x}{k_a}\right]\right\},$$

$$M(x) = \frac{x-a}{3}(P_1 + 2P_2),$$

$$Q(x) = \frac{1}{3}(P_1 + 2P_2), \quad \text{for } 2a/3 \le x \le a.$$

**3.11:** In this Exercise, we denote $k_0 = k_0^*/6EI$, $k_a = k_a^*/6EI$, and $K = 6a^2 k_0 k_a EI$.

**a)**

$$w(x) = \frac{q_0}{360EI}\left\{x^6 - 3ax(x^4 + a^4) + \frac{5a^2}{k_0 k_a}[x(k_a - k_0) - ak_a(1 - k_0 x^3)]\right\};$$

**b)**

$$w(x) = \frac{M_0}{K}\left\{[k_a(x-a) + axk_0k_a(x^2 - a^2) + k_0 x] + \frac{3}{4}a^3 x k_0 k_a\right\} + \frac{q_0}{24EI}\left\{x(x-a)(x^2 - ax - a^2) + \frac{2}{k_0 k_a}[ak_a + x(k_0 - k_a)]\right\}, \quad \text{for } x \le a/2;$$

$$w(x) = \frac{M_0}{K}\left\{[k_a(x-a) + (1 + axk_0(x - 2a)) + k_0 x] + \frac{3}{4}a^3(x-a)k_0k_a\right\} + \frac{q_0}{24EI}\left\{x(x-a)(x^2 - ax - a^2) + \frac{2}{k_0 k_a}[ak_a + x(k_0 - k_a)]\right\}, \quad \text{for } x \ge a/2.$$

**c)**

$$w(x) = \frac{M_0}{K}\left\{[k_a(x-a) + axk_0k_a(x^2 - a^2) + k_0 x] + \frac{3}{4}a^3 x k_0 k_a\right\} - \frac{P_0 a}{64K}\left\{3k_a[16(x-a) + axk_0(16(x^2 - a^2) + 9a^2)] - 16xk_0\right\}, \quad \text{for } x \le a/4;$$

$$w(x) = \frac{M_0}{K}\left\{[k_a(x-a) + axk_0k_a(x^2 - a^2) + k_0 x] + \frac{3}{4}a^3 x k_0 k_a\right\} + \frac{P_0 a}{64K}\left\{k_a(x-a)[k_0 a(16(x-a)^2 - 15a^2) - 48] + 16xk_0\right\},$$

for $a/4 \le x \le a/2$;

$$w(x) = \frac{M_0}{K}\left\{[k_a(x-a) + (1 + axk_0(x-2a)) + k_0x] + \frac{3}{4}a^3(x-a)k_0k_a\right\} + \frac{P_0a}{64K}\{k_a(x-a)[k_0a(16(x-a)^2 - 15a^2) - 48] + 16xk_0\},$$

for $a/2 \le x \le a$.

**d)** $$w(x) = \frac{q_0}{24ak_0}\{\sqrt{\pi}[axk_0(2x^2+3)(\mathrm{erf}(x) - \mathrm{erf}(a)) - 2(x\,\mathrm{erf}(x) - a\,\mathrm{erf}(a))] + \frac{2}{ak_a}[ak_a - (k_0+k_a)x - k_0k_aax(x^2+1)](e^{-a^2}-1) + 2ak_0(x^2+1)e^{-x^2} - 2ak_0(3x^2 - 2ax + 1)\}$$

where erf$(x)$ represents a special function which is referred to [4, 9, 32, 53, 60] as the *error function* defined as

$$\mathrm{erf}(x) = \frac{2}{\sqrt{\pi}}\int_0^x e^{-t^2}\,dt$$

The error function is well tabulated and can readily be valuated in an environment of every contemporary mathematics or statistics software. This makes erf$(x)$ convenient for practical use as an elementary function.

Note that in Exercises 3.12 through 3.15 the parameter $k$ is introduced for compactness as $k = (k_0/4EI)^{1/4}$.

**3.12:**

**a)** $$w(x) = \frac{q_0}{8k^6EI}\{e^{k(x-a)}[k(k + a(1+ka))\cos k(x-a) + (1+ka)\sin k(x-a)] - [k(k+b(1+kb))\cos k(x-b) + (1+kb)\sin k(x-b)]e^{k(x-b)}\}, \quad \text{for } x \le a;$$

$$w(x) = -\frac{q_0}{8k^6EI}\left\{e^{k(a-x)}[k(k - a(1-ka))\cos k(x-a) + (1-ka)\sin k(x-a)] + [k(k+b(1+kb))\cos k(x-b) + (1+kb)\sin k(x-b)]e^{k(x-b)} - 2k^2(x^2+1)\right\},$$

for $a \le x \le b$;

$$w(x) = \frac{q_0}{8k^6EI}\{[k(k + b(kb - 1)) \cos k(x - b)$$
$$+ (kb - 1) \sin k(x - b)]e^{-k(x-b)}$$
$$- [k(k + b(ka - 1)) \cos k(x - a)$$
$$+ (ka - 1) \sin k(x - a)]e^{-k(x-a)}\}, \quad x \geq b.$$

**b)** $$w(x) = \frac{P_1}{8k^3EI}e^{k(x-a)}[\cos k(x - a) - \sin k(x - a)]$$
$$+ \frac{P_2}{8k^3EI}e^{k(x-b)}[\cos k(x - b) - \sin k(x - b)], \quad \text{for } x \leq a;$$
$$w(x) = \frac{P_1}{8k^3EI}e^{k(a-x)}[\cos k(x - a) + \sin k(x - a)]$$
$$+ \frac{P_2}{8k^3EI}e^{k(x-b)}[\cos k(x - b) - \sin k(x - b)],$$
for $a \leq x \leq b$;
$$w(x) = \frac{P_1}{8k^3EI}e^{k(a-x)}[\cos k(x - a) + \sin k(x - a)]$$
$$+ \frac{P_2}{8k^3EI}e^{k(b-x)}[\cos k(x - b) + \sin k(x - b)], \quad \text{for } x \geq b.$$

**c)** $$w(x) = \frac{q_0}{8k^4EI}[e^{k(x-a_1)} \cos k(x - a_1) - e^{k(x-a_2)} \cos k(x - a_2)]$$
$$+ \frac{P_0}{8k^3EI}e^{k(x-b)}[\cos k(x - b) - \sin k(x - b)],$$
for $x \leq a_1$;
$$w(x) = \frac{q_0}{8k^4EI}[2 - e^{-k(x-a_1)} \cos k(x - a_1)$$
$$- e^{k(x-a_2)} \cos k(x - a_2)]$$
$$+ \frac{P_0}{8k^3EI}e^{k(x-b)}[\cos k(x - b) - \sin k(x - b)],$$
for $a_1 \leq x \leq a_2$;
$$w(x) = \frac{q_0}{8k^4EI}[e^{-k(x-a_2)} \cos k(x - a_2) - e^{-k(x-a_1)} \cos k(x - a_1)]$$
$$+ \frac{P_0}{8k^3EI}e^{k(x-b)}[\cos k(x - b) - \sin k(x - b)],$$
for $a_2 \leq x \leq b$;
$$w(x) = \frac{q_0}{8k^4EI}[e^{-k(x-a_2)} \cos k(x - a_2) - e^{-k(x-a_1)} \cos k(x - a_1)]$$
$$+ \frac{P_0}{8k^3EI}e^{-k(x-b)}[\cos k(x - b) + \sin k(x - b)],$$
for $x \geq b$.

**d)** $$w(x) = \frac{1}{4k^2EI}[M_1 e^{k(x-a)} \sin k(x-a) + M_2 e^{k(x-b)} \sin k(x-b)], \quad x \le a;$$

$$w(x) = \frac{1}{4k^2EI}[M_1 e^{k(a-x)} \sin k(x-a) + M_2 e^{k(x-b)} \sin k(x-b)], \quad a \le x \le b;$$

$$w(x) = \frac{1}{4k^2EI}[M_1 e^{k(a-x)} \sin k(x-a) + M_2 e^{k(b-x)} \sin k(x-b)], \quad x \ge b.$$

**3.13:**

**a)** $$w(x) = \frac{\cosh kx}{2k^2EI}[M_1 e^{-ka} \sin k(x-a) + M_2 e^{-kb} \sin k(x-b)], \quad x \le a;$$

$$w(x) = \frac{M_1}{2k^2EI} e^{-kx} \sin k(x-a) \cosh ka + \frac{M_2}{2k^2EI} e^{-kb} \sin k(x-b) \cosh kx, \quad \text{for } a \le x \le b;$$

$$w(x) = \frac{e^{-kx}}{2k^2EI}[M_1 \sin k(x-a) \cosh ka + M_2 \sin k(x-b) \cosh kb], \quad x \le b.$$

**b)** $$w(x) = \frac{P_0}{8k^3EI}\{e^{k(x-a)}[\cos k(x-a) - \sin k(x-a)] - e^{-k(x+a)}[\sin k(x+a) - 2\cos kx \cos ka - \cos k(x-a)]\} + \frac{M_0}{2k^2EI} e^{-kb} \sin k(x-b) \cosh kx, \quad \text{for } x \le a;$$

$$w(x) = \frac{P_0}{8k^3EI}\{e^{k(a-x)}[\cos k(x-a) - \sin k(a-x)] - e^{-k(x+a)}[\sin k(x+a) - 2\cos kx \cos ka - \cos k(x-a)]\} + \frac{M_0}{2k^2EI} e^{-kb} \sin k(x-b) \cosh kx, \quad \text{for } a \le x \le b;$$

$$w(x) = \frac{P_0}{8k^3EI}\{e^{k(a-x)}[\cos k(x-a) - \sin k(a-x)] - e^{-k(x+a)}[\sin k(x+a) - 2\cos kx \cos ka - \cos k(x-a)]\} + \frac{M_0}{2k^2EI} e^{-kx} \sin k(x-b) \cosh kb, \quad \text{for } x \ge a.$$

**3.14:**

$$g(x,s) = \frac{1}{8k^3} \begin{cases} e^{k(x-s)}[\sin k(x-s) - \cos k(x-s)] \\ \quad +e^{-k(x+s)}[\sin k(x+s) + \cos k(x+s)], & \text{for } x \le s \\ e^{k(s-x)}[\sin k(s-x) - \cos k(s-x)] \\ \quad +e^{-k(x+s)}[\sin k(x+s) + \cos k(x+s)], & \text{for } s \le x \end{cases}$$

where $k = (k_0/4EI)^{1/4}$.

**3.15:**

**a)** $$w(x) = \frac{M_0}{2k^2EI}e^{-ka}\sin k(x-a)\cosh kx + \frac{P_0}{8k^3EI}\{e^{k(x-b)}[\cos k(x-b) - \sin k(x-b)] - e^{-k(x+b)}[\sin k(x+b) + \cos k(x+b)]\},$$

for $x \le a$;

$$w(x) = \frac{M_0}{2k^2EI}e^{-kx}\sin k(x-a)\cosh ka + \frac{P_0}{8k^3EI}\{e^{k(x-b)}[\cos k(x-b) - \sin k(x-b)] - e^{-k(x+b)}[\sin k(x+b) + \cos k(x+b)]\},$$

for $a \le x \le b$;

$$w(x) = \frac{M_0}{2k^2EI}e^{-kx}\sin k(x-a)\cosh ka + \frac{P_0}{8k^3EI}\{e^{k(b-x)}[\cos k(x-b) + \sin k(x-b)] - e^{-k(x+b)}[\sin k(x+b) + \cos k(x+b)]\},$$

for $x \ge b$.

**b)** $$w(x) = \frac{\cosh kx}{2k^3EI}[M_1e^{-ka}\sin k(x-a) + M_2e^{-kb}\sin k(x-b)], \quad x \le a;$$

$$w(x) = \frac{1}{2k^3EI}[M_1e^{-kx}\cosh ka\sin k(x-a) + M_2e^{-kb}\cosh kx\sin k(x-b)], \quad \text{for } a \le x \le b;$$

$$w(x) = \frac{e^{-kx}}{2k^3EI}[M_1\cosh ka\sin k(x-a) + M_2\cosh kb\sin k(x-b)], \quad x \ge b.$$

## Chapter 4

**4.1:**

**a)**
$$w(x) = \frac{q_0}{12m^4}\left\{mx[6(ma+b)^2 + mx(3(ma+b) + m(3a-x))]\right.$$
$$\left. \times\, 6(ma+b)^2(mx+b)\ln\left(\frac{b}{mx+b}\right)\right\}.$$

**b)**
$$w(x) = \frac{P_1}{2m^3}\left\{2(ma_1+b)(mx+b)\ln\left(\frac{b}{mx+b}\right) + mx[mx + 2(ma_1+b)]\right\}$$
$$+ \frac{P_2}{2m^3}\left\{2(ma_2+b)(mx+b)\ln\left(\frac{b}{mx+b}\right) + mx[mx + 2(ma_2+b)]\right\},$$

for $x \le a_1$;

$$w(x) = \frac{P_1}{2m^3}\left\{2(ma_1+b)(mx+b)\ln\left(\frac{b}{ma_1+b}\right) + ma_1[ma_1 + 2(mx+b)]\right\}$$
$$+ \frac{P_2}{2m^3}\left\{2(ma_2+b)(mx+b)\ln\left(\frac{b}{mx+b}\right) + mx[mx + 2(ma_2+b)]\right\},$$

$a_1 \le x \le a_2$;

$$w(x) = \frac{P_1}{2m^3}\left\{2(ma_1+b)(mx+b)\ln\left(\frac{b}{ma_1+b}\right) + ma_1[ma_1 + 2(mx+b)]\right\}$$
$$+ \frac{P_2}{2m^3}\left\{2(ma_2+b)(mx+b)\ln\left(\frac{b}{ma_2+b}\right) + ma_2[ma_2 + 2(mx+b)]\right\},$$

$x \ge a_2$.

**c)**
$$w(x) = \frac{M_1}{m^2}\left[mx + (mx+b)\ln\left(\frac{b}{mx+b}\right)\right]$$
$$+ \frac{M_2}{m^2}\left[mx + (mx+b)\ln\left(\frac{b}{mx+b}\right)\right], \quad \text{for } x \le a_1;$$
$$w(x) = \frac{M_1}{m^2}\left[ma_1 + (mx+b)\ln\left(\frac{b}{ma_1+b}\right)\right]$$
$$+ \frac{M_2}{m^2}\left[mx + (mx+b)\ln\left(\frac{b}{mx+b}\right)\right], \quad \text{for } a_1 \le x \le a_2;$$
$$w(x) = \frac{M_1}{m^2}\left[ma_1 + (mx+b)\ln\left(\frac{b}{ma_1+b}\right)\right]$$
$$+ \frac{M_2}{m^2}\left[ma_2 + (mx+b)\ln\left(\frac{b}{ma_2+b}\right)\right], \quad \text{for } x \ge a_2.$$

**4.2:**

$$g(x,s)=\frac{1}{2m}\begin{cases}2xs[1+\ln(a/s)]-2a(x+s)+(a^2+s^2), & x\le s\\ 2xs[1+\ln(a/x)]-2a(x+s)+(a^2+x^2), & s\le x\end{cases}$$

**4.3:**

$$\begin{aligned}w(x)&=\frac{P_0}{2m}\left[2a_1x\left(1+\ln\left(\frac{a}{a_1}\right)\right)-2a(x+a_1)+(a^2+a_1^2)\right]\\&\quad+\frac{M_0}{m}\left[x\ln\left(\frac{a}{a_2}\right)+(a_2-a)\right],\quad \text{for } x\le a_1;\\ w(x)&=\frac{P_0}{2m}\left[2a_1x\left(1+\ln\left(\frac{a}{x}\right)\right)-2a(x+a_1)+(a^2+x^2)\right]\\&\quad+\frac{M_0}{m}\left[x\ln\left(\frac{a}{a_2}\right)+(a_2-a)\right],\quad \text{for } a_1\le x\le a_2;\\ w(x)&=\frac{P_0}{2m}\left[2a_1x\left(1+\ln\left(\frac{a}{x}\right)\right)-2a(x+a_1)+(a^2+x^2)\right]\\&\quad+\frac{M_0}{m}\left[x\ln\left(\frac{a}{x}\right)+(x-a)\right],\quad \text{for } x\ge a_2.\end{aligned}$$

**4.4:**

$$g(x,s)=\frac{1}{mb^3}\begin{cases}[b(x+s)-2]-b^2xs+e^{-bx}[b(x-s)+2], & x\le s\\ [b(x+s)-2]-b^2xs+e^{-bs}[b(s-x)+2], & s\le x\end{cases}$$

**4.5:**

$$\begin{aligned}w(x)&=\frac{P_0}{mb^3}\left\{\left[b\left(x+\frac{a}{4}\right)-2\right]-\frac{1}{4}ab^2x+e^{-bx}\left[b\left(x-\frac{a}{4}\right)+2\right]\right\}\\&\quad-\frac{q_0a}{8mb^3}\{4+(bx-1)(3ab-4)+[3ab-4(bx+2)]e^{-bx}\},\\&\quad\text{for } x\le a/4;\\ w(x)&=\frac{P_0}{mb^3}\left\{\left[b\left(x+\frac{a}{4}\right)-2\right]-\frac{1}{4}ab^2x+e^{-ab/4}\left[b\left(\frac{a}{4}-x\right)+2\right]\right\}\\&\quad-\frac{q_0a}{8mb^3}\{4+(bx-1)(3ab-4)+[3ab-4(bx+2)]e^{-bx}\},\\&\quad\text{for } a/4\le x\le a/2;\end{aligned}$$

$$w(x) = \frac{P_0}{mb^3}\left\{\left[b\left(x+\frac{a}{4}\right)-2\right]-\frac{1}{4}ab^2x+e^{-ab/4}\left[b\left(\frac{a}{4}-x\right)+2\right]\right\}$$
$$-\frac{q_0}{8mb^4}\{4[b^2(x-a)^2+2(3+2b(x-a))]e^{-bx}$$
$$+4[2(bx-2)-(ab+2)]e^{-ab/2}+ab[3ab(bx-1)-4(bx-2)]\},$$
for $x \geq a/2$.

**4.6:**

**a)**
$$w_1(x) = \frac{q_0x^2(x-a)}{120aEI}[(x-2a)^2+10a^2],$$
$$M_1(x) = \frac{q_0}{30aEI}[5(x-a)^3+2a^2(6x-a)],$$
$$Q_1(x) = \frac{q_0}{10aEI}[5(x-a)^2+4a^2], \quad \text{for } 0 \leq x \leq a;$$
$$w_2(x) = \frac{q_0(x-a)}{120aEI}[(x-a)^4-10a^2(x-2a)^2+21a^4],$$
$$M_2(x) = \frac{q_0(x+a)}{6aEI}(x-2a)^2, \quad Q_2(x) = \frac{q_0x}{2aEI}(x-2a), \quad \text{for } a \leq x \leq 2a.$$

**b)**
$$w_1(x) = \frac{1}{48EI}(x^2(x-a)[q_1(2x-3a)+6q_2a],$$
$$M_1(x) = \frac{1}{8EI}[q_1(4x^2-5ax+a^2)+2q_2a(3x-a)],$$
$$Q_1(x) = \frac{1}{8EI}[q_1(8x-5a)+6q_2a], \quad \text{for } 0 \leq x \leq a;$$
$$w_2(x) = \frac{(a-x)}{48EI}[q_1a^3-q_2(2x^3-14ax^2+34a^2x-16a^3)],$$
$$M_2(x) = \frac{q_2}{2EI}(x-2a)^2, \quad Q_2(x) = \frac{q_2}{EI}(x-2a), \quad \text{for } a \leq x \leq 2a.$$

**c)**
$$w_1(x) = \frac{q_0x^2(a-x)}{720a^2EI}(2x^3-10ax^2-10a^2x-15a^3),$$
$$M_1(x) = -\frac{q_0}{24a^2EI}(2x^4-8ax^3-a^3x+a^4),$$
$$Q_1(x) = -\frac{q_0}{24a^2EI}(8x^3-24ax^2-a^3), \quad \text{for } 0 \leq x \leq a;$$
$$w_2(x) = \frac{q_0(a-x)}{720a^2EI}(2x^5-10ax^4-10a^2x^3+150a^3x^2-330a^4x+165a^5),$$
$$M_2(x) = \frac{q_0(x+2a)}{12a^2EI}(2a-x)^3,$$
$$Q_2(x) = -\frac{q_0(x+a)}{3a^2EI}(x-2a)^2, \quad \text{for } a \leq x \leq 2a.$$

In Exercises 4.7 through 4.9, we use notations: $R = EI_1 + EI_2$, $R_1 = EI_1 R$, and $R_2 = EI_2 R$.

**4.7:**

**a)**

$$w_1(x) = \frac{x(x+a)}{720R_1}\{R[q_0(2x^4 - 2ax^3 + 2a^2x^2 + 12a^3x) + q_2(30x^2 + 45ax)] \\ + q_1[EI_1(6x^3 - 6ax^2 - 7a^2x + 28a^3) + EI_2(6x^3 - 6ax^2 - 21a^2x)]\},$$

$$M_1(x) = \frac{(x+a)}{120R_1}\{R[q_0(10x^3 - 10ax^2 + 10a^2x + 4a^3) + q_2(60x + 15a)] \\ + q_1[EI_1(20x^2 - 20ax + 7a^2) + EI_2(20x^2 - 20ax - 7a^2)]\},$$

$$Q_1(x) = \frac{1}{120R_1}\{R[q_0(40x^3 + 14a^3) + q_2(120x + 75a)] \\ + q_1[EI_1(60x^2 - 13a^2) + EI_2(60x^2 - 27a^2)]\}, \quad \text{for } -a \le x \le 0;$$

$$w_2(x) = \frac{x(x-a)}{720R_2}\{R[q_0(2x^4 + 2ax^3 + 2a^2x^2 - 12a^3x) + q_2(30x^2 - 45ax)] \\ + q_1[EI_1(6x^3 + 6ax^2 - 21a^2x) + EI_2(6x^3 + 6ax^2 - 7a^2x - 28a^3)]\},$$

$$M_2(x) = \frac{(x-a)}{120R_2}\{R[q_0(10x^3 + 10ax^2 + 10a^2x - 4a^3) + q_2(60x - 15a)] \\ + q_1[EI_1(20x^2 + 20ax - 7a^2) + EI_2(20x^2 + 20ax + 7a^2)]\},$$

$$Q_2(x) = \frac{1}{120R_2}\{R[q_0(40x^3 - 14a^3) + q_2(120x - 75a)] \\ + q_1[EI_1(60x^2 - 27a^2) + EI_2(60x^2 - 13a^2)]\}, \quad \text{for } 0 \le x \le a.$$

**4.8:** The entries $g_{ij}(x, s)$, $(i, j = 1, 2)$ of the influence matrix are defined in the form:

$$g_{11}(x,s) = \frac{1}{12a^3R_1}\begin{cases} s(a+x)^2[EI_1(3a^2x + 2as^2 - xs^2) \\ \quad +2EI_2s(as - 3ax - 2xs)], & \text{for } -a \le x \le s \le 0 \\ x(a+s)^2[EI_1(3a^2s + 2ax^2 - sx^2) \\ \quad +2EI_2x(ax - 3as - 2xs)], & \text{for } -a \le s \le x \le 0 \end{cases}$$

$$g_{12}(x,s) = \frac{1}{6a^3R}xs(a-s)^2(a+x)^2, \quad \text{for } -a \le x \le 0 \le s \le a$$

$$g_{21}(x,s) = \frac{1}{6a^3R}xs(a-x)^2(a+s)^2, \quad \text{for } -a \le s \le 0 \le x \le a$$

$$g_{22}(x,s) = \frac{1}{12a^3R_2}\begin{cases} x(a-s)^2[2EI_1x(3as - ax - 2xs) \\ \quad +EI_2(3a^2s - 2ax^2 - x^2s)], & \text{for } 0 \le x \le s \le a \\ s(a-x)^2[2EI_1s(3ax - as - 2xs) \\ \quad +EI_2(3a^2x - 2as^2 - xs^2)], & \text{for } 0 \le s \le x \le a \end{cases}$$

**4.9:**

**a)**

$$w_1(x) = \frac{1}{240R} ax((x+a)^2(2aq_0 + 5q_1),$$

$$M_1(x) = \frac{a}{120R}(3x+2a)(2aq_0 + 5q_1),$$

$$Q_1(x) = \frac{a}{40R}(2aq_0 + 5q_1), \quad \text{for } -a \le x \le 0;$$

$$w_2(x) = \frac{x(x-a)^2}{240R_2}\{2q_0[Rx(x+2a) + EI_2a^2] + 5q_1(2Rx + EI_a)\},$$

$$M_2(x) = \frac{1}{120R_2}\{2q_0[EI_1(10x^3 - 9xa^2 + 2a^3) + 2EI_2x(5x^2 - 3a^2)]$$
$$+ 5q_1[2EI_1(6x^2 - 6ax + a^2) + 3EI_2x(4x - 3a)]\},$$

$$Q_2(x) = \frac{1}{40R_2}\{2q_0[EI_1(10x^2 - 3a^2) + 2EI_2(5x^2 - a^2)]$$
$$+ 5q_1[4EI_1(2x - a) + EI_2(8x - 3a)]\}, \quad \text{for } 0 \le x \le a.$$

**b)**

$$w_1(x) = \frac{x(x+a)^2}{120R_1}\{q_0[Rx(x-2a) + 2EI_1a^2] + 5q_1Rx\},$$

$$M_1(x) = \frac{1}{60R_1}\{q_0[EI_1(10x^3 - 3a^2x + 2a^3) + EI_2(10x^3 - 9a^2x - 2a^3)]$$
$$+ 5q_1R(6x^2 + 6ax + a^2)\},$$

$$Q_1(x) = \frac{1}{20R_1}\{10q_1R(2x + a) + q_0[EI_1(10x^2 - a^2)$$
$$+ EI_2(10x^2 - 3a^2)]\}, \quad \text{for } -a \le x \le 0;$$

$$w_2(x) = \frac{x(x-a)^2}{120R_2}\{q_0[Rx(x+2a) + 2EI_1a^2] + 5q_1Rx\},$$

$$M_2(x) = \frac{1}{60R_2}\{q_0[EI_1(10x^3 - 9a^2x + 2a^3) + EI_2(10x^3 - 3a^2x - 2a^3)]$$
$$+ 5q_1R(6x^2 - 6ax + a^2)\},$$

$$Q_2(x) = \frac{1}{20R_2}\{10q_1R(2x - a) + q_0[EI_1(10x^2 - 3a^2)$$
$$+ EI_2(10x^2 - a^2)]\}, \quad \text{for } 0 \le x \le a.$$

In exercises 4.10 and 4.11, we denote $R_0 = 4EI_1 + 3EI_2$, $R_1 = EI_1R_0$, and $R_2 = EI_2R_0$.

**4.10:** The entries $g_{ij}(x, s)$, $(i, j = 1, 2)$ of the influence matrix are defined in the form:

$$g_{11}(x,s) = -\frac{1}{6a^3R_1}\begin{cases} s(a+x)^2[2EI_1(s^2x-3a^2x-2as^2) \\ \quad +3EI_2s(2xs+3ax-as)], & \text{for } -a\le x\le s\le 0 \\ x(a+s)^2[2EI_1(x^2s-3a^2s-2ax^2) \\ \quad +3EI_2x(2sx+3as-ax)], & \text{for } -a\le s\le x\le 0 \end{cases}$$

$$g_{12}(x,s) = \frac{1}{2a^3R_0}xs(2a-s)(a-s)(a+x)^2, \quad \text{for } -a\le x\le 0\le s\le a$$

$$g_{21}(x,s) = \frac{1}{2a^3R_0}xs(x-a)(x-2a)(a+s)^2, \quad \text{for } -a\le s\le 0\le x\le a$$

$$g_{22}(x,s) = \frac{1}{6a^3R_2}\begin{cases} x(s-a)[2EI_1x(2a^2x+2axs-xs^2-6a^2s+3as^2) \\ \quad +3EI_2a^2(x^2-2as+s^2)], & \text{for } 0\le x\le s\le a \\ s(x-a)[2EI_1s(2a^2s+2axs-x^2s-6a^2x+3ax^2) \\ \quad +3EI_2a^2(s^2-2ax+x^2)], & \text{for } 0\le s\le x\le a \end{cases}$$

**4.11:**

**a)**

$$w_1(x) = \frac{x(x+a)^2}{24R_1}\{q_1[2EI_1(2x-a)+3xEI_2]+q_2aEI_1\},$$

$$M_1(x) = \frac{1}{4R_1}\{q_1[2EI_1x(4x+3a)+EI_1(6x^2+6ax+a^2)] + q_2EI_1a(2a+3x)\},$$

$$Q_1(x) = \frac{1}{4R_1}\{2q_1[EI_1(8x+3a)+3EI_2(2x+a)]+3q_2aEI_1\}, \quad -a\le x\le 0;$$

$$w_2(x) = \frac{x(x-a)^2}{24R_2} \times\{q_1EI_2a(2a-x)+q_2[2EI_1x(2x-3a)+3EI_2(x^2-ax-a^2)]\},$$

$$M_2(x) = \frac{1}{4R_2}(a-x)\{q_1aEI_2-2q_2[EI_1(4x-a)+3x]\},$$

$$Q_2(x) = \frac{1}{4R_2}\{-q_1aEI_2+2q_2[EI_1(8x-5a)+3EI_2(2x-a)]\}, \quad 0\le x\le a.$$

**b)**
$$w_1(x) = \frac{q_0 a}{15\pi^4 R_1}\left\{EI_1\left[\pi(1+60\pi)x(x+a)^2 + 240(a^2-x^2))\right.\right.$$
$$\left. - 1920a^3\sin^2\left(\frac{\pi x}{4a}\right)\right] + 480x(x^2-3a^2)$$
$$\left. + 360EI_2\left[x^2((4-\pi)(x+a)-a) - 4\sin^2\left(\frac{\pi x}{4a}\right)\right]\right\},$$
$$M_1(x) = \frac{2q_0 a}{15\pi^4 R_1}\left\{EI_1\left[180(\pi^2-4\pi+8)x + \pi^4(3x+2a)\right.\right.$$
$$\left. + 240a\pi^2\sin^2\left(\frac{\pi x}{4a}\right)\right]$$
$$\left. + 90EI_2\left[12(4-\pi)x - 4(\pi-6)a - a\pi^2\cos\left(\frac{\pi x}{2a}\right)\right]\right\},$$
$$Q_1(x) = \frac{2q_0 a}{5\pi^4 R_1}\left\{EI_1\left[60(\pi^2-4\pi+8) + \pi^4 + 20\pi^3\sin\left(\frac{\pi x}{2a}\right)\right]\right.$$
$$\left. + 15EI_2\left[24(4-\pi) + \pi^3\sin\left(\frac{\pi x}{2a}\right)\right]\right\}, \quad \text{for } -a \le x \le 0;$$
$$w_2(x) = \frac{q_0}{120a\pi^4 R_2}x(x-a)(x-2a)\{4EI_1\pi^4 x(2a-x)$$
$$+ EI_2[240a^2(\pi^2+4\pi+24) + \pi^4(4a^2+6ax-3x^2)]\},$$
$$M_2(x) = \frac{q_0(a-x)}{30a\pi^4 R_2}\{4EI_1\pi^4(5x^2-10ax+2a^2)$$
$$+ 15EI_2[a^2(576-24\pi^2-96\pi) + \pi^4 x(x-2a)]\};$$
$$Q_2(x) = \frac{q_0}{10a\pi^4 R_2}\{4EI_1\pi^4(5x^2-10ax+4a^2)$$
$$+ 5EI_2[120a^2(24-4\pi-\pi^2) + \pi^4(3x^2-6ax+2a^2)]\},$$
for $0 \le x \le a$;

**4.12:** The entries $g_{ij}(x, s)$, $(i, j = 1, 2, 3)$ of the influence matrix are defined in the form:

$$g_{11}(x,s) = \frac{1}{12a^3 EI_1}\begin{cases} x^2(a-s)[s(x-3a)(s-2a) - 2a^2 x], \\ \qquad \text{for } 0 \le x \le s \le a \\ s^2(a-x)[x(s-3a)(x-2a) - 2a^2 s], \\ \qquad \text{for } 0 \le s \le x \le a \end{cases}$$

$$g_{12}(x,s) = \frac{1}{4aEI_1}x^2(a-s)(a-x), \quad \text{for } 0 \le x \le a \le s \le 2a;$$

$$g_{13}(x,s) = \frac{1}{4aEI_1}x^2(a-s)(a-x), \quad \text{for } 0 \le x \le a,\ 2a \le s \le 3a$$

$$g_{21}(x,s) = \frac{1}{4aEI_1}s^2(a-s)(a-x), \quad \text{for } 0 \le s \le a \le x \le 2a$$

$$g_{22}(x,s) = \frac{1}{12EI_1}\begin{cases}(a-x)[(2x-a)(a-3s)+2x^2], & \text{for } a \le x \le s \le 2a\\ (a-s)[(2s-a)(a-3x)+2s^2], & \text{for } a \le s \le x \le 2a\end{cases}$$

$$g_{23}(x,s) = \frac{1}{12EI_1}(a-x)[(2x-a)(a-3s)+2x^2], \quad \text{for } a \le x \le 2a \le s \le 3a;$$

$$g_{31}(x,s) = \frac{1}{4aEI_1}s^2(a-s)(a-x), \quad \text{for } 2a \le x \le 3a,\ 0 \le s \le a$$

$$g_{32}(x,s) = \frac{1}{12EI_1}(s-a)[(a-3x)(a-2s)-2s^2], \quad \text{for } a \le s \le 2a \le x \le 3a$$

$$g_{33}(x,s) = -\frac{1}{12EI_1EI_2}\begin{cases}2EI_1[x^2(x-3s)+12(xs-ax-as)+16a^3]\\ \quad +3aEI_2[7a(x+s)-5xs], & \text{for } 2a \le x \le s \le 3a\\ 2EI_1[s^2(s-3x)+12(xs-as-ax)+16a^3]\\ \quad +3aEI_2[7a(x+s)-5xs], & \text{for } 2a \le s \le x \le 3a\end{cases}$$

**4.13:** The entries $g_{ij}(x,s)$, $(i, j = 1, 2, 3)$ of the influence matrix are defined in the form:

$$g_{11}(x,s) = -\frac{1}{6EI}\begin{cases}(a-s)(s^2+as-4a^2+5ax-3xs), & \text{for } 0 \le x \le s \le a\\ (a-x)(x^2+ax-4a^2+5as-3xs), & \text{for } 0 \le s \le x \le a\end{cases}$$

$$g_{12}(x,s) = -\frac{1}{6aEI}(a-s)(2a-s)(3a-s)(x-a), \quad \text{for } 0 \le x \le a \le s \le 2a;$$

$$g_{13}(x,s) = \frac{a}{6EI}(2a-s)(x-a), \quad \text{for } 0 \le x \le a,\ 2a \le s \le 3a$$

$$g_{21}(x,s) = -\frac{1}{6aEI}(a-s)(x-a)(x-2a)(x-3a), \quad \text{for } 0 \le s \le a \le x \le 2a$$

$$g_{22}(x,s) = -\frac{1}{6aEI}\begin{cases}(x-a)(2a-s)[(x+s-2a)^2+2x(a-s)],\\ \quad a \le x \le s \le 2a\\ (s-a)(2a-x)[(s+x-2a)^2+2s(a-x)],\\ \quad a \le s \le x \le 2a\end{cases}$$

$$g_{23}(x,s) = -\frac{x}{6aEI}(x-a)(x-2a)(2a-s), \quad \text{for } a \le x \le 2a \le s \le 3a;$$

$$g_{31}(x,s) = \frac{a}{6EI}(a-s)(x-2a), \quad \text{for } 2a \le x \le 3a,\ 0 \le s \le a$$

$$g_{32}(x,s) = \frac{s}{6aEI}(a-s)(2a-s)(x-2a), \quad \text{for } a \le s \le 2a \le x \le 3a$$

$$g_{33}(x,s) = -\frac{1}{6EI}\begin{cases}(x-2a)(x^2+2ax-3xs+4as-4a^2), \\ \quad 2a \le x \le s \le 3a \\ (s-2a)(s^2+2as-3xs+4ax-4a^2), \\ \quad 2a \le s \le x \le 3a\end{cases}$$

## Chapter 5

**5.1:** $$G(x,y;s,t) = \frac{2}{\pi}\sum_{n=1}^{\infty}\frac{1+\nu|x-s|}{4\nu^3}e^{-\nu|x-s|}\sin\nu y\sin\nu t, \qquad \nu = \frac{n\pi}{b}$$

**5.2:**

**b)** $$G(x,y;s,t) = \frac{1}{4b}\sum_{n=1}^{\infty}\frac{g_n(x,s)}{\nu^3(1+2\nu^2a^2-\cosh 2\nu a)}\sin\nu y\sin\nu t,$$

where $\nu = n\pi/b$ and the coefficient $g_n(x,s)$ is defined as

$$\begin{aligned} g_n(x,s) = {} & 2\sinh\nu x\{e^{\nu s}[(\nu s-1)(1+2\nu a-e^{-2\nu a})-2\nu^2a^2] \\ & - e^{-\nu s}[(\nu s+1)(1-2\nu a-e^{2\nu a})+2\nu^2a^2]\} \\ & + \nu x e^{-\nu x}\{e^{\nu s}[1+2\nu a+4\nu^2a(a-s)] \\ & + e^{-\nu s}[(1+2\nu s)(1-e^{2\nu a})-2\nu a]\} \\ & - \nu x e^{\nu x}\{e^{nus}[(2\nu s-1)(1-e^{-2\nu a})-2\nu a] \\ & - e^{-\nu s}[1-2\nu a+4\nu^2a(a-s)-e^{2\nu a}]\}, \quad \text{for } x \le s \end{aligned}$$

The expression for $g_n(x,s)$ that is valid for $x \ge s$ can be obtained from the above by interchanging $x$ with $s$;

**c)** $$G(x,y;s,t) = \frac{1}{b}\sum_{n=1}^{\infty}\frac{g_n(x,s)}{\nu^3(\sinh 2\nu a-2\nu a)}\sin\nu y\sin\nu t, \qquad \nu = \frac{n\pi}{b},$$

where the coefficient $g_n(x,s)$ is defined, for $x \le s$, as

$$\begin{aligned} g_n(x,s) = {} & \nu x\cosh\nu x[2\nu(s-a)\cosh\nu s-\sinh\nu s-\sinh\nu(s-2a)] \\ & - \sinh\nu x[\nu s\cosh\nu(s-2a)-\sinh\nu(s-2a) \\ & + \nu(s-2a)\cosh\nu s+(2\nu^2a(s-a)-1)\sinh\nu s] \end{aligned}$$

from which the expression for $g_n(x,s)$ that is valid for $x \ge s$ can be obtained by interchanging $x$ with $s$.

In Exercises 5.3–5.5, we use the notations: $\mu = m\pi/a$ and $\nu = n\pi/b$.

**5.3:** $$w(x, y) = -\frac{4Q_0}{abD}\sum_{m=1}^{\infty}\sum_{n=1}^{\infty}\frac{\sin\mu x \sin\nu y}{\mu\nu(\mu^2+\nu^2)^2} \times (\cos\mu a_2 - \cos\mu a_1)(\cos\nu b_2 - \cos\nu b_1);$$

$$M_x(x, y) = \frac{4Q_0}{ab}\sum_{m=1}^{\infty}\sum_{n=1}^{\infty}\frac{(\mu^2+\sigma\nu^2)\sin\mu x \sin\nu y}{\mu\nu(\mu^2+\nu^2)^2} \times (\cos\mu a_2 - \cos\mu a_1)(\cos\nu b_2 - \cos\nu b_1);$$

$$M_y(x, y) = \frac{4Q_0}{ab}\sum_{m=1}^{\infty}\sum_{n=1}^{\infty}\frac{(\nu^2+\sigma\mu^2)\sin\mu x \sin\nu y}{\mu\nu(\mu^2+\nu^2)^2} \times (\cos\mu a_2 - \cos\mu a_1)(\cos\nu b_2 - \cos\nu b_1);$$

**5.4:**

**a)** $$w(x, y) = -\frac{4Q_0}{abD}\sum_{m=1}^{\infty}\sum_{n=1}^{\infty}\frac{\sin\mu x \sin\nu y}{\mu\nu(\mu^2+\nu^2)^2} \times \left(1 - \cos\frac{\nu b}{2}\right)(1 - \cos\mu a);$$

**b)** $$w(x, y) = -\frac{2Q_0}{abD}\sum_{m=1}^{\infty}\sum_{n=1}^{\infty}\frac{\sin\mu x \sin\nu y}{\mu^2\nu^2(\mu^2+\nu^2)^2} \times \left(2\sin\frac{\mu a}{2} - \mu a\cos\frac{\mu a}{2}\right)(\sin\nu b - \nu b\cos\nu b);$$

**c)** $$w(x, y) = -\frac{4Q_0}{abD}\sum_{m=1}^{\infty}\sum_{n=1}^{\infty}\frac{\sin\mu x \sin\nu y}{\mu^3\nu^3(\mu^2+\nu^2)^2} \times (2 - \mu a\sin\mu a - 2\cos\mu a)(2 - \nu b\sin\nu b - 2\cos\nu b);$$

**5.5:**

**a)** $$w(x, y) = -\frac{4Q_0}{abD}\sum_{m=1}^{\infty}\sum_{n=1}^{\infty}\frac{\sin\mu x \sin\nu y}{\mu\nu[(\mu^2+\nu^2)^2+\lambda]} \times \left(1 - \cos\frac{\mu a}{2}\right)(1 - \cos\nu b);$$

**b)** $$w(x, y) = -\frac{4Q_0}{abD}\sum_{m=1}^{\infty}\sum_{n=1}^{\infty}\frac{\sin\mu x \sin\nu y}{\mu\nu[(\mu^2+\nu^2)^2+\lambda]} \times \left(1 - \cos\frac{\mu a}{2}\right)\left(1 - \cos\frac{\nu b}{2}\right);$$

**c)** $$w(x, y) = -\frac{4Q_0}{abD}\sum_{m=1}^{\infty}\sum_{n=1}^{\infty}\frac{\sin\mu x\sin\nu y}{\mu^2\nu^2[(\mu^2+\nu^2)^2+\lambda]}$$
$$\times(\sin\mu a - \mu a\cos\mu a)(\sin\nu b - \nu b\cos\nu b);$$

**5.8:** $$u(x, \varphi) = -\frac{48a^4(1-\sigma^2)Z_0}{Ehl\varphi_0}\sum_{m=1}^{\infty}\sum_{n=1}^{\infty}\frac{[\widetilde{\mu}^2(1+\sigma)-(\widetilde{\mu}^2+\nu^2)]}{\nu\Delta}$$
$$\times\cos\mu x\sin\nu\varphi\left(\cos\frac{\mu l}{2}-1\right)(\cos\nu\varphi_0-1);$$
$$v(x, \varphi) = -\frac{48a^4(1-\sigma^2)Z_0}{Ehl\varphi_0}\sum_{m=1}^{\infty}\sum_{n=1}^{\infty}\frac{[\widetilde{\mu}^2(1+\sigma)+(\widetilde{\mu}^2+\nu^2)]}{\widetilde{\mu}\Delta}$$
$$\times\cos\mu x\sin\nu\varphi\left(\cos\frac{\mu l}{2}-1\right)(\cos\nu\varphi_0-1);$$
$$w(x, \varphi) = \frac{48a^4(1-\sigma^2)Z_0}{Ehl\varphi_0}\sum_{m=1}^{\infty}\sum_{n=1}^{\infty}\frac{(\widetilde{\mu}^2+\nu^2)^2}{\widetilde{\mu}\nu\Delta}$$
$$\times\cos\mu x\sin\nu\varphi\left(\cos\frac{\mu l}{2}-1\right)(\cos\nu\varphi_0-1),$$

where we use the following notations:

$$\mu = \frac{m\pi}{l}, \quad \nu = \frac{n\pi}{\varphi_0}, \quad \widetilde{\mu} = \mu a$$

and $\Delta = 12\widetilde{\mu}^4a^2(1-\sigma^2)+h^2(\widetilde{\mu}^2+\nu^2)^4$.

# References

1. Abramovitz, M. and Stegun, I., Handbook of Mathematical Functions, National Bureau of Standards, Washington, D.C., 1972.

2. Agarwal, R.P., *Boundary Value Problems for Higher Order Differential Equations*, World Scientific, Singapore, 1985.

3. Amdursky, V. and Ziv, A., On the numerical solution of stiff linear systems of the oscillatory type, *SIAM J. of Applied Mathematics*, **33**, 593–606, 1977.

4. Andrews, L.C., *Special Functions for Engineers and Applied Mathematicians*, Macmillan, New York, 1985.

5. Atkinson, K., *An Introduction to Numerical Analysis*, John Wiley, New York, 1989.

6. Babushka, I., Prager M. and Vitasek, E., *Numerical Processes in Differential Equations*, John Wiley, New York, 1966.

7. Banerjee, P.K. and Butterfield, R., *Boundary Element Method in Engineering Science*, McGraw-Hill, London, 1981.

8. Barton, G., *Elements of Green's Functions and Propagation*, Clarendon Press, Oxford, 1989.

9. Beck, J.V., Cole, K.D., Haji-Sheikh, A. and Litkouhi, B., *Heat Conduction Using Green's Functions*, Hemisphere, Publishing Corp., London–Washington, D.C.–Philadelphia, 1992.

10. Berger, J.R., Boundary element analysis of anisotropic bimaterials with special Green's functions, *Engineering Analysis with Boundary Elements*, **14**(2), 123–131, 1994.

11. Bergman, S. and Schiffer, M., *Kernel Functions and Elliptic Differential Equations in Mathematical Physics*, Academic Press, New York, 1953.

12. Biggs, J.M., *Introduction to Structural Engineering Analysis and Design*, Prentice-Hall, Englewood Cliffs, New Jersey, 1986.

13. Blevins, R.D., *Formulas for Natural Frequency and Mode Shape*, Van Nostrand Reinhold, New York, 1979.

14. Boley, B.A., A method for the construction of Green's functions, *Quarterly Journal of Applied Mathematics*, **14**, 249–257, 1956.

15. Boyce, W.E. and DiPrima, R.C., *Elementary Differential Equations and Boundary Value Problems*, John Wiley, New York, 1997.

16. Brebbia, C.A., *The Boundary Element Method for Engineers*, Pentech Press/Halstead Press, London–New York, 1978.

17. Butkovsky, A.G., *Green's Functions and Transfer Functions Handbook*, Translation from Russian, Halstead Press, New York, 1982.

18. Chen-To, Tai, *Dyadic Green's Functions in Electromagnetic Theory*, IEEE Press, New York, 1994.

19. Collatz, L., *Numerical Treatment of Differential Equations*, Springer-Verlag, Berlin, 1960.

20. Courant, R. and Hilbert, D., *Methods of Mathematical Physics*, Interscience, New York, 1953.

21. Crandall, S.H., Dahl, N.C. and Lardner, T.J., *An Introduction to the Mechanics of Solids*, McGraw-Hill, New York, 1972.

22. Cruse, T.A., *Boundary Element Analysis in Computational Fracture Mechanics*, Kluwer Academic Publisher, Dordrecht, 1988.

23. Davis, P.W., *Differential Equations*, Prentice Hall, New Jersey, 1999.

24. Dolgova, I.M. and Melnikov, Yu.A., Construction of Green's functions and matrices for equations and systems of elliptic type, *Translation from Russian PMM* (*J. Appl. Math. Mech.*), **42**, 740–746, 1978.

25. Duffy, D., *Green's Functions with Applications*, CRC Press, Boca Raton, 2001.

26. Economou, E.N., *Green's Functions in Quantum Physics*, Springer-Verlag, Berlin, 1983.

27. Elias, Z.M., *Theory and Methods of Structural Analysis*, John Wiley, New York, 1986.

28. Flugge, W., *Stresses in Shells*, Springer-Verlag, Berlin–Heidelberg–New York, 1973.

29. Gavelya, S.P., On one method of construction of Green's matrices for joint shells, *Reports of the Ukrainian Academy of Sciences*, Ser. A, 12, 1107–1111 (in Russian), 1969.

30. Gradstein, I.S. and Ryzhik, I.M., *Tables of Integrals, Series and Products*, Academic Press, New York, 1980.

31. Greenberg, M.D., *Application of Green's Functions in Science and Engineering*, Prentice Hall, Englewood Cliffs, New Jersey, 1971.

32. Haberman, R., *Elementary Applied Partial Differential Equations*, Prentice Hall, Englewood Cliffs, New Jersey, 1987.

33. Hasebe, N., Jun Quin and Yizhou Chen, Fundamental solutions for half plane with an oblique edge crack, *Engineering Analysis with Boundary Elements*, **17**(4), 263–267, 1996.

34. Hwu, C. and Yen, W., Green's functions of two-dimensional anisotropic plane containing an elliptic hole, *Int. J. of Solids and Structures*, **27**, 1705–1719, 1991.

35. Irschik, H. and Ziegler, F., Application of the Green's function method to thin elastic polygonal plates, *Acta Mechanica*, **39**, 1980.

36. Kantorovich, L.V. and Krylov, V.I., *Approximate Methods of Higher Analysis*, Interscience, New York, 1964.

37. Kupradze, V.D., *Potential Method in the Theory of Elasticity*, Davey, New York, 1965.

38. Lebedev, N.N., Skal'skaya, I.P. and Uflyand, Ya.S., *Problems of Mathematical Physics*, Pergamon Press, New York, 1966.

39. Marshall, S.L., A rapidly convergent modified Green's function for Laplace equation in a rectangular region, *Proc. of Royal Society*, London, **155**, 1739–1766, 1999.

40. Melnikov, Yu.A., Computation of heat potentials, *Proc. of the V Ukrainian Conference for Graduate Students in Mathematics*, Kiev, Ukrainian Academy of Sciences Publishers, 221–222 (in Russian), 1970.

41. Melnikov, Yu.A., Some applications of the Green's function method in mechanics, *Int. J. of Solids and Structures*, **13**, 1045–1058, 1977.

42. Melnikov, Yu.A., *Green's Functions and Matrices for Elliptic Equations and Systems*, Dnepropetrovsk State University Publishers, Dnepropetrovsk (in Russian), 1991.

43. Melnikov, Yu.A., *Green's Functions in Applied Mechanics*, Computational Mechanics Publications, Southampton–Boston, 1995.

44. Melnikov, Yu.A., Green's function formalism extended to systems of differential equations posed on graphs, *J. of Engineering Mathematics*, **34**, 369–386, 1998.

45. Melnikov, Yu.A., *Influence Functions and Matrices*, Marcel Dekker, New York–Basel, 1999.

46. Melnikov, Y.A., An alternative construction of Green's functions for the two-dimensional heat equation, *Engineering Analysis with Boundary Elements*, **24**, 467–475, 2000.

47. Melnikov, Y.A., Influence functions of a point source for perforated compound plates with facial convection, *J. of Engineering Mathematics*, **49**, 253–270, 2004.

48. Melnikov, Yu.A. and Bobylyov, Ye.A., Green's function method solution of the Reissner's plate problem, *Engineering Analysis with Boundary Elements*, **17**, 255–262, 1996.

49. Melnikov, Yu.A. and Koshnarjova, V.A., Green's matrices and 2-D elastopotentials for external Boundary value problems, *Applied Mathematical Modelling*, **18**, 161–167, 1994.

50. Melnikov, Y.A. and Melnikov, M.Y., Computability of series representations for Green's functions in a rectangle, *Engineering Analysis with Boundary Elements*, **30**, 774–780, 2006.

51. Melnikov, Y.A. and Sheremet, V.D., Some new results on the bending of a circular plate subject to a transverse point force, *Mathematics and Mechanics of Solids*, **6**, 29–45, 2001.

52. Melnikov, Yu.A. and Titarenko, S.A., On a new approach to 2-D optimal shape design, *Int. J. Numer. Methods in Engineering*, **36**, 2017–2030, 1993.

53. Mikhlin, S.G., *Linear Equations of Mathematical Physics*, Holt, Rinehart and Winston, New York, 1967.

54. Morse, P.M. and Feshbach, H., *Methods of Theoretical Physics*, McGraw-Hill, New York–Toronto–London, 1953.

55. Olsen, G.A., *Elements of Mechanics of Materials*, Prentice Hall, Englewood Cliffs, New Jersey, 1982.

56. Reissner, E., The effect of transverse shear deformation on the bending of elastic plates, *ASME J. of Applied Mechanics*, **12**, 69–77, 1945.

57. Roach, G.F., *Green's Functions*, Cambridge University Press, New York, 1982.

58. Shames, I.H., *Mechanics of Deformable Solids*, Prentice Hall, Englewood Cliffs, New Jersey, 1964.

59. Sheremet, V.D., *Handbook of Green's Functions and Matrices*, WIT Press, Southampton–Boston, 2002.

60. Smirnov, V.I., *A Course of Higher Mathematics*, Pergamon Press, Oxford–New York, 1964.

61. Stakgold, I., *Green's Functions and Boundary Value Problems*, John Wiley, New York, 1980.

62. Tewary, V.K., Elastic Green's function for a bimaterial composite solid containing a free surface normal to the interface, *J. of Materials Research*, **6**, 2592–2608, 1991.

63. Timoshenko, S. and Gere, J.M., *Theory of Elastic Stability*, McGraw-Hill, New York, 1961.

64. Timoshenko, S. and Goodier, J.N., *Theory of Elasticity*, McGraw-Hill, New York, 1970.

65. Timoshenko, S.P. and Woinowsky-Krieger, S., *Theory of Plates and Shells*, McGraw-Hill, New York, 1976.

66. Wempner, G., *Mechanics of Solids*, RWS Publishing Company, Boston, 1995.

67. Wright, D.J., *Introduction to Linear Algebra*, McGraw-Hill, Boston–Toronto, 1999.

68. Zill D.G. and Cullen, M.R., *Differential Equations with Boundary-Value Problems*, PWS Publishing Company, Boston, 1993.

# Index

**M**